Birkhäuser

Frontiers in Mathematics

This series is designed to be a repository for up-to-date research results which have been prepared for a wider audience. Graduates and postgraduates as well as scientists will benefit from the latest developments at the research frontiers in mathematics and at the "frontiers" between mathematics and other fields like computer science, physics, biology, economics, finance, etc. All volumes are online available at SpringerLink.

Edmond W. H. Lee

Advances in the Theory of Varieties of Semigroups

Birkhäuser

Edmond W. H. Lee
Department of Mathematics
Nova Southeastern University
Fort Lauderdale, FL, USA

ISSN 1660-8046 ISSN 1660-8054 (electronic)
Frontiers in Mathematics
ISBN 978-3-031-16496-5 ISBN 978-3-031-16497-2 (eBook)
https://doi.org/10.1007/978-3-031-16497-2

This book is published under the imprint Birkhäuser, www.birkhauser-science.com by the registered company Springer Nature Switzerland AG
The registered company address is: Gewerbestrasse 11, 6330 Cham, Switzerland

To my mother and late father

Preface

> *In order to guide research and organize knowledge, we group algebras into varieties. This way of classification has been so successful that it has no serious competitor.*
>
> —McKenzie, McNulty, and Taylor (1987)

Recall that a *variety* is a nonempty class of algebras that is closed under the formation of homomorphic images, subalgebras, and arbitrary direct products. According to Birkhoff's theorem (1935), varieties are precisely equationally defined classes of algebras and hence can be investigated, in principle, by both semantic and syntactic methods. The theory of varieties is one of the most central topics of universal algebra; some standard references include the monographs of Burris and Sankappanavar (1981), Grätzer (1979), and McKenzie et al. (1987).

A *semigroup* is a nonempty set endowed with an associative binary operation. The algebraic theory of semigroups dates back to Suschkewitsch (1928) and has grown into an independent branch of algebra after the work of Clifford (1933, 1941), Dubreil (1941), Rees (1940), and many others. Refer to Hollings (2009) for a historical overview of the early development of the theory. Since the 1960s, varieties of semigroups—which include periodic varieties of groups—have been intensely investigated. Varieties of semigroups endowed with additional operations have also received much attention; see, for example, Neumann (1967) for varieties of groups, Petrich (1984) for varieties of inverse semigroups, and Petrich and Reilly (1999) for varieties of completely regular semigroups. Over the years, several surveys have been written on various aspects of varieties, such as equational properties and the finite basis problem (Araújo et al. 2019; Gupta and Krasilnikov 2003; Shevrin and Volkov 1985; Taylor 1979; Volkov 2000, 2001), structural properties of semigroups in varieties (Shevrin and Martynov 1985; Shevrin and Sukhanov 1989), lattices of varieties (Aĭzenshtat and Boguta 1979; Evans 1971; Gusev et al. 2022; Shevrin et al. 2009; Vernikov 2015), and algorithmic problems (Kharlampovich and Sapir 1995).

The present monograph is a revised version of my Doctor of Sciences thesis, recently defended at the National Research University Higher School of Economics (2020a). The primary focus is on equational properties of varieties of three types of algebras—semigroups, involution semigroups, and monoids—with an eye on highlighting counterintuitive differences between them. The syntactic method is the predominant tool of investigation. The monograph is suitable for researchers in universal algebra; some knowledge of semigroup theory may be helpful but not essential.

An overview of the main results and a general survey of the relevant literature are presented in Chap. 1. After developing some preliminaries in Chap. 2, detailed justifications of the main results are given in Chaps. 3–14, which are grouped into three parts.

Part I (Chaps. 3–6) is concerned with semigroups. Major results here include a positive solution to the finite basis problem for all aperiodic Rees–Suschkewitsch varieties and a complete description of all those that are Cross, finitely generated, or small (Chap. 3); a complete description of pseudo-simple identities that are hereditarily finitely based (Chap. 4); a sufficient condition under which a semigroup is non-finitely based (Chap. 5); and a sufficient condition under which a non-finitely based finite algebra has no irredundant identity bases (Sect. 6.1). An infinite class of non-finitely based finite semigroups without irredundant identity bases is also exhibited (Chap. 6).

Part II (Chaps. 7–10) is dedicated to exploring, with respect to equational properties, interrelations between an involution semigroup and its semigroup reduct and interrelations between different involution semigroups sharing the same semigroup reduct. The non-finitely based finite semigroups from Chap. 6 are endowed with different involutions to produce the first examples of finite involution semigroups with an infinite irredundant identity basis (Chap. 7) and the first examples of finitely based finite involution semigroups with a non-finitely based semigroup reduct (Chap. 8). Three finite involution semigroups sharing a common semigroup reduct are then constructed so that one has a finite identity basis, one has an infinite irredundant identity basis, and one has no irredundant identity bases (Sect. 9.1). These examples demonstrate that an involution semigroup and its semigroup reduct can satisfy very different equational properties. On the other hand, it is proved that for any involution semigroup that generates a variety with some semilattice admitting a nontrivial involution, the non-finite basis property is inherited from its semigroup reduct (Chap. 10). An infinite class of finite involution semigroups is also exhibited with the property that as involution semigroups, the varieties they generate constitute two incomparable infinite chains, while as semigroups, the varieties they generate constitute a single infinite chain (Sect. 9.2).

Part III (Chaps. 11–14) is devoted to monoids. The existence and uniqueness of two maximal hereditarily finitely based overcommutative varieties of monoids are first established (Chap. 11). This facilitates a description of hereditarily finitely based varieties of monoids from several large classes. Varieties of aperiodic monoids with central idempotents that are Cross or inherently non-finitely generated are then completely characterized (Chap. 12), and the description of Cross varieties is extended to a certain class of aperiodic monoids with commuting idempotents (Chap. 13). Examples are also given to demonstrate the instability of the finite basis property under certain operations; specifically, the direct product of a certain finitely based finite monoid with any noncommutative group of finite exponent is shown to be non-finitely based (Sect. 14.1) and a recipe is given to construct the first examples of non-finitely based finite semigroups that generate finitely based monoids (Sect. 14.2).

A list of symbols and an index are provided at the end for the reader's convenience.

I am very grateful to the anonymous reviewers and the following colleagues for valuable comments and suggestions: Lev Beklemishev, Alexei Belov, Igor Dolinka, Sergey Gusev, Marcel Jackson, Alexei Krasilnikov, Jian Rong Li, Stuart Margolis, Mario Petrich, Norman Reilly, John Rhodes, Olga Sapir, Boris Vernikov, Mikhail Volkov, and Wen Ting Zhang.

Marcel Jackson and Mikhail Volkov have significantly influenced my work. For many years, they have regularly shared knowledge and insight without reservation. I thank them for their generosity and friendship.

I also thank my wife Wendy, the manager and love of my life, for being very supportive throughout the time I spent on this monograph and its earlier thesis version.

Most importantly, I thank the LORD my God for guidance, provision, and especially aligning several crucial events that were beyond my control; otherwise, writing this monograph would not have been possible.

Fort Lauderdale, FL, USA E. W. H. Lee
June 2022

Contents

Historical Overview and Main Results 1

1.1 Important Varietal Properties

In the theory of varieties of algebras, the investigation of several finiteness properties is highly relevant and popular: a variety is *finitely based* if its equational theory is finitely axiomatizable, *finitely generated* if it is generated by a finite algebra, and *small* if it contains finitely many subvarieties. As observed by M.V. Sapir (1991), these three properties are independent in the sense that a variety that satisfies any two properties need not satisfy the third. Since an algebra and the variety it generates satisfy the same identities, it is unambiguous to say that an algebra satisfies a certain equational property if the variety it generates satisfies the same property. For instance, an algebra is *finitely based* if the variety it generates is finitely based.

To distinguish varieties of the three types of algebras considered in the present monograph, let $\mathscr{V}_{\mathsf{sem}}\mathfrak{C}$, $\mathscr{V}_{\mathsf{inv}}\mathfrak{C}$, and $\mathscr{V}_{\mathsf{mon}}\mathfrak{C}$ denote the variety of semigroups generated by $\mathfrak{C}$, the variety of involution semigroups generated by $\mathfrak{C}$, and the variety of monoids generated by $\mathfrak{C}$, respectively. Let $\mathbf{0}$ denote the trivial variety regardless of algebra type.

1.1.1 Finite Basis Problem

The *finite basis problem*—the question of which algebras are finitely based—is one of the most prominent research problems in universal algebra. Apart from being very natural by itself, this problem also has several interesting and unexpected connections with other topics of theoretical and practical importance, for example, feasible algorithms for membership in certain classes of formal languages (see Almeida 1994) and classical number-theoretic conjectures, such as the twin prime conjecture, Goldbach's conjecture, and the odd perfect number conjecture (Perkins 1989). One approach to the finite basis

E. W. H. Lee, *Advances in the Theory of Varieties of Semigroups*, Frontiers in Mathematics, https://doi.org/10.1007/978-3-031-16497-2_1

problem is to classify non-finitely based algebras, and an effective method is to locate algebras that satisfy a stronger property: an algebra in a locally finite variety is *inherently non-finitely based* if every locally finite variety containing it is non-finitely based. In particular, a finite algebra is non-finitely based if the variety it generates contains some inherently non-finitely based algebra. A finite algebra that is not inherently non-finitely based is *weakly finitely based*.

1.1.2 Hereditary Finite Basis Property

A finitely based variety that satisfies the stronger property of having only finitely based subvarieties is *hereditarily finitely based*, and an algebra is *hereditarily finitely based* if it generates a hereditarily finitely based variety. For varieties of associative algebras, the hereditary finite basis property is also known as the *Specht property* in honor of Specht (1950), who importantly asked if every variety of associative algebras over a field of characteristic zero is finitely based. This question was affirmatively answered by Kemer (1988). Refer to Belov (2010) for a survey of results related to the Specht property for rings and some other general algebras.

Since there can only be countably many finite sets of identities up to renaming of variables, a hereditarily finitely based variety of algebras with finitely many operations, such as semigroups and monoids, can contain at most countably many subvarieties. Hereditarily finitely based varieties are closely related to minimal non-finitely based varieties, also commonly called *limit varieties*. Indeed, since Zorn's lemma implies that each non-finitely based variety must contain some limit subvariety, a variety is hereditarily finitely based if and only if it excludes limit subvarieties.

1.1.3 Cross Varieties

A variety is *Cross* if it is finitely based, finitely generated, and small. Being Cross is a hereditary property since every subvariety of each Cross variety is Cross (see MacDonald and Vaughan-Lee 1978). Consequently, every Cross variety is hereditarily finitely based. Minimal non-Cross varieties are also called *almost Cross varieties*. As in the case of non-finitely based varieties possessing limit subvarieties, it follows from Zorn's lemma that each non-Cross variety contains some almost Cross subvariety. Therefore, a variety is Cross if and only if it excludes almost Cross subvarieties.

1.2 Varieties Generated by Completely 0-Simple Semigroups

A semigroup is *simple* if it has no proper ideals. A semigroup with zero element is *0-simple* if it has no proper nonzero ideals. A *completely (0-)simple* semigroup is a (0-)simple semigroup S that contains some *primitive* idempotent, that is, a nonzero idempotent e such that for any nonzero idempotent $f \in S$, the implication

$$ef = fe = f \quad \Longrightarrow \quad e = f$$

holds; see Howie (1995, Sect. 3.2).

The class of completely 0-simple semigroups was one of the first classes of semigroups to be studied, in the groundbreaking work of Rees (1940) and Suschkewitsch (1928), and remains one of the most important classes of semigroups. Rees's theorem—representing each completely 0-simple semigroup as a Rees matrix semigroup $\mathcal{M}^0(I, G, \Lambda; P)$ over some group G with normalized $\Lambda \times I$ matrix P (see Howie 1995, Theorem 3.2.3)—has been essential to the development of semigroup theory throughout the decades. As each finite semigroup can be obtained from finite completely 0-simple semigroups by a sequence of ideal extensions, the role that finite completely 0-simple semigroups play in semigroup theory is comparable to the role that finite simple groups play in group theory. Naturally, varieties generated by completely 0-simple semigroups deserve special consideration.

1.2.1 Rees–Suschkewitsch Varieties

For each $n \geq 1$, let $\mathbf{RS}_n$ denote the variety generated by all completely 0-simple semigroups over groups of exponent dividing n. Following Kublanovskiĭ (2011), semigroups in $\mathbf{RS}_n$ are called *Rees–Suschkewitsch semigroups*, and subvarieties of $\mathbf{RS}_n$ are called *Rees–Suschkewitsch varieties*. One important result concerning Rees–Suschkewitsch varieties is the finite basis property for $\mathbf{RS}_n$.

Theorem 1.1 (Hall et al. 1997; Mashevitzky 1991) *For each $n \geq 1$, the identities*

$$x^{n+2} \approx x^2, \quad (xy)^{n+1}x \approx xyx, \quad x(yx)^n zx \approx xzx(yx)^n$$

constitute an identity basis for the variety $\mathbf{RS}_n$.

However, not all Rees–Suschkewitsch varieties are finitely based, as non-finitely based examples exist in abundance (Lee and Reilly 2008; Mashevitzky 1999), some of which are varieties of completely simple semigroups (Auinger and Szendrei 1999; Kad'ourek 2018; Mashevitzky 1984) and many of which are varieties of groups (see Gupta and Krasilnikov 2003).

Rees–Suschkewitsch varieties have received much attention since the publication of Theorem 1.1 in Hall et al. (1997). Lee (2002, 2004, 2006, 2007d) studied Rees–Suschkewitsch varieties that are *aperiodic* in the sense that they exclude nontrivial groups. The *intersection problem*—the problem of computing a finite semigroup that generates the intersection of two given finitely generated varieties—was considered by Lee and Reilly (2006), and a solution to the problem was found for varieties of central completely simple semigroups. Lee and Reilly (2008) also investigated Rees–Suschkewitsch varieties with certain centrality conditions, while Reilly (2009) examined those with certain orthodoxy conditions.

One huge obstacle—and a unique characteristic—in the study of Rees–Suschkewitsch varieties is that not all of them are generated by completely 0-simple semigroups, whence Rees's theorem is not always applicable. Rees–Suschkewitsch varieties that are generated by completely simple or completely 0-simple semigroups are said to be *exact*. Kublanovskiĭ et al. (2008) exhibited some necessary conditions and some sufficient conditions related to the exactness of a Rees–Suschkewitsch variety. Reilly (2008b) then refined these conditions into necessary and sufficient conditions, thus giving a complete characterization of exact Rees–Suschkewitsch varieties.

1.2.2 The Varieties $\mathbf{A}_2$ and $\mathbf{B}_2$

A semigroup is *aperiodic* if all its subgroups are trivial. Two very important aperiodic Rees–Suschkewitsch semigroups are the completely 0-simple semigroups

$$\mathcal{A}_2 = \langle \mathsf{a}, \mathsf{e} \mid \mathsf{a}^2 = 0,\ \mathsf{aea} = \mathsf{a},\ \mathsf{e}^2 = \mathsf{eae} = \mathsf{e} \rangle$$
$$\text{and } \mathcal{B}_2 = \langle \mathsf{a}, \mathsf{b} \mid \mathsf{a}^2 = \mathsf{b}^2 = 0,\ \mathsf{aba} = \mathsf{a},\ \mathsf{bab} = \mathsf{b} \rangle;$$

see Table 1.1. These semigroups can also be given as the Rees matrix semigroups

$$\mathcal{A}_2 = \mathcal{M}^0\left(\{1,2\}, \{1\}, \{1,2\}; \begin{bmatrix} 1 & 1 \\ 1 & 0 \end{bmatrix}\right)$$
$$\text{and } \mathcal{B}_2 = \mathcal{M}^0\left(\{1,2\}, \{1\}, \{1,2\}; \begin{bmatrix} 1 & 0 \\ 0 & 1 \end{bmatrix}\right)$$

Table 1.1 Multiplication tables of $\mathcal{A}_2$ and $\mathcal{B}_2$

$\mathcal{A}_2$	0	a	ae	ea	e
0	0	0	0	0	0
a	0	0	0	a	ae
ae	0	a	ae	a	ae
ea	0	0	0	ea	e
e	0	ea	e	ea	e

$\mathcal{B}_2$	0	a	ab	ba	b
0	0	0	0	0	0
a	0	0	0	a	ab
ab	0	a	ab	0	0
ba	0	0	0	ba	b
b	0	ba	b	0	0

over the trivial group $\{1\}$ or as the matrix semigroups

$$\mathcal{A}_2 = \left\{ \begin{bmatrix} 0 & 0 \\ 0 & 0 \end{bmatrix}, \begin{bmatrix} 1 & 0 \\ 0 & 0 \end{bmatrix}, \begin{bmatrix} 0 & 1 \\ 0 & 0 \end{bmatrix}, \begin{bmatrix} 1 & 0 \\ 1 & 0 \end{bmatrix}, \begin{bmatrix} 0 & 1 \\ 0 & 1 \end{bmatrix} \right\}$$

$$\text{and } \mathcal{B}_2 = \left\{ \begin{bmatrix} 0 & 0 \\ 0 & 0 \end{bmatrix}, \begin{bmatrix} 1 & 0 \\ 0 & 0 \end{bmatrix}, \begin{bmatrix} 0 & 1 \\ 0 & 0 \end{bmatrix}, \begin{bmatrix} 0 & 0 \\ 1 & 0 \end{bmatrix}, \begin{bmatrix} 0 & 0 \\ 0 & 1 \end{bmatrix} \right\}$$

under usual matrix multiplication.

Remark 1.2

(i) The subscript 2 of the symbols $\mathcal{A}_2$ and $\mathcal{B}_2$ refers to the size of the 2×2 sandwich matrices in their Rees matrix representations. The semigroup $\mathcal{A}_2$ is idempotent generated, while the semigroup $\mathcal{B}_2$, being both completely 0-simple and inverse, is a *Brandt semigroup*.

(ii) In general, Brandt semigroups are precisely semigroups isomorphic to a Rees matrix semigroup $\mathcal{M}^0(I, G, I; E)$ over some group G and index set I with $I \times I$ identity matrix E; see Howie (1995, Theorem 5.1.8).

Let $\mathbf{A_2}$ and $\mathbf{B_2}$ denote the varieties generated by the semigroups $\mathcal{A}_2$ and $\mathcal{B}_2$, respectively; these varieties are aperiodic, exact, and finitely based (Reilly 2008a; Trahtman 1994). By comparing the identity bases for $\mathbf{RS}_n$ and $\mathbf{A_2}$ in Theorem 1.1 and Lemma 3.1, it is easily seen that $\mathbf{RS}_1 = \mathbf{A_2}$. Therefore, the variety $\mathbf{RS}_1$ is finitely generated by $\mathcal{A}_2$. In general, not every $\mathbf{RS}_n$ is finitely generated because it contains the Burnside variety of all groups of exponent dividing n, which is not locally finite for many values of n. Now since $\mathbf{A_2} = \mathbf{RS}_1$ is the largest aperiodic Rees–Suschkewitsch variety, the inclusion $\mathbf{B_2} \subseteq \mathbf{A_2}$ follows; alternatively, this inclusion holds because the semigroup $\mathcal{B}_2$ *divides* the direct product $\mathcal{A}_2 \times \mathcal{A}_2$ in the sense that $\mathcal{B}_2$ is a homomorphic image of some subsemigroup of $\mathcal{A}_2 \times \mathcal{A}_2$ (Reilly 2007).

Remark 1.3

(i) A finite identity basis for the variety $\mathbf{B_2}$ was first established by Trahtman (1981a). Later, Reilly (2008a) discovered a gap in the proof of this result and provided an independent proof to confirm its correctness. Another proof of the same identity basis for $\mathbf{B_2}$ can also be found in Lee and Volkov (2007).

(ii) Volkov (1985) demonstrated how an identity basis for the variety generated by a Brandt semigroup $\mathcal{M}^0(I, G, I; E)$ over any group G of finite exponent can be constructed from an identity basis for G. However, the proof of this result is incomplete since it uses a version of Trahtman's flawed argument (1981a) that is responsible for the gap mentioned in part (i). This problem was recently remedied in Volkov (2019) by an independent proof of the result.

(iii) Refer to Lemmas 3.1 and 3.2(ii) for explicit finite identity bases for the varieties $\mathbf{A}_2$ and $\mathbf{B}_2$.

Besides being important examples of Rees–Suschkewitsch varieties, $\mathbf{A}_2$ and $\mathbf{B}_2$ also play other roles in the theory of varieties of semigroups. For instance, the variety $\mathbf{A}_2$ coincides with the class of semigroups that are *2-testable* in the sense that they satisfy any identity formed by a pair of words that begin with the same variable, end with the same variable, and share the same set of factors of length two (Trahtman 1999); and periodic varieties that exclude $\mathbf{B}_2$ are precisely those that consist of bands of Archimedean semigroups (Sapir and Sukhanov 1981). Refer to the surveys by Shevrin et al. (2009), Shevrin and Volkov (1985), and Volkov (2001) for more information on the varieties $\mathbf{A}_2$ and $\mathbf{B}_2$.

1.2.3 Aperiodic Rees–Suschkewitsch Varieties

Most results overviewed in Sect. 1.2.1 are concerned with exact varieties. Substantial progress in the understanding of Rees–Suschkewitsch varieties thus requires the exploration of those that are not exact. It is logical to begin this task with aperiodic Rees–Suschkewitsch varieties, that is, subvarieties of $\mathbf{A}_2 = \mathbf{RS}_1$. Further, since every known example of non-finitely based Rees–Suschkewitsch variety contains nontrivial groups, it is also natural to question if the variety $\mathbf{A}_2$ contains some non-finitely based subvariety.

Problem 1.4 Investigate the lattice $\mathfrak{L}(\mathbf{A}_2)$ of subvarieties of $\mathbf{A}_2$. Specifically, determine whether or not the variety $\mathbf{A}_2$ is hereditarily finitely based.

Although the variety $\mathbf{A}_2$ contains only 13 exact subvarieties (Reilly 2007), a complete solution to Problem 1.4 will be highly nontrivial because it follows from Volkov (1989a) that the lattice $\mathfrak{L}(\mathbf{A}_2)$ contains an isomorphic copy of every finite lattice; a variety having such a lattice of subvarieties is said to be *finitely universal*. In particular, the variety $\mathbf{A}_2$ contains infinitely many non-exact subvarieties.

Problems related to the cardinality of the lattice $\mathfrak{L}(\mathbf{A}_2)$ were also posed by Jackson (2000) in his study of varieties with continuum many subvarieties.

Problem 1.5 (Jackson 2000, Questions 4.6 and 4.8; Volkov 2001, Problem 6.5)

(i) Does the variety $\mathbf{A}_2$ contain continuum many subvarieties?
(ii) Does the variety $\mathbf{B}_2$ contain continuum many subvarieties?
(iii) Is the variety $\mathbf{B}_2$ hereditarily finitely based?

Note that Problem 1.5 would be completely solved if the variety $\mathbf{A}_2$ is shown to be hereditarily finitely based. On the contrary, as noted by Volkov (2001), a negative solution to part (iii) of this problem would clarify the structure of hereditarily finitely based finite semigroups.

The subsemigroups $\mathcal{A}_0 = \mathcal{A}_2 \backslash \{\mathsf{e}\}$ and $\mathcal{B}_0 = \mathcal{B}_2 \backslash \{\mathsf{b}\}$ of $\mathcal{A}_2$ and $\mathcal{B}_2$, respectively, are crucial to the description of subvarieties of $\mathbf{A_2}$. These subsemigroups can be given by the presentations

$$\mathcal{A}_0 = \langle \mathsf{e}, \mathsf{f} \mid \mathsf{e}^2 = \mathsf{e},\ \mathsf{f}^2 = \mathsf{f},\ \mathsf{ef} = 0 \rangle$$

$$\text{and } \mathcal{B}_0 = \langle \mathsf{a}, \mathsf{e}, \mathsf{f} \mid \mathsf{af} = \mathsf{ea} = \mathsf{a},\ \mathsf{e}^2 = \mathsf{e},\ \mathsf{f}^2 = \mathsf{f},\ \mathsf{ef} = \mathsf{fe} = 0 \rangle;$$

see Table 1.2. The varieties $\mathbf{A_0}$ and $\mathbf{B_0}$ generated by $\mathcal{A}_0$ and $\mathcal{B}_0$, respectively, are non-exact Rees–Suschkewitsch varieties. Edmunds (1980) proved that all varieties generated by a semigroup of order four, and so also $\mathbf{A_0}$ and $\mathbf{B_0}$, are finitely based. It also follows from Volkov (1989a) that $\mathbf{A_0}$ and $\mathbf{B_0}$ are finitely universal. Apart from these results, not much was known about non-exact Rees–Suschkewitsch varieties prior to the turn of the millennium. The first detailed study of non-exact subvarieties of $\mathbf{A_2}$ was performed in Lee (2002), where subvarieties of $\mathbf{A_0}$ were described and the structure of the lattice $\mathfrak{L}(\mathbf{A_0})$ was given modulo certain varieties that satisfy permutation identities. In particular, the variety $\mathbf{A_0}$ is hereditarily finitely based and $\mathbf{B_0}$ is its unique maximal subvariety.

For any subvariety $\mathbf{V}$ of $\mathbf{A_2}$, the unique largest subvariety of $\mathbf{A_2}$ that excludes $\mathbf{V}$, if it exists, is denoted by $\overline{\mathbf{V}}$. The existence of the varieties $\overline{\mathbf{A_2}}$, $\overline{\mathbf{B_2}}$, and $\overline{\mathbf{A_0}}$ follows from Escada (2000), Sapir and Sukhanov (1981), and Torlopova (1984), respectively. Therefore, for any $\mathbf{V} \in \{\mathbf{A_2}, \mathbf{B_2}, \mathbf{A_0}\}$, the intersection $\mathbf{V} \cap \overline{\mathbf{V}}$ coincides with the unique maximal subvariety of $\mathbf{V}$; in particular, $\overline{\mathbf{A_2}}$ is the unique maximal subvariety of $\mathbf{A_2}$, so that $\mathfrak{L}(\mathbf{A_2}) = \mathfrak{L}(\overline{\mathbf{A_2}}) \cup \{\mathbf{A_2}\}$. After the initial work of Lee (2002), some results concerning the varieties $\overline{\mathbf{A_2}}$ and $\overline{\mathbf{B_2}}$ were established. Lee (2004) exhibited explicit identity bases for the varieties $\overline{\mathbf{A_2}}$ and $\overline{\mathbf{B_2}}$ and proved that $\mathbf{B_2} \cap \overline{\mathbf{B_2}} = \mathbf{B_0}$. It turned out that the varieties $\mathbf{A_0}$ and $\mathbf{B_2}$ share the same proper subvarieties—with $\mathbf{B_0}$ being the unique maximal subvariety of both $\mathbf{A_0}$ and $\mathbf{B_2}$—whence the variety $\mathbf{B_2}$ is hereditarily finitely based; see Lee (2004, Sects. 3 and 4) for more details. Consequently, parts (ii) and (iii) of Problem 1.5 were solved. The question of whether or not the equality $\overline{\mathbf{A_2}} = \overline{\mathbf{B_2}} \vee \mathbf{B_2}$ holds was raised by Lee (2004) and affirmatively answered by Volkov (2005) shortly after. Lee (2007d) demonstrated the existence of the variety $\overline{\mathbf{B_0}}$ and proved that it is hereditarily finitely based. Other results on the variety $\overline{\mathbf{B_2}}$ were also established by Lee (2006) and Reilly (2008a), while further information on the lattice $\mathfrak{L}(\mathbf{A_2})$ and its elements was found by Lee and Volkov (2007).

Table 1.2 Multiplication tables of $\mathcal{A}_0$ and $\mathcal{B}_0$

$\mathcal{A}_0$	0	fe	f	e
0	0	0	0	0
fe	0	0	0	fe
f	0	fe	f	fe
e	0	0	0	e

$\mathcal{B}_0$	0	a	e	f
0	0	0	0	0
a	0	0	0	a
e	0	a	e	0
f	0	0	0	f

The investigation of the lattice $\mathfrak{L}(\mathbf{A}_2)$ is continued in Chap. 3. This chapter has three main goals, the first of which is concerned with the finite basis problem. In Sect. 3.1, it is shown that the lattice $\mathfrak{L}(\overline{\mathbf{A}}_2)$ is the disjoint union of five intervals:

$$[\mathbf{A}_0 \vee \mathbf{B}_2, \overline{\mathbf{A}}_2], \quad [\mathbf{A}_0, \overline{\mathbf{B}}_2], \quad [\mathbf{B}_2, \overline{\mathbf{A}}_0], \quad [\mathbf{B}_0, \overline{\mathbf{A}}_0 \cap \overline{\mathbf{B}}_2], \quad \mathfrak{L}(\overline{\mathbf{B}}_0). \tag{1.1}$$

The varieties in each of these intervals are then shown in Sect. 3.2 to be finitely based, thus solving the finite basis problem for all subvarieties of $\mathbf{A}_2$.

Theorem 1.6 *The variety* $\mathbf{A}_2$ *is hereditarily finitely based.*

Consequently, the lattice $\mathfrak{L}(\mathbf{A}_2)$ is countably infinite and Problem 1.5 is completely solved.

The second goal of Chap. 3 is to examine the structure of the lattice $\mathfrak{L}(\mathbf{A}_2)$. Since the variety $\overline{\mathbf{B}}_0$ is finitely universal, it is impossible to give an explicit description of the lattice $\mathfrak{L}(\overline{\mathbf{B}}_0)$. In contrast, it is shown in Sect. 3.3 that the first four intervals in (1.1) are each isomorphic to a direct product of two countably infinite chains. It follows that the first four intervals in (1.1), together with $\mathbf{A}_2$, form the infinite distributive subinterval $[\mathbf{B}_0, \mathbf{A}_2]$ of $\mathfrak{L}(\mathbf{A}_2)$. Of particular interest is the subinterval $[\mathbf{B}_2, \mathbf{A}_2]$ of $[\mathbf{B}_0, \mathbf{A}_2]$; see Fig. 1.1. It is counterintuitive that this interval contains infinitely many non-finitely generated varieties even though its endpoints $\mathbf{B}_2$ and $\mathbf{A}_2$ are generated by $\mathcal{B}_2$ and $\mathcal{A}_2$, two semigroups with similar multiplication tables; see Table 1.1.

The third and last goal of Chap. 3 is to investigate subvarieties of $\mathbf{A}_2$ that are Cross, finitely generated, or small. A characterization of these varieties is given in Sect. 3.4. To this end, *diverse identities*—identities of the form $x_1x_2\cdots x_n \approx \mathbf{w}$ that are not permutation identities—and the variety $\mathbf{H}_{\mathsf{com}}$ defined by the identity system $\{xyx \approx y^2, xy \approx yx\}$ are required.

Theorem 1.7 *The following statements on any subvariety* $\mathbf{V}$ *of* $\mathbf{A}_2$ *are equivalent*:

(a) $\mathbf{V}$ *is Cross*;
(b) $\mathbf{V}$ *is small*;
(c) $\mathbf{V}$ *is finitely generated and* $\mathbf{B}_0 \nsubseteq \mathbf{V}$;
(d) $\mathbf{V}$ *satisfies some diverse identity*;
(e) $\mathbf{H}_{\mathsf{com}} \nsubseteq \mathbf{V}$.

It follows that every small subvariety of $\mathbf{A}_2$ is Cross, the variety $\mathbf{H}_{\mathsf{com}}$ is the unique almost Cross subvariety of $\mathbf{A}_2$, and the non-small subvarieties of $\mathbf{A}_2$ constitute the interval $[\mathbf{H}_{\mathsf{com}}, \mathbf{A}_2]$.

The Cross subvarieties of $\mathbf{A}_2$ and the finitely generated subvarieties of $\mathbf{A}_2$ are also shown to constitute incomplete sublattices of $\mathfrak{L}(\mathbf{A}_2)$. In general, however, Cross varieties, finitely generated varieties, and small varieties do not form sublattices of the lattice $\mathfrak{L}_{\mathsf{sem}}$

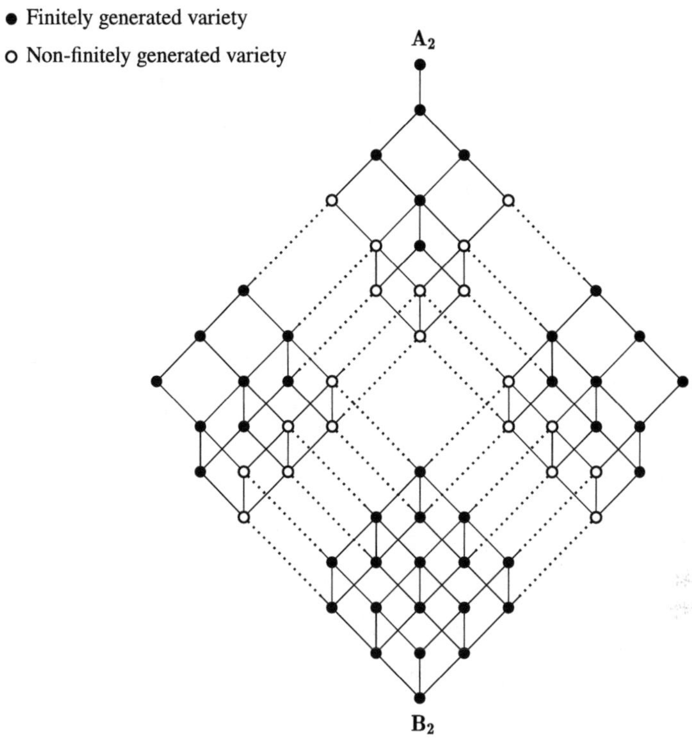

Fig. 1.1 The interval $[\mathbf{B}_2, \mathbf{A}_2]$

of all varieties of semigroups; see, for instance, M.V. Sapir (1991) and Shevrin et al. (2009).

1.2.4 Rees–Suschkewitsch Varieties Containing Nontrivial Groups

A next step in the study of Rees–Suschkewitsch varieties is to consider those that contain nontrivial groups. In this case, the situation is much more complicated. For example, Mashevitzky (1999) presented an infinite chain $\mathbf{V}_1 \supset \mathbf{V}_2 \supset \mathbf{V}_3 \supset \cdots$ of finitely generated exact varieties in $\mathfrak{L}(\mathbf{RS}_2)$ that are alternately finitely based and non-finitely based, that is, $\mathbf{V}_k$ is finitely based if and only if k is even. Further, Lee and Reilly (2008) exhibited a continuum of ultimately incomparable infinite chains of finitely generated exact varieties that are alternately finitely based and non-finitely based.

For each $n \geq 1$, let $\mathbf{Z}_n$ denote the variety generated by the cyclic group

$$\mathcal{Z}_n = \langle \mathsf{g} \mid \mathsf{g}^n = 1 \rangle$$

of order n. Even though the varieties $\mathbf{A}_2$ and $\mathbf{Z}_n$ are hereditarily finitely based, their varietal join $\mathbf{A}_2 \vee \mathbf{Z}_n$ is non-finitely based for all $n \geq 2$ (Volkov 1989b). It follows that $\mathbf{A}_2$ is a maximal hereditarily finitely based Rees–Suschkewitsch variety. Techniques similar to Chap. 3 were employed to exhibit an infinite class of limit Rees–Suschkewitsch varieties.

Theorem 1.8 (Lee and Volkov 2011) *The varieties $\mathbf{A}_2 \vee \mathbf{Z}_p$, where p ranges over the primes, are precisely all limit varieties generated by completely 0-simple semigroups over commutative groups.*

It is unknown if any limit variety is generated by completely 0-simple semigroups over noncommutative groups; see Lee and Volkov (2011, Sect. 6). Prior to the discovery in Theorem 1.8, the only other limit varieties of semigroups with an explicit description were given by Pollák (1989), M.V. Sapir (1991), and Volkov (1982).

More comprehensive results on Rees–Suschkewitsch varieties containing nontrivial groups—some of which generalize results in Chap. 3—were also established by Kublanovskiĭ (2011). It turns out that limit varieties do not exist among non-exact Rees–Suschkewitsch varieties.

1.2.5 Aperiodic Rees–Suschkewitsch Monoids

For any semigroup S, let S^1 denote the monoid obtained by adjoining a new unit element $\mathtt{1}$ to S. It is reasonable to call a monoid of the form S^1, where S is any Rees–Suschkewitsch semigroup, a *Rees–Suschkewitsch monoid*. Finite identity bases for varieties generated by some small Rees–Suschkewitsch monoids, such as $\mathcal{A}_0^1$ and $\mathcal{B}_0^1$, have been found by Edmunds (1977, 1980) and Tishchenko (1980). Subvarieties of some varieties generated by an aperiodic Rees–Suschkewitsch monoid were considered in Lee (2007c, 2008b, 2009b).

Aperiodic Rees–Suschkewitsch semigroups are precisely semigroups in the variety $\mathbf{A}_2$ and so by Theorem 1.6, generate hereditarily finitely based varieties. However, not every variety generated by an aperiodic Rees–Suschkewitsch monoid is finitely based; for example, the varieties $\mathscr{V}_{\mathsf{sem}}\{\mathcal{A}_2^1\}$ and $\mathscr{V}_{\mathsf{sem}}\{\mathcal{B}_2^1\}$ are not only non-finitely based but also inherently non-finitely based (M.V. Sapir 1987b). Based on this result, the following dichotomy on the equational property of the variety generated by an aperiodic Rees–Suschkewitsch monoid was established.

Theorem 1.9 (Lee 2011b) *For any aperiodic Rees–Suschkewitsch semigroup S, the variety $\mathscr{V}_{\mathsf{sem}}\{S^1\}$ is finitely based if and only if $\mathcal{B}_2 \notin \mathscr{V}_{\mathsf{sem}}\{S\}$. More specifically, the variety $\mathscr{V}_{\mathsf{sem}}\{S^1\}$ is hereditarily finitely based if $\mathcal{B}_2 \notin \mathscr{V}_{\mathsf{sem}}\{S\}$; otherwise, it is inherently non-finitely based.*

Recall from Fig. 1.1 that infinitely many varieties of aperiodic Rees–Suschkewitsch semigroups are non-finitely generated. But the situation for varieties generated by aperiodic Rees–Suschkewitsch monoids is much simpler.

Theorem 1.10 (Lee 2012c) *There exist only* 28 *varieties generated by aperiodic Rees–Suschkewitsch monoids, all of which are finitely generated and precisely* 19 *of which are finitely based.*

1.3 Hereditary Finite Basis Property

1.3.1 Hereditarily Finitely Based Identities

The hereditary finite basis property, when satisfied by a variety $\mathbf{V}$ of semigroups, imposes a number of restrictions on the lattice $\mathfrak{L}(\mathbf{V})$ of subvarieties of $\mathbf{V}$. Most notably, as observed earlier in Sect. 1.1.2, a hereditarily finitely based variety of semigroups contains at most countably many subvarieties. Further, a variety $\mathbf{V}$ is hereditarily finitely based if and only if $\mathbf{V}$ is finitely based and $\mathfrak{L}(\mathbf{V})$ satisfies the descending chain condition. Not all finitely based varieties are hereditarily finitely based. For instance, for any $n \geq 2$, the variety $\mathbf{RS}_n$ is finitely based by Theorem 1.1, but it is not hereditarily finitely based because its subvariety $\mathbf{A_2} \vee \mathbf{Z}_n$ is non-finitely based (Volkov 1989b). Finitely based varieties that contain continuum many subvarieties are also not hereditarily finitely based; some examples of finite semigroups generating such varieties can be found in Edmunds et al. (2010), Jackson (2000), Jackson and Lee (2018), and Zhang and Luo (2008).

The hereditary finite basis property of a variety, say with identity basis Σ, also imposes a restriction on any identity system containing Σ. Specifically, an identity system Σ defines a hereditarily finitely based variety if and only if every identity system containing Σ is equivalent to some finite identity system; in this case, the identity system Σ is said to be *hereditarily finitely based*. An identity is *hereditarily finitely based* if it forms a hereditarily finitely based identity system on its own. The first examples of hereditarily finitely based identity systems are due to Perkins (1969): any nontrivial permutation identity and any periodicity identity form a hereditarily finitely based identity system. Since the variety **Com** of all commutative semigroups (Perkins 1969) and the variety **Idem** of all idempotent semigroups (Birjukov 1970; Fennemore 1970; Gerhard 1970) are hereditarily finitely based, the commutativity identity $xy \approx yx$ and the idempotency identity $x^2 \approx x$ are hereditarily finitely based.

An important motivation to study hereditarily finitely based identity systems is that they provide easily verifiable sufficient conditions for the finite basis property. Hereditarily finitely based identities, due to their simplicity, naturally attract the most interest. Pollák (1975) initiated the investigation of hereditarily finitely based identities and, for over a decade, contributed immensely to their identification. In Pollák (1979b), hereditarily finitely based identities are classified into the following four types:

(T1) one side of the identity is *simple*, that is, it contains any variable at most once;
(T2) one side of the identity is of the form xyx;
(T3) one side of the identity is of the form $xyzx$ and the other side is of the form x^2;
(T4) one side of the identity is of the form

$$x_1^{e_1}x_2^{e_2}\cdots x_m^{e_m}$$

with $e_1, e_2, \ldots, e_m \in \{1, 2\}$ and the other side is of the form

$$x_1x_2\cdots x_{k-1}yx_kyx_{k+1}x_{k+2}\cdots x_n$$

with $n \geq 2$ and $k \in \{1, 2, \ldots, n\}$. (Note that in the extreme cases $k = 1$ and $k = n$, the above word becomes $yx_1yx_2x_3\cdots x_n$ and $x_1x_2\cdots x_{n-1}yx_ny$, respectively.)

All identities in (T3) are hereditarily finitely based (Pollák 1979a), while a description of hereditarily finitely based identities from (T4) follows from Pollák (1986) and Pollák and Volkov (1985).

The description of hereditarily finitely based identities from (T1) and (T2) is, however, less straightforward because there are two cases depending on whether or not the identity in question is *homotypical* in the sense that it is formed by a pair of words sharing the same set of variables. As explained by Volkov (2002, page 881), it is infeasible to completely classify non-homotypical hereditarily finitely based identities from (T1) and (T2) since doing so would require as a special case a description of all hereditarily finitely based periodic varieties of groups—a problem which has remained unsolved since the 1970s. Nevertheless, Pollák (1982) classified non-homotypical hereditarily finitely based identities from (T1) and (T2) modulo simple semigroups and completely described all homotypical hereditarily finitely based identities from (T2). Regarding homotypical hereditarily finitely based identities from (T1), only permutation identities have been described: a permutation identity is hereditarily finitely based if and only if it is formed by a pair of words that begin or end with different variables (Pollák and Volkov 1985).

For more information on Pollák's accomplishments in hereditarily finitely based identities and semigroup varieties in general, refer to the detailed survey of his work (Volkov 2002).

1.3.2 Pseudo-Simple Hereditarily Finitely Based Identities

A word is *almost simple* if it contains one variable twice and any other variable at most once. An identity is *almost simple* if it is formed by a pair of almost simple words; it is *pseudo-simple* if it is formed by a simple word and an almost simple word. Pollák and Volkov (1985) published a complete description of almost simple hereditarily finitely based identities, some consequences of which contributed to the classification of hereditarily finitely based identities described in the previous subsection. The similarity between almost simple identities and pseudo-simple identities led to several questions.

Problem 1.11 (Pollák and Volkov 1985)

(i) Which pseudo-simple identities are hereditarily finitely based?
(ii) Is the homotypical pseudo-simple identity $x^2y \approx xy$ hereditarily finitely based?

Since the idempotency identity $x^2 \approx x$—the simplest homotypical pseudo-simple identity—is hereditarily finitely based (Birjukov 1970; Fennemore 1970; Gerhard 1970), the next simplest homotypical pseudo-simple identity $x^2y \approx xy$ and its dual $xy^2 \approx xy$ are logical choices for investigation.

A solution to Problem 1.11(i) clearly constitutes a nontrivial step toward a more complete description of hereditarily finitely based identities from (T1). But as remarked by Pollák (1979b, page 447), it "looks rather hopeless" to obtain such a description, and his pessimism seemed well founded because there was no progress on either part of Problem 1.11 for many years.

It was not until 15 years after Problem 1.11 was published that a solution to part (ii) was found: the identity $x^2y \approx xy$ defines a variety with continuum many subvarieties and so is not hereditarily finitely based (Kad'ourek 2000). It follows that the identity

$$x_1x_2\cdots x_\ell y^2 z_1z_2\cdots z_r \approx x_1x_2\cdots x_\ell y z_1z_2\cdots z_r, \tag{1.2}$$

where $\ell, r \geq 0$ and $(\ell, r) \neq (0, 0)$, is also not hereditarily finitely based. It turns out that this result can be extended to a complete solution to Problem 1.11(i). In Chap. 4, it is shown that every pseudo-simple identity that is not of the form (1.2) is hereditarily finitely based. Consequently, a pseudo-simple identity is hereditarily finitely based if and only if it is not of the form (1.2).

The results on hereditarily finitely based identities surveyed in this and the previous subsections are summarized in Table 1.3 (see Volkov 2002, Table 1). What remains to be investigated are homotypical hereditarily finitely based identities from (T1) that are neither permutation identities nor pseudo-simple identities.

Problem 1.12 Classify homotypical hereditarily finitely based identities of the form

$$x_1x_2\cdots x_n \approx \mathbf{w},$$

where $\mathbf{w}$ is any word of length at least $n+2$.

Table 1.3 Current results on the classification of hereditarily finitely based identities

Type	Homotypical identities	Non-homotypical identities
(T1)	Classification only for pseudo-simple identities (Chap. 4) and permutation identities (Pollák and Volkov 1985)	Classification modulo simple semigroups (Pollák 1982)
(T2)	Complete classification (Pollák 1982)	
(T3)	Always hereditarily finitely based (Pollák 1979a)	
(T4)	Complete classification (Pollák 1986; Pollák and Volkov 1985)	

1.3.3 Minimal Non-Finitely Based Semigroups

Since non-finitely based finite semigroups exist, there exists one with the smallest possible order; such a semigroup is said to be *minimal non-finitely based*. The identification of all minimal non-finitely based semigroups was a project that employed many hereditarily finitely based identity systems.

The discovery by Perkins (1966) of the non-finitely based semigroup $\mathcal{B}_2^1$ of order six naturally focused much attention upon the finite basis problem for smaller semigroups. This problem, explicitly posed by Tarski (1968), attracted the interest of Bol'bot (1979), Edmunds (1977, 1980), Karnofsky (1970), Perkins (1969), Simel'gor (1978), Tishchenko (1980), and Trahtman (1981a,b). A complete solution was eventually announced by Trahtman (1983).

Theorem 1.13 *Every semigroup of order five or less is finitely based.*

Trahtman's proof of this result—the publication of which was delayed for a few years (1991)—invoked some hereditarily finitely based identity systems that were available at that time in addition to establishing the finite basis property for several sporadic cases.

Recently, Lee (2013b) generalized Theorem 1.13 by proving that up to isomorphism and anti-isomorphism, every semigroup of order five or less, with the possible exception of the monoid

$$\mathcal{M}_5 = \langle \mathsf{a}, \mathsf{e} \mid \mathsf{a}^2\mathsf{e} = \mathsf{a}^2,\ \mathsf{e}^2 = \mathsf{ea} = \mathsf{e} \rangle \cup \{1\},$$

is hereditarily finitely based; see Table 1.4.

Problem 1.14 (Lee 2013b, Problem 1.4; See Also Edmunds et al. 2010, Question 4.2) Is the semigroup $\mathcal{M}_5$ hereditarily finitely based?

Table 1.4 Multiplication table of $\mathcal{M}_5$

$\mathcal{M}_5$	a^2	a	ae	1	e
a^2	a^2	a^2	a^2	a^2	a^2
a	a^2	a^2	a^2	a	ae
ae	ae	ae	ae	ae	ae
1	a^2	a	ae	1	e
e	e	e	e	e	e

Table 1.5 Multiplication tables of $\mathcal{A}_2^{g}$ and $\mathcal{L}_3$

$\mathcal{A}_2^{g}$	0	a	ae	ea	e	g
0	0	0	0	0	0	g
a	0	0	0	a	ae	g
ae	0	a	ae	a	ae	g
ea	0	0	0	ea	e	g
e	0	ea	e	ea	e	g
g	g	g	g	g	g	0

$\mathcal{L}_3$	0	fef	ef	fe	e	f
0	0	0	0	0	0	0
fef	0	0	0	0	0	fef
ef	0	0	0	0	0	ef
fe	0	0	fef	0	fe	fef
e	0	0	ef	0	e	ef
f	0	fef	fef	fe	fe	f

This problem currently remains open. But regardless of its solution, due to Theorem 1.13, minimal non-finitely based semigroups are of order six, and the monoids $\mathcal{A}_2^1$ and $\mathcal{B}_2^1$ are such examples (M.V. Sapir 1987b). Two further examples were more recently discovered: the semigroup $\mathcal{A}_2^{g}$ obtained by adjoining a new element g to $\mathcal{A}_2$, where multiplication involving g is given by $g\mathcal{A}_2 = \mathcal{A}_2 g = \{g\}$ and $g^2 = 0$, and the $\mathscr{J}$-trivial semigroup

$$\mathcal{L}_3 = \langle e, f \mid e^2 = e,\ f^2 = f,\ efe = 0\rangle;$$

see Table 1.5. (Note that the zero element 0 of $\mathcal{A}_2$ is no longer a zero element in $\mathcal{A}_2^{g}$.) These semigroups can also be given as the 3×3 matrix semigroups

$$\mathcal{A}_2^{g} = \left\{ \begin{bmatrix} 1&0&0\\0&0&0\\0&0&0 \end{bmatrix}, \begin{bmatrix} 1&0&0\\0&1&0\\0&0&0 \end{bmatrix}, \begin{bmatrix} 1&0&0\\0&0&1\\0&0&0 \end{bmatrix}, \begin{bmatrix} 1&0&0\\0&1&0\\0&1&0 \end{bmatrix}, \begin{bmatrix} 1&0&0\\0&0&1\\0&0&1 \end{bmatrix}, \begin{bmatrix} -1&0&0\\0&0&0\\0&0&0 \end{bmatrix} \right\}$$

$$\text{and } \mathcal{L}_3 = \left\{ \begin{bmatrix} 0&0&0\\0&0&0\\0&0&0 \end{bmatrix}, \begin{bmatrix} 0&1&0\\0&0&0\\0&0&0 \end{bmatrix}, \begin{bmatrix} 0&0&1\\0&0&0\\0&0&0 \end{bmatrix}, \begin{bmatrix} 0&1&0\\0&1&0\\0&0&0 \end{bmatrix}, \begin{bmatrix} 0&0&1\\0&0&1\\0&0&0 \end{bmatrix}, \begin{bmatrix} 1&0&0\\0&0&1\\0&0&1 \end{bmatrix} \right\}$$

under usual matrix multiplication. The semigroup $\mathcal{A}_2^{g}$ was discovered by Volkov and is non-finitely based by either Mashevitzky (1983) or Volkov (1989b); it generates the limit variety $\mathbf{A_2} \vee \mathbf{Z_2}$ in Theorem 1.8. The semigroup $\mathcal{L}_3$ is non-finitely based by either Lee (2012a) or Zhang and Luo (2011).

Remark 1.15 The subscript 3 of the symbol $\mathcal{L}_3$ refers to the length of the product efe on the left side of the relation efe $= 0$ in the presentation of $\mathcal{L}_3$. By replacing efe in this relation with the alternating products ef, efe, efef, ... of arbitrary length, the $\mathcal{J}$-trivial semigroups $\mathcal{L}_2, \mathcal{L}_3, \mathcal{L}_4, \ldots$ are obtained. It is clear that $\mathcal{L}_2 = \mathcal{A}_0$, but the semigroups $\mathcal{L}_3, \mathcal{L}_4, \mathcal{L}_5, \ldots$ have not previously been investigated.

Two semigroups are *equivalent* if they are either isomorphic or anti-isomorphic. There exist 1,309 nonequivalent semigroups of order up to five. But there are 15,973 nonequivalent semigroups of order six, among which 1,373 are monoids and 14,600 are non-unital (Distler and Kelsey 2009, 2014). In view of these large numbers, it is natural to question the existence of a minimal non-finitely based semigroup that is not equivalent to any of $\mathcal{A}_2^1$, $\mathcal{B}_2^1$, $\mathcal{A}_2^{\mathsf{g}}$, and $\mathcal{L}_3$.

Problem 1.16 Identify all minimal non-finitely based semigroups.

The solution to this problem obviously requires complete knowledge of the finite basis property of all semigroups of order six. But significant work in this direction has not been undertaken, perhaps due to the sheer number of semigroups involved.

In 2007, I initiated the investigation of Problem 1.16 after Volkov (email communication, 3 Apr 2006) informed me of the then new example of minimal non-finitely based semigroup $\mathcal{A}_2^{\mathsf{g}}$—the third example following $\mathcal{B}_2^1$ and $\mathcal{A}_2^1$. Over the next few years, Li and Zhang joined the project. During this period, the semigroup $\mathcal{L}_3$ was found by me and shown by Zhang to be non-finitely based. The discovery of a fourth non-finitely based semigroup of order six was highly unanticipated at that time since the project was extremely close to completion: the finite basis property of only two of the 15,973 nonequivalent semigroups of order six remained to be determined, and $\mathcal{L}_3$ was one of them. Shortly after, Problem 1.16 was completely solved and the result was announced in Lee et al. (2012).

Theorem 1.17 *Up to isomorphism, the semigroups* $\mathcal{A}_2^1$, $\mathcal{B}_2^1$, $\mathcal{A}_2^{\mathsf{g}}$, *and* $\mathcal{L}_3$ *are precisely all minimal non-finitely based semigroups.*

Identification by anti-isomorphism is immaterial since each of the four minimal non-finitely based semigroups is anti-isomorphic to itself.

The complete proof of Theorem 1.17 was published in two parts: $\mathcal{A}_2^1$ and $\mathcal{B}_2^1$ were first shown to be the only non-finitely based monoids of order six (Lee and Li 2011), then $\mathcal{A}_2^{\mathsf{g}}$ and $\mathcal{L}_3$ were shown to be the only non-finitely based non-unital semigroups of order six (Lee and Zhang 2015). Establishing these results required many hereditarily finitely based identity systems in addition to the examination of numerous sporadic cases.

1.4 Non-Finite Basis Property

1.4.1 Finite Basis Problem for Finite Semigroups

Finite members from well-studied classes of algebras such as groups (Oates and Powell 1964), associative rings (Kruse 1973; L'vov 1973), Lie rings (Bahturin and Ol'shanskiĭ 1975), and lattices (McKenzie 1970) are finitely based. But as documented in Sect. 1.3.3, not all finite semigroups are finitely based. The finite basis problem has been settled for all semigroups of order up to six but remains open in general for larger semigroups.

Problem 1.18 Classify all finitely based finite semigroups.

The algorithmic version of Problem 1.18, posed by Tarski (1968) for finite algebras, is commonly known as *Tarski's finite basis problem* for finite semigroups.

Problem 1.19 (Shevrin and Volkov 1985, Question 8.3; Volkov 2001, Problem 2.1) Is there an algorithm that when given a finite semigroup, decides whether or not it is finitely based?

Problem 1.19 presently remains open. In general, however, Tarski's finite basis problem for finite algebras is undecidable (McKenzie 1996).

Since consideration of Problem 1.18 began in the 1960s, a few classes of finitely based semigroups have been found. Besides finite groups (Oates and Powell 1964), some other important examples include

- commutative semigroups (Perkins 1969);
- periodic permutative semigroups (Perkins 1969), in particular, nilpotent semigroups;
- idempotent semigroups (Birjukov 1970; Fennemore 1970; Gerhard 1970);
- Brandt semigroups over a finitely based group of finite exponent (Volkov 2019), in particular, finite Brandt semigroups;
- finite central locally orthodox completely regular semigroups (Kad'ourek 2021), in particular, finite orthodox completely regular semigroups (Rasin 1982) and finite central simple semigroups (Shevrin and Volkov 1985, Theorem 20.3);
- aperiodic Rees–Suschkewitsch semigroups (Chap. 3).

Other positive contributions toward solving Problem 1.18 include some hereditarily finitely based identity systems—such as those used to establish Theorems 1.13 and 1.17—and a few other sufficient conditions for the finite basis property (Lee and Li 2011; Lee and Zhang 2015; O.B. Sapir 2015a).

1.4.2 Establishing the Non-Finite Basis Property

Despite years of work, overall progress on solving Problem 1.18 has remained "a collection of isolated theorems rather than a unified theory" (Volkov 2001, page 190). On the other hand, techniques on establishing the non-finite basis property have flourished in a few directions. The following is a brief description of three popular methods and a comparison of their ranges of applicability. Refer to Volkov (2001) for a detailed survey of other methods and related results obtained over the past few decades.

1.4.2.1 Critical Rees Matrix Semigroups

For any finite semigroup S, let $\mathscr{V}^{(n)}_{\mathsf{sem}}\{S\}$ denote the variety defined by identities of S that involve at most n distinct variables. It is clear that the inclusions

$$\mathscr{V}^{(1)}_{\mathsf{sem}}\{S\} \supseteq \mathscr{V}^{(2)}_{\mathsf{sem}}\{S\} \supseteq \mathscr{V}^{(3)}_{\mathsf{sem}}\{S\} \supseteq \cdots \supseteq \mathscr{V}_{\mathsf{sem}}\{S\}$$

hold. Since $\mathscr{V}^{(n)}_{\mathsf{sem}}\{S\}$ is finitely based for all $n \geq 1$ (Birkhoff 1935), the semigroup S is finitely based if and only if $\mathscr{V}^{(n)}_{\mathsf{sem}}\{S\} = \mathscr{V}_{\mathsf{sem}}\{S\}$ for some $n \geq 1$. It follows that a finite semigroup S is non-finitely based if and only if for all $n \geq 1$, the inclusion $\mathscr{V}^{(n)}_{\mathsf{sem}}\{S\} \supseteq \mathscr{V}_{\mathsf{sem}}\{S\}$ is proper, that is, there exists some semigroup $S_n \in \mathscr{V}^{(n)}_{\mathsf{sem}}\{S\} \backslash \mathscr{V}_{\mathsf{sem}}\{S\}$, called a *critical semigroup* for S.

Over the years, several types of critical semigroups have been constructed to exhibit non-finitely based examples of inverse semigroups (Kleĭman 1979), restriction semigroups (Jones 2013), and pseudovarieties (Cowan and Reilly 1995; Kleĭman 1982; Repnitskiĭ and Volkov 1998; Rhodes and Steinberg 2005; Trotter and Volkov 1996; Volkov 1998). Within the class of all semigroups, the most prominent type of critical semigroups has been Rees matrix semigroups. Kad'ourek (2018), Mashevitzky (1983, 1999, 2007, 2012), and Volkov (1989b) have constructed critical Rees matrix semigroups to present several examples of non-finitely based finite semigroups. In particular, the following sufficient condition for the non-finite basis property is a representative result.

Theorem 1.20 (Volkov 1989b; See Also Volkov 2001, Theorem 4.7) *Let S be any semigroup and S_E be the subsemigroup of S generated by its idempotents. Suppose that $\mathcal{A}_2 \in \mathscr{V}_{\mathsf{sem}}\{S\}$ and that some group exists in $\mathscr{V}_{\mathsf{sem}}\{S\} \backslash \mathscr{V}_{\mathsf{sem}}\{S_E\}$. Then S is non-finitely based.*

The extensive list of semigroups to which Theorem 1.20 is applicable includes

- the semigroup $\mathsf{M}_n(\mathfrak{F})$ of $n \times n$ matrices over any finite field $\mathfrak{F}$ with $n \geq 2$;
- the semigroup $\mathcal{T}_n$ of transformations of $\{1, 2, \ldots, n\}$ with $n \geq 3$;

- the semigroup $\mathcal{BR}_n$ of binary relations on $\{1, 2, \ldots, n\}$ with $n \geq 2$;
- the direct product $\mathcal{A}_2 \times G$ of the semigroup $\mathcal{A}_2$ with any nontrivial finite group G, in particular, the semigroup $\mathcal{A}_2 \times \mathcal{Z}_p$ that generates the limit variety $\mathbf{A_2} \vee \mathbf{Z}_p$ in Theorem 1.8;
- any Rees matrix semigroup $\mathcal{M}^0(I, G, \Lambda; P)$ such that the matrix P has a 2×2 submatrix with precisely one zero entry and $G \notin \mathscr{V}_{\mathsf{sem}}\{G_P\}$, where G_P is the subgroup of G generated by the nonzero entries of P.

Critical Rees matrix semigroups have also been constructed to provide non-finitely based examples of varieties (Sapir and Volkov 1994) and pseudovarieties (Volkov 1995, 1996).

1.4.2.2 Inherently Non-Finitely Based Finite Semigroups

An *isoterm* for a semigroup is a word that cannot be paired with another word to form a nontrivial identity for the semigroup. M.V. Sapir (1987b) presented an elegant characterization of inherently non-finitely based finite semigroups using *Zimin's words* $\mathbf{z}_1, \mathbf{z}_2, \mathbf{z}_3, \ldots$ defined by $\mathbf{z}_1 = x_1$ and $\mathbf{z}_n = \mathbf{z}_{n-1} x_n \mathbf{z}_{n-1}$ for all $n \geq 2$ (Zimin 1982).

Theorem 1.21 (M.V. Sapir 1987b, 2014) *A finite semigroup S is inherently non-finitely based if and only if for all $n \leq |S|^3$, the word $\mathbf{z}_n$ is an isoterm for S.*

Using this theorem, M.V. Sapir (1987b) deduced that the minimal non-finitely based semigroups $\mathcal{A}_2^1$ and $\mathcal{B}_2^1$ are inherently non-finitely based; the other two minimal non-finitely based semigroups $\mathcal{A}_2^{\mathsf{g}}$ and $\mathcal{L}_3$, however, are not inherently non-finitely based because they satisfy the identity $xyxzx \approx xzxyx$, so that the word $\mathbf{z}_3 = x_1x_2x_1x_3x_1x_2x_1$ is not an isoterm. It follows that a finite semigroup is non-finitely based if the variety it generates contains either $\mathcal{A}_2^1$ or $\mathcal{B}_2^1$. By considering the matrix representations of $\mathcal{A}_2$ and $\mathcal{B}_2$ in Sect. 1.2.2, the monoids $\mathcal{A}_2^1 = \mathcal{A}_2 \cup \left\{ \left[\begin{smallmatrix} 1 & 0 \\ 0 & 1 \end{smallmatrix}\right] \right\}$ and $\mathcal{B}_2^1 = \mathcal{B}_2 \cup \left\{ \left[\begin{smallmatrix} 1 & 0 \\ 0 & 1 \end{smallmatrix}\right] \right\}$ are clearly embeddable in $\mathsf{M}_n(\mathfrak{F})$ for each $n \geq 2$, whence these matrix semigroups are all inherently non-finitely based. Other similar examples include

- the semigroup $\mathsf{TrM}_n(\mathfrak{B})$ of upper triangular $n \times n$ matrices over the Boolean semiring $\mathfrak{B}$ with $n \geq 3$ (Li and Luo 2011; Volkov and Gol'dberg 2004);
- the semigroup $\mathcal{T}_n$ with $n \geq 3$ (M.V. Sapir 1987b);
- the semigroup $\mathcal{OT}_n$ of order-preserving transformations of $\{1, 2, \ldots, n\}$ with $n \geq 3$;
- the semigroup $\mathcal{T}_{n,k}$ of transformations of $\{1, 2, \ldots, n\}$ of rank up to k with $n > k \geq 3$.

The monoid $\mathcal{A}_2^1$ is embeddable in the last two semigroups because

$$\mathcal{A}_2 = \langle \mathsf{a}, \mathsf{e} \rangle \cong \begin{cases} \left\langle \begin{pmatrix} 1\ 2\ 3\ 4\ 5 \cdots n \\ 1\ 1\ 2\ 2\ 2 \cdots 2 \end{pmatrix}, \begin{pmatrix} 1\ 2\ 3\ 4\ 5 \cdots n \\ 1\ 3\ 3\ 3\ 3 \cdots 3 \end{pmatrix} \right\rangle \subset \mathcal{OT}_n & \text{with } n \geq 3, \\ \left\langle \begin{pmatrix} 1\ 2\ 3\ 4\ 5 \cdots n \\ 1\ 1\ 2\ 1\ 1 \cdots 1 \end{pmatrix}, \begin{pmatrix} 1\ 2\ 3\ 4\ 5 \cdots n \\ 1\ 3\ 3\ 1\ 1 \cdots 1 \end{pmatrix} \right\rangle \subset \mathcal{T}_{n,k} & \text{with } n > k \geq 3. \end{cases}$$

M.V. Sapir (1987a) also provided an efficient algorithmic description of inherently non-finitely based finite semigroups. In contrast, there is no algorithm that can decide if a finite groupoid is inherently non-finitely based (McKenzie 1996). Using the result of M.V. Sapir (1987a), Jackson (2002) deduced a description of all inherently non-finitely based finite semigroups that are minimal with respect to division; since every inherently non-finitely based finite semigroup has at least one minimal inherently non-finitely based divisor, this provides another algorithmic characterization of inherently non-finitely based finite semigroups.

The aforementioned result of M.V. Sapir (1987a) also enabled Volkov and Gol'dberg (2003) to prove that the semigroup $\mathsf{TrM}_n(\mathfrak{F})$ of upper triangular $n \times n$ matrices over any finite field $\mathfrak{F}$ is inherently non-finitely based if and only if $n \geq 4$ and $|\mathfrak{F}| \geq 3$. Unlike all other inherently non-finitely based examples listed above, the variety generated by $\mathsf{TrM}_n(\mathfrak{F})$ excludes both $\mathcal{A}_2^1$ and $\mathcal{B}_2^1$.

1.4.2.3 Syntactic Method

Recall from Sect. 1.4.2.1 that a finite semigroup S is non-finitely based if and only if the exclusions $\mathscr{V}^{(1)}_{\mathsf{sem}}\{S\}, \mathscr{V}^{(2)}_{\mathsf{sem}}\{S\}, \mathscr{V}^{(3)}_{\mathsf{sem}}\{S\}, \ldots \nsubseteq \mathscr{V}_{\mathsf{sem}}\{S\}$ hold, and these exclusions can be confirmed by the presence of critical semigroups for S. Another method to establish these exclusions is by syntactic arguments involving identities satisfied by S. This can be achieved by locating an infinite list $\sigma_1, \sigma_2, \sigma_3, \ldots$ of identities satisfied by S, called *critical identities* of S, with the property that for each $n \geq 1$, there exists some sufficiently large k such that σ_k is not deducible from the identities of S that involve at most n distinct variables. This approach brought forth the first sufficient condition for the non-finite basis property.

Theorem 1.22 (Perkins 1969) *Suppose that S is any monoid with isoterms $xyzyx$ and $xzyxy$ such that S satisfies the identities*

$$x\left(\prod_{i=1}^{m} y_i\right)x\left(\prod_{i=m}^{1} y_i\right) \approx x\left(\prod_{i=m}^{1} y_i\right)x\left(\prod_{i=1}^{m} y_i\right), \quad m \in \{2, 3, 4, \ldots\} \tag{1.3}$$

but violates the identities $x^2y \approx yx^2$ and $xyxy \approx xy^2x$. Then S is non-finitely based.

In the proof of Theorem 1.22 (Perkins 1969, Sect. 3), the identities (1.3) serve as critical identities for the semigroup S. Recently, O.B. Sapir (2015b) generalized this theorem by showing that the assumption for S to violate the identity $x^2y \approx yx^2$ is superfluous.

By applying Theorem 1.22, Perkins (1969) proved that the Brandt monoid $\mathcal{B}_2^1$ is non-finitely based. In fact, for any monoid M that satisfies the identities (1.3), such as any commutative monoid, the direct product $\mathcal{B}_2^1 \times M$ is also non-finitely based.

Perkins (1969) also constructed another example of a non-finitely based finite semigroup. For any set $\mathscr{W}$ of words over a countably infinite alphabet $\mathscr{A}$, let $\mathcal{R}_\mathsf{Q}\mathscr{W}$ denote the Rees quotient of the free monoid $\mathscr{A}_\varnothing^+ = \mathscr{A}^+ \cup \{\varnothing\}$ over the ideal of all words that are not factors of any word in $\mathscr{W}$. Then the monoid

$$\mathcal{P}_{25} = \mathcal{R}_\mathsf{Q}\{xyxy, xyzyx, xzyxy, x^2z\}$$

of order 25 is non-finitely based by Theorem 1.22.

Rees quotients of $\mathscr{A}_\varnothing^+$ constitute a large source of examples for the investigation of the finite basis problem and of varieties of monoids; their importance will be apparent in Part III, where varieties of monoids are the main object of study. As a possible approach to Tarski's finite basis problem for finite semigroups (Problem 1.19), M.V. Sapir (2014) suggested specializing the investigation to the class of Rees quotients of $\mathscr{A}_\varnothing^+$.

Problem 1.23 (M.V. Sapir 2014, Problem 3.10.10; Shevrin and Volkov 1985, Question 7.1; Volkov 2001, Problem 4.1) Is there an algorithm that when given a finite set $\mathscr{W}$ of words, decides whether or not the semigroup $\mathcal{R}_\mathsf{Q}\mathscr{W}$ is finitely based?

A negative solution to this open problem clearly implies a negative solution to Problem 1.19.

Since the pioneering work of Perkins (1969), the study of Rees quotients of $\mathscr{A}_\varnothing^+$ has attracted much interest; see, for example, Jackson (1999, 2001, 2005a,b), Jackson and Sapir (2000), Li and Luo (2015), Li et al. (2013), and O.B. Sapir (1997, 2000, 2015b, 2016, 2019). These numerous investigations resulted in the discovery of many sufficient conditions similar to Theorem 1.22 and recipes to construct explicit examples of finitely based and non-finitely based semigroups. A partial positive solution to Problem 1.23 was also found.

Theorem 1.24 (O.B. Sapir 2000, 2016) *For any word* $\mathbf{w}$ *with at most two non-simple variables, it is decidable if the monoid* $\mathcal{R}_\mathsf{Q}\{\mathbf{w}\}$ *is finitely based. Specifically, for any word* $\mathbf{w} \in \{x, y\}^+$ *that begins with* x*, the monoid* $\mathcal{R}_\mathsf{Q}\{\mathbf{w}\}$ *is finitely based if and only if* $\mathbf{w}$ *is either* x^m, $x^m y^n$, *or* $x^m y x^n$ *for some* $m, n \geq 1$.

It follows that the monoid $\mathcal{R}_\mathsf{Q}\{xyxy\}$ of order nine is non-finitely based; in fact, it is the smallest Rees quotient of $\mathscr{A}_\varnothing^+$ that is non-finitely based (Jackson and Sapir 2000). Li and Luo (2015) also described certain words $\mathbf{w} \in \{x, y, z\}^+$ for which $\mathcal{R}_{\mathcal{Q}}\{\mathbf{w}\}$ is finitely based.

The syntactic method has also been used to establish other results not involving Rees quotients of $\mathscr{A}_{\varnothing}^+$; these include some sufficient conditions for the non-finite basis property (see Chen et al. 2016; Zhang and Luo 2016) and examples of non-finitely based finite semigroups, such as

- the monoid $\mathcal{A}_2^1$ (Trahtman 1987);
- the semigroup $\mathcal{L}_3 \times \mathcal{Z}_n$ with $n \geq 1$ (Lee 2012a);
- the monoid $\mathcal{L}_\ell^1$ with $\ell \geq 3$ (Mikhailova and Sapir 2018; O.B. Sapir 2018; Zhang 2013);
- the semigroup $\mathcal{E}_n$ of extensive transformations of $\{1, 2, \ldots, n\}$ with $n \geq 5$ (Gol'dberg 2007);
- the semigroup $\mathcal{PEI}_n$ of partial extensive injective transformations of $\{1, 2, \ldots, n\}$ with $n \geq 3$ (Hu et al. 2015).

A sufficient condition for the non-finite basis property that is found by the syntactic method would typically require a semigroup to satisfy some assumptions beyond the satisfaction of an infinite list of critical identities. In many cases—Theorem 1.22 being a prime example—these assumptions involve either isoterms, violation of identities, or both. Further, most if not all sufficient conditions syntactically found so far are only applicable to monoids.

In Chap. 5, the syntactic method is employed to establish a new sufficient condition for the non-finite basis property that is different in several aspects.

Theorem 1.25 *Suppose that S is any semigroup that satisfies the identities*

$$x^n y_1^n y_2^n \cdots y_m^n x^n \approx x^n y_m^n y_{m-1}^n \cdots y_1^n x^n, \quad m \in \{2, 3, 4, \ldots\} \tag{1.4}$$

for some fixed $n \geq 2$ and that $\mathcal{L}_3 \in \mathscr{V}_{\mathsf{sem}}\{S\}$. Then S is non-finitely based.

One big difference between Theorem 1.25 and many previous sufficient conditions is that its formulation does not require any isoterms, although the condition $\mathcal{L}_3 \in \mathscr{V}_{\mathsf{sem}}\{S\}$ implicitly implies that every simple word is an isoterm for S. Another difference is that the theorem applies to only non-unital semigroups. Indeed, any monoid M that satisfies the identities (1.4) has commuting idempotents, but this implies the exclusion $\mathcal{L}_3 \notin \mathscr{V}_{\mathsf{sem}}\{M\}$ because the two nonzero idempotents in $\mathcal{L}_3$ do not commute.

The statement of Theorem 1.25 is also simpler than quite a number of previous sufficient conditions since it requires only the single assumption $\mathcal{L}_3 \in \mathscr{V}_{\mathsf{sem}}\{S\}$ in addition to the satisfaction of critical identities. Further, it is shown that for any semigroup S that is uniformly periodic, say S satisfies the identity $x^{2n} \approx x^n$ for some $n \geq 2$, the inclusion $\mathcal{L}_3 \in \mathscr{V}_{\mathsf{sem}}\{S\}$ holds if and only if S violates the identity $(x^n y^n x^n)^{n+1} \approx x^n y^n x^n$.

Consequently, for uniformly periodic semigroups, Theorem 1.25 can be stated entirely with satisfaction and violation of identities.

Theorem 1.26 *Any semigroup that satisfies the identities*

$$x^{2n} \approx x^n, \quad x^n y_1^n y_2^n \cdots y_m^n x^n \approx x^n y_m^n y_{m-1}^n \cdots y_1^n x^n, \quad m \in \{2,3,4,\ldots\}$$

for some fixed $n \geq 2$ but violates the identity $(x^n y^n x^n)^{n+1} \approx x^n y^n x^n$ is non-finitely based.

It is easily verified that the $\mathscr{J}$-trivial semigroup $\mathcal{L}_3$ is non-finitely based by Theorem 1.25. In fact, this theorem can be used to solve the finite basis problem for all of the semigroups introduced in Remark 1.15:

$$\mathcal{L}_\ell = \langle \mathsf{e}, \mathsf{f} \mid \mathsf{e}^2 = \mathsf{e},\ \mathsf{f}^2 = \mathsf{f},\ \underbrace{\mathsf{efefe}\cdots}_{\text{length } \ell} = 0 \rangle.$$

Specifically, the semigroup $\mathcal{L}_2 = \mathcal{A}_0$ is finitely based (Edmunds 1980), while the semigroups $\mathcal{L}_3, \mathcal{L}_4, \mathcal{L}_5, \ldots$ are non-finitely based by Theorem 1.25. It follows that for any $\ell \geq 3$ and any semigroup T that satisfies the identities (1.4), the direct product $\mathcal{L}_\ell \times T$ is non-finitely based. In particular, the direct product

$$\mathcal{L}_{\ell,n} = \mathcal{L}_\ell \times \mathcal{Z}_n$$

is non-finitely based for all $\ell \geq 3$ and $n \geq 1$. On the other hand, the semigroup $\mathcal{L}_{2,n}$ is finitely based by either Kublanovskiĭ (2011) or Lee and Volkov (2011).

Recently, O.B. Sapir (2015b, Lemma 2.5) established a method that can be used to deduce the proofs (but not the statements) of all sufficient conditions in the present subsubsection.

1.4.2.4 A Comparison of the Three Methods

For the three methods of establishing the non-finite basis property that were surveyed in Sects. 1.4.2.1–1.4.2.3, their ranges of applicability are displayed in Fig. 1.2 (see Volkov 2001, Fig. 3). A semigroup appearing in a box indicates that it can be shown to be non-finitely based by the method labeling the box. The vertical dashed line separates aperiodic semigroups from semigroups with nontrivial subgroups. The only semigroup in this figure that has not been introduced is the symmetric group $\mathcal{S}ym_n$ over n symbols. For each $n \geq 3$, the direct product $\mathcal{R}_Q\{xyx\} \times \mathcal{S}ym_n$ is non-finitely based by Theorem 1.69, a result that is established in Sect. 14.1 by the syntactic method.

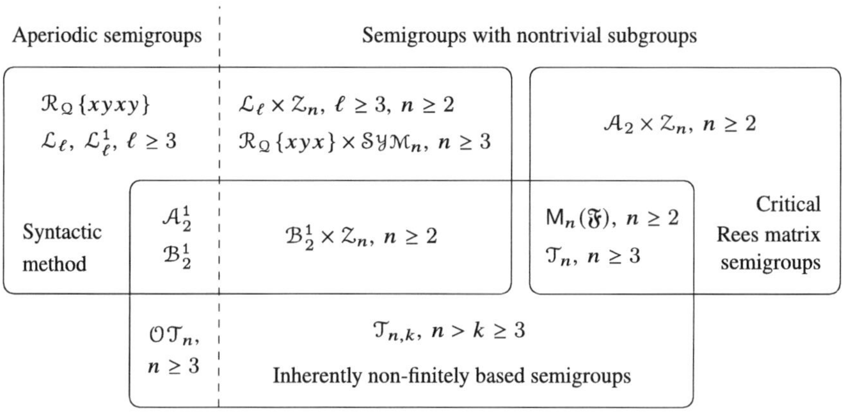

Fig. 1.2 The ranges of applicability of the three standard methods

1.4.3 Irredundant Identity Bases

An identity basis for an algebra is *irredundant* if each of its proper subsets fails to be an identity basis for the algebra. Every finitely based algebra has some irredundant identity basis—simply eliminate redundant identities from any finite identity basis, one by one, until no redundancies exist. The contemplation of irredundant identity bases was due to the presumption that a finite algebra without finite identity bases might at least have an irredundant one. However, this optimism was refuted by subsequent examples of non-finitely based finite semigroups without irredundant identity bases (Mashevitzky 1983; M.V. Sapir 1991).

On the other hand, it was unknown if a non-finitely based finite semigroup can have an infinite irredundant identity basis, and the existence of such a semigroup has been questioned since the 1970s; see Shevrin and Volkov (1985, Question 8.6) and Volkov (2001, Problem 2.6). This question remained open until after the turn of the millennium, when Jackson (2005a) established some sufficient conditions for a semigroup to possess an infinite irredundant identity basis and applied them to exhibit a few finite examples, the smallest of which is the monoid $\mathcal{R}_{\mathcal{Q}}\{xyxy\}$ of order nine.

The investigation is continued in Chap. 6 with the property of algebras not having any irredundant identity bases. The chapter begins by establishing in Sect. 6.1 the first and presently only known sufficient condition under which a non-finitely based finite algebra of any signature type has no irredundant identity bases. Then in Sects. 6.2–6.5, an explicit identity basis is found for the non-finitely based semigroup $\mathcal{L}_{3,n} = \mathcal{L}_3 \times \mathcal{Z}_n$ for any $n \geq 1$.

Theorem 1.27 *The identities*

$$x^{n+2} \approx x^2, \quad x^{n+1}yx^{n+1} \approx xyx, \quad xhykxty \approx yhxkytx, \tag{1.5a}$$

$$x\left(\prod_{i=1}^{m}(y_ih_iy_i)\right)x \approx x\left(\prod_{i=m}^{1}(y_ih_iy_i)\right)x, \quad m \in \{2,3,4,\ldots\} \tag{1.5b}$$

constitute an identity basis for the semigroup $\mathcal{L}_{3,n}$.

Many examples of non-finitely based finite semigroups have been discovered since the first published examples (Perkins 1969), but identity bases for most of them have not been explicitly given. Prior to $\mathcal{L}_{3,n}$, the only non-finitely based finite semigroups whose identity bases have been explicitly found are the Rees–Suschkewitsch semigroups $\mathcal{A}_2 \times \mathcal{Z}_n$ with $n \geq 2$ (Lee and Volkov 2011) and a certain monoid (Jackson 2005a, Proposition 4.1).

The explicit identity basis for the semigroup $\mathcal{L}_{3,n}$ is easily shown to be redundant. In fact, using the sufficient condition from Sect. 6.1, it is shown in Sect. 6.6 that $\mathcal{L}_{3,n}$ has no irredundant identity bases. This result is not too surprising on its own, but it will play a vital role in Part II in the construction of some counterintuitive examples of involution semigroups.

The identity basis (1.5) for $\mathcal{L}_{3,n}$ also allows one to easily check if a finite semigroup belongs to the variety $\mathscr{V}_{\mathsf{sem}}\{\mathcal{L}_{3,n}\}$ even though it is non-finitely based.

Theorem 1.28 (Lee 2018c) *A nontrivial finite semigroup* S *belongs to the variety* $\mathscr{V}_{\mathsf{sem}}\{\mathcal{L}_{3,n}\}$ *if and only if* S *satisfies both the identities* (1.5a) *and the identity from* (1.5b) *with* $m = |S|$.

1.5 Varieties of Involution Semigroups

A unary operation * on a semigroup S is an *involution* if S satisfies the identities

$$(x^\star)^\star \approx x \text{ and } (xy)^\star \approx y^\star x^\star.$$

An *involution semigroup* is a pair $\langle S, {}^\star\rangle$, where the *reduct* S is a semigroup with involution *. Commutative semigroups $\langle S, {}^{\mathsf{tr}}\rangle$ with the trivial involution ^{tr}, groups $\langle G, {}^{-1}\rangle$ with inversion $^{-1}$, and the matrix semigroups $\langle \mathsf{M}_n(\mathfrak{F}), {}^T\rangle$ with transposition ^{T} are examples of involution semigroups.

Any involution on a finite semigroup S, when treated as a permutation of S, is a product of disjoint transpositions. Therefore, when defining an involution on a finite semigroup, it suffices to specify only the nontrivial transpositions. For instance, the Rees quotient $\mathcal{R}_Q\{xyxy, yxyx\}$ admits the involution * defined by

$$\begin{aligned} &\mathtt{0}^\star = \mathtt{0}, \quad x^\star = x, \quad (xy)^\star = yx, \quad (xyx)^\star = xyx, \quad (xyxy)^\star = yxyx, \\ &\mathtt{1}^\star = \mathtt{1}, \quad y^\star = y, \quad (yx)^\star = xy, \quad (yxy)^\star = yxy, \quad (yxyx)^\star = xyxy; \end{aligned}$$

this involution can be given by the transpositions $(xy)^\star = yx$ and $(xyxy)^\star = yxyx$.

Example 1.29 Some small involution semigroups include

$$\begin{aligned} &\langle \mathcal{A}_0^1, {}^\star\rangle \text{ with } \mathsf{e}^\star = \mathsf{f}, \qquad \langle \mathcal{A}_2^1, {}^\star\rangle \text{ with } (\mathsf{ae})^\star = \mathsf{ea}, \qquad \langle \mathcal{A}_2^{\mathsf{g}}, {}^\star\rangle \text{ with } (\mathsf{ae})^\star = \mathsf{ea}, \\ &\langle \mathcal{L}_3, {}^\star\rangle \text{ with } (\mathsf{ef})^\star = \mathsf{fe}, \qquad \langle \mathcal{B}_2^1, {}^\star\rangle \text{ with } (\mathsf{ab})^\star = \mathsf{ba}, \qquad \langle \mathcal{B}_2^1, {}^{-1}\rangle \text{ with } \mathsf{a}^{-1} = \mathsf{b}. \end{aligned}$$

Note that the last two involution semigroups in Example 1.29 share the same semigroup reduct. In general, it is possible for a semigroup to admit two or more involutions. For instance, if $\mathfrak{R} \in \{1, 2, \ldots, n\}$ is a *square root of unity modulo n* in the sense that $\mathfrak{R}^2 \equiv 1 \pmod{n}$, then endowing the cyclic group $\mathcal{Z}_n$ with the unary operation $x \mapsto x^{\mathfrak{R}}$ results in an involution semigroup, denoted by $\langle \mathcal{Z}_n, {}^{\mathfrak{R}}\rangle$. In particular, $\langle \mathcal{Z}_n, {}^1\rangle = \langle \mathcal{Z}_n, {}^{\mathsf{tr}}\rangle$ for all $n \geq 1$ and $\langle \mathcal{Z}_n, {}^{n-1}\rangle = \langle \mathcal{Z}_n, {}^{-1}\rangle$ for all $n \geq 3$. Conversely, any involution semigroup with semigroup reduct $\mathcal{Z}_n$ is isomorphic to $\langle \mathcal{Z}_n, {}^{\mathfrak{R}}\rangle$ for some $\mathfrak{R}$ from

$$\mathsf{sq}(n) = \{\mathfrak{R} \in \{1, 2, \ldots, n\} \mid \mathfrak{R}^2 \equiv 1 \pmod{n}\},$$

and the number of such involution semigroups is

$$|\mathsf{sq}(n)| = \begin{cases} 2^{r+1} & \text{if } n \equiv 0 \pmod{8}, \\ 2^{r-1} & \text{if } n \equiv \pm 2 \pmod{8}, \\ 2^r & \text{otherwise,} \end{cases}$$

where r is the number of distinct prime factors of n (Hage and Harju 2009, Theorem 2).

An *inverse semigroup* is an involution semigroup $\langle S, {}^\star\rangle$ that satisfies the identities

$$xx^\star x \approx x \quad \text{and} \quad xx^\star yy^\star \approx yy^\star xx^\star.$$

Examples of inverse semigroups include the inverse Brandt monoid $\langle \mathcal{B}_2^1, {}^{-1}\rangle$ and all involution groups $\langle G, {}^{-1}\rangle$, while involution semigroups such as $\langle \mathsf{M}_n(\mathfrak{F}), {}^T\rangle$ and the first five involution semigroups in Example 1.29 are not inverse semigroups.

1.5.1 Equational Properties of Involution Semigroups

The finite basis problem for inverse semigroups was examined in the 1970s shortly after the problem was considered for semigroups (Problem 1.18). The initial work of Kleĭman

(1977, 1979) was followed by a long lapse of over a decade until the investigations by Kad'ourek (1992, 2003) and M.V. Sapir (1993). In short, with respect to the finite basis problem, inverse semigroups have received much less attention than semigroups. Although the situation with involution semigroups is similar, the topic has gained much interest lately; see, for example, Ashikmin et al. (2015), Auinger et al. (2012a,b), Auinger et al. (2014), Auinger et al. (2015), Dolinka (2010), Gao et al. (2020a,b, 2021, 2022), Han et al. (2021), Jackson and Volkov (2010), Volkov (2015), Zhang and Luo (2020), Zhang et al. (2017), and Zhang et al. (2020). The investigations so far have predominantly been concerned with non-finitely based involution semigroups.

Some techniques developed to tackle the finite basis problem for semigroups have been successfully adapted to involution semigroups. A common way to establish the non-finite basis property is the critical semigroups method. Critical Rees matrix involution semigroups have been constructed by Auinger et al. (2012b) to produce many examples of non-finitely based involution semigroups. One of their most useful results is an involution version of Volkov's $\mathcal{A}_2$-theorem (Theorem 1.20), the statement of which requires the Rees matrix semigroup

$$\mathcal{A}_3 = \mathcal{M}^0\left(\{1, 2, 3\}, \{1\}, \{1, 2, 3\}; \begin{bmatrix} 1 & 1 & 1 \\ 1 & 1 & 0 \\ 1 & 0 & 1 \end{bmatrix}\right)$$

of order ten with involution * given by $(i, 1, j)^\star = (j, 1, i)$ and $0^\star = 0$.

Theorem 1.30 (Auinger et al. 2012b, Theorem 2.2) *Let $\langle S, {}^\star \rangle$ be any involution semigroup and $\langle S_H, {}^\star \rangle$ be the involution subsemigroup of $\langle S, {}^\star \rangle$ generated by the elements in $H = \{xx^\star \mid x \in S\}$. Suppose that $\langle \mathcal{A}_3, {}^\star \rangle \in \mathscr{V}_{\mathsf{inv}}\{\langle S, {}^\star \rangle\}$ and that some group exists in $\mathscr{V}_{\mathsf{inv}}\{\langle S, {}^\star \rangle\} \backslash \mathscr{V}_{\mathsf{inv}}\{\langle S_H, {}^\star \rangle\}$. Then $\langle S, {}^\star \rangle$ is non-finitely based.*

This theorem has been applied to establish the non-finite basis property of the matrix involution semigroup $\langle \mathsf{M}_2(\mathfrak{F}), {}^T \rangle$ over any finite field $\mathfrak{F}$ of odd characteristic (Auinger et al. 2012b, Theorem 3.8) and a few involution semigroups of partitions (Auinger et al. 2012a, Theorem 2.12).

By constructing certain critical inverse semigroups of partial transformations, Kleĭman (1979) proved that the inverse Brandt monoid $\langle \mathcal{B}_2^1, {}^{-1} \rangle$ is non-finitely based. Kad'ourek (1992) generalized this result by showing that $\langle \mathcal{B}_2^1, {}^{-1} \rangle$ has no irredundant identity bases even though it is unknown whether or not its reduct $\mathcal{B}_2^1$ has such an identity basis. This leads to the following related open problem.

Problem 1.31 Is there a finite involution semigroup with an infinite irredundant identity basis?

Kad'ourek (2003) also established a stronger equational property of $\langle \mathcal{B}_2^1, {}^{-1}\rangle$: any finite inverse semigroup $\langle S, {}^{-1}\rangle$ with solvable subgroups such that $\langle \mathcal{B}_2^1, {}^{-1}\rangle \in \mathscr{V}_{\mathsf{inv}}\{\langle S, {}^{-1}\rangle\}$ is non-finitely based. However, unlike its semigroup reduct $\mathcal{B}_2^1$, which is inherently non-finitely based (M.V. Sapir 1987b), the inverse semigroup $\langle \mathcal{B}_2^1, {}^{-1}\rangle$ is weakly finitely based. In fact, every finite inverse semigroup is weakly finitely based (M.V. Sapir 1993). Nevertheless, Auinger et al. (2012b) generalized one half of M.V. Sapir's characterization of inherently non-finitely based finite semigroups (Theorem 1.21) to involution semigroups.

Theorem 1.32 (Auinger et al. 2012b, Theorem 2.3; Dolinka 2010, Theorem 1.1) *Let $\langle S, {}^\star\rangle$ be any involution semigroup in some locally finite variety. Suppose that every Zimin's word $\mathbf{z}_n$ is an isoterm for $\langle S, {}^\star\rangle$. Then $\langle S, {}^\star\rangle$ is inherently non-finitely based.*

It follows that $\langle \mathcal{A}_2^1, {}^\star\rangle$, $\langle \mathcal{B}_2^1, {}^\star\rangle$, and several involution semigroups of matrices over either the Boolean semiring or various finite fields are inherently non-finitely based (Auinger et al. 2012b). It is unknown if the converse of Theorem 1.32 holds, but Dolinka (2010) has shown that if $\langle S, {}^\star\rangle$ is any inherently non-finitely based finite involution semigroup such that $\langle \mathcal{B}_2^1, {}^{-1}\rangle \notin \mathscr{V}_{\mathsf{inv}}\{\langle S, {}^\star\rangle\}$, then every Zimin's word is an isoterm for $\langle S, {}^\star\rangle$.

The motivation for investigating involution semigroups is the same as that for semigroups—to study natural generalizations of groups—only that the involution is meant to describe the intricate symmetries of groups more accurately. Although the aforementioned differences in the equational properties satisfied by the involution semigroups $\langle \mathcal{B}_2^1, {}^{-1}\rangle$ and $\langle \mathcal{B}_2^1, {}^\star\rangle$ and their reduct $\mathcal{B}_2^1$ are undesirable for the grand scheme of generalization, they inspire several interesting venues of exploration. Most notably, since $\langle \mathcal{B}_2^1, {}^\star\rangle$ is inherently non-finitely based while $\langle \mathcal{B}_2^1, {}^{-1}\rangle$ is not, the following general problem is relevant.

Problem 1.33 Are there two involution semigroups, both sharing the same semigroup reduct, such that one is finitely based and one is not?

On the other hand, it is not too surprising that $\mathcal{B}_2^1$ is inherently non-finitely based while $\langle \mathcal{B}_2^1, {}^{-1}\rangle$ is not, due to the existence of the following examples:

(E1) finitely based involution semigroups $\langle S, {}^\star\rangle$ with a non-finitely based reduct S;
(E2) non-finitely based involution semigroups $\langle S, {}^\star\rangle$ with a finitely based reduct S.

Refer to the survey by Volkov (2001, Sect. 2) for more details. However, unlike $\langle \mathcal{B}_2^1, {}^{-1}\rangle$, the examples of (E1) and (E2) given in this survey are infinite. The first finite example of (E2) can be deduced from Jackson and Volkov (2010, Proposition 6.3). Recently, Lee (2019a) demonstrated how other finite examples of (E2) can be found and exhibited one of order 18. The smallest example of (E2) currently known is the involution semigroup

$\langle \mathcal{A}_0^1, {}^\star \rangle$ of order five (Gao et al. 2020a). Regarding finite examples of (E1), as observed by Auinger et al. (2014, Sect. 1), none had yet been found.

Problem 1.34 Is there a finitely based finite involution semigroup with a non-finitely based semigroup reduct?

It turns out that the direct product $\langle \mathcal{L}_3, {}^\star \rangle \times \langle \mathcal{Z}_n, {}^{\mathfrak{R}} \rangle$, where $n \geq 1$ and $\mathfrak{R} \in \mathsf{sq}(n)$, provides examples that affirmatively answer Problems 1.31, 1.33, and 1.34. In view of the abbreviation $\mathcal{L}_{3,n} = \mathcal{L}_3 \times \mathcal{Z}_n$, it is unambiguous to write

$$\langle \mathcal{L}_{3,n}, {}^{\star\mathfrak{R}} \rangle = \langle \mathcal{L}_3, {}^\star \rangle \times \langle \mathcal{Z}_n, {}^{\mathfrak{R}} \rangle,$$

where the involution ${}^{\star\mathfrak{R}}$ on $\mathcal{L}_{3,n}$ is given by $(x, y) \mapsto (x^\star, y^{\mathfrak{R}})$. Note that if $\mathfrak{R} > 1$ in $\langle \mathcal{L}_{3,n}, {}^{\star\mathfrak{R}} \rangle$, then it is necessary that $n \geq 3$.

In Chap. 7, the involution semigroup $\langle \mathcal{L}_{3,n}, {}^{\star\mathfrak{R}} \rangle$ is shown to possess an infinite irredundant identity basis if $\mathfrak{R} > 1$, thereby positively solving Problem 1.31. Further, an involution subsemigroup of $\langle \mathcal{L}_{3,3}, {}^{\star 2} \rangle$ of order eight with an infinite irredundant identity basis is also exhibited; in comparison, recall that the monoid $\mathcal{R}_{\mathcal{Q}}\{xyxy\}$ of order nine is presently the smallest semigroup known to have an infinite irredundant identity basis (Jackson 2005a). In contrast, the involution semigroup $\langle \mathcal{L}_{3,n}, {}^{\star 1} \rangle$ is shown in Chap. 8 to be finitely based. Hence, the finite basis problem for the involution semigroup $\langle \mathcal{L}_{3,n}, {}^{\star\mathfrak{R}} \rangle$ has a complete solution for all values of $n \geq 1$ and $\mathfrak{R} \in \mathsf{sq}(n)$.

Theorem 1.35 *The involution semigroup $\langle \mathcal{L}_{3,n}, {}^{\star\mathfrak{R}} \rangle$ has a finite identity basis if $\mathfrak{R} = 1$; otherwise, it has an infinite irredundant identity basis. Consequently, $\langle \mathcal{L}_{3,n}, {}^{\star\mathfrak{R}} \rangle$ has an irredundant identity basis regardless of the values of n and $\mathfrak{R}$.*

Recall from Sect. 1.4.2 that the semigroup $\mathcal{L}_{3,n}$ is non-finitely based. Hence, Problem 1.34 is positively solved by the finitely based involution semigroup $\langle \mathcal{L}_{3,n}, {}^{\star 1} \rangle$. In particular, the involution semigroup $\langle \mathcal{L}_3, {}^\star \rangle \cong \langle \mathcal{L}_{3,1}, {}^{\star 1} \rangle$ of order six is finitely based while its reduct $\mathcal{L}_3$ is not. Further, for any $n \geq 3$ and $\mathfrak{R} > 1$, the non-finitely based involution semigroup $\langle \mathcal{L}_{3,n}, {}^{\star\mathfrak{R}} \rangle$ and the finitely based involution semigroup $\langle \mathcal{L}_{3,n}, {}^{\star 1} \rangle$ share the same reduct $\mathcal{L}_{3,n}$. Therefore, Problem 1.33 is also positively solved.

In general, an algebra can possess either a finite identity basis, an infinite irredundant identity basis, or no irredundant identity bases. The positive solution to Problem 1.33 prompts the following intriguing and more refined problem: can finite involution semigroups from all three types share the same semigroup reduct? The involution semigroup $\langle \mathcal{L}_{3,n}, {}^{\star\mathfrak{R}} \rangle$ does not provide an example because by Theorem 1.35, it always has an irredundant identity basis—what is lacking is an involution semigroup without irredundant identity bases. Nevertheless, the involution semigroup $\langle \mathcal{L}_{3,n}, {}^{\star\mathfrak{R}} \rangle$ and its reduct $\mathcal{L}_{3,n}$ play a crucial role in addressing the problem.

Recall from Sect. 1.4.3 that the semigroup $\mathcal{L}_{3,n}$ has no irredundant identity bases. It so happens that a construction of Crvenković et al. (2000), when applied to $\mathcal{L}_{3,n}$, produces an involution semigroup $\langle S, {}^\star \rangle$ with no irredundant identity bases such that the semigroup reduct S is isomorphic to the 0-direct union $\mathcal{L}_{3,n} \uplus \mathcal{L}_{3,n}$ of two copies of $\mathcal{L}_{3,n}$. In contrast, involution semigroups with semigroup reduct $\mathcal{L}_{3,n} \uplus \mathcal{L}_{3,n}$ having either a finite identity basis or an infinite irredundant identity basis can easily be constructed from $\langle \mathcal{L}_{3,n}, {}^{\star\mathfrak{R}} \rangle$. Consequently, the aforementioned refined problem has an affirmative solution.

Theorem 1.36 *There exist three finite involution semigroups, all sharing the same semigroup reduct, such that one has a finite identity basis, one has an infinite irredundant identity basis, and one has no irredundant identity bases.*

Details on the construction of the three finite involution semigroups in Theorem 1.36 are given in Sect. 9.1.

1.5.2 Lattice of Varieties of Involution Semigroups

The lattice $\mathfrak{L}_{\mathsf{sem}}$ of varieties of semigroups has been intensely investigated since the 1970s. Some varieties forming sublattices of $\mathfrak{L}_{\mathsf{sem}}$ that are well understood by now include varieties of commutative semigroups (Kisielewicz 1994); *overcommutative varieties*, that is, varieties containing **Com** (Vernikov 1999, 2001; Volkov 1994); periodic varieties of completely regular semigroups (Polák 1985, 1987, 1988); and aperiodic Rees–Suschkewitsch varieties (Chap. 3). Results concerning the variety **Idem** of all idempotent semigroups have been the most complete and satisfactory. In particular, the variety **Idem** is almost Cross and the lattice $\mathfrak{L}$(**Idem**) is distributive and of width three (Birjukov 1970; Fennemore 1970; Gerhard 1970); see Fig. 1.3. Refer to the surveys by Evans (1971) and Shevrin et al. (2009) for more results on the lattice $\mathfrak{L}_{\mathsf{sem}}$.

The lattice $\mathfrak{L}_{\mathsf{inv}}$ of varieties of involution semigroups has also been investigated since the 1970s, but much less is known about its structure compared with that of the lattice $\mathfrak{L}_{\mathsf{sem}}$. Although only a few small regions of the lattice $\mathfrak{L}_{\mathsf{inv}}$ have been considered, the results obtained so far are sufficient to demonstrate that the lattices $\mathfrak{L}_{\mathsf{inv}}$ and $\mathfrak{L}_{\mathsf{sem}}$ do not bear much resemblance to each other. For instance, the variety **Idem*** of all involution idempotent semigroups is not almost Cross (Adair 1982) and the lattice $\mathfrak{L}$(**Idem***) is non-modular and of infinite width (Dolinka 2000a,b); cf. Fig. 1.3. Refer to the survey by Crvenković and Dolinka (2002) for more information on the lattice $\mathfrak{L}$(**Idem***) and to the monograph of Petrich (1984) for more information on the lower regions of the lattice of varieties of inverse semigroups. Other differences between the lattices $\mathfrak{L}_{\mathsf{inv}}$ and $\mathfrak{L}_{\mathsf{sem}}$ can also be seen by comparing results related to the atoms of $\mathfrak{L}_{\mathsf{inv}}$ (Dolinka 2001; Fajtlowicz 1971) and of $\mathfrak{L}_{\mathsf{sem}}$ (Evans 1971; Kalicki and Scott 1955).

An involution semigroup $\langle S, {}^\star \rangle$ is *singular* if S does not admit an involution different from *. Nonsingular involution semigroups are the main culprits for many of the

Fig. 1.3 The lattice $\mathfrak{L}(\mathbf{IDEM})$

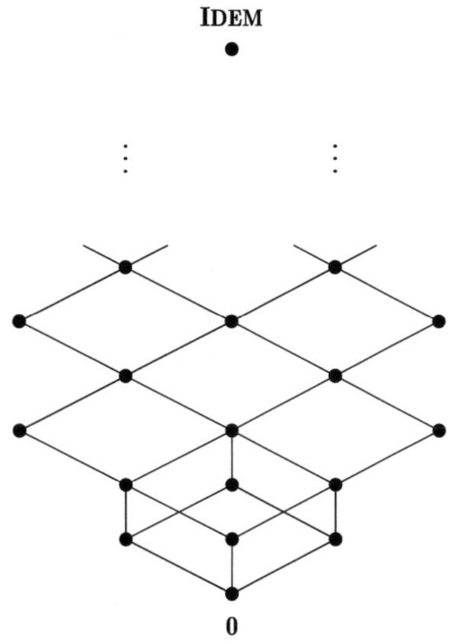

aforementioned differences between the lattices $\mathfrak{L}_{\mathsf{inv}}$ and $\mathfrak{L}_{\mathsf{sem}}$. More specifically, for any involution semigroups $\langle S, {}^\star \rangle$ and $\langle T, {}^\star \rangle$, the implication

$$\mathscr{V}_{\mathsf{sem}}\{S\} \subseteq \mathscr{V}_{\mathsf{sem}}\{T\} \quad \Longrightarrow \quad \mathscr{V}_{\mathsf{inv}}\{\langle S, {}^\star \rangle\} \subseteq \mathscr{V}_{\mathsf{inv}}\{\langle T, {}^\star \rangle\} \tag{1.6}$$

does not hold in general. For instance, for each $n \geq 3$, the nonsingular involution groups $\langle \mathbb{Z}_n, {}^{\mathsf{tr}} \rangle$ and $\langle \mathbb{Z}_n, {}^{-1} \rangle$ serve as a counterexample pair for (1.6) since they generate incomparable varieties.

Problem 1.37 Does the implication (1.6) hold for singular involution semigroups?

It turns out that for each $\ell \geq 2$, the semigroup $\mathcal{L}_\ell$ admits a unique involution *, and the singular involution semigroups $\{\langle \mathcal{L}_\ell, {}^\star \rangle \mid \ell = 2, 3, 4, \ldots\}$ provide infinitely many counterexample pairs to negatively solve Problem 1.37.

Theorem 1.38 *The varieties*

$$\mathscr{V}_{\mathsf{sem}}\{\mathcal{L}_2\} \subset \mathscr{V}_{\mathsf{sem}}\{\mathcal{L}_3\} \subset \mathscr{V}_{\mathsf{sem}}\{\mathcal{L}_4\} \subset \cdots$$

constitute a single infinite chain in the lattice $\mathfrak{L}_{\mathsf{sem}}$, *while the varieties*

$$\mathscr{V}_{\mathsf{inv}}\{\langle \mathcal{L}_2, {}^\star \rangle\} \subset \mathscr{V}_{\mathsf{inv}}\{\langle \mathcal{L}_4, {}^\star \rangle\} \subset \mathscr{V}_{\mathsf{inv}}\{\langle \mathcal{L}_6, {}^\star \rangle\} \subset \cdots$$
$$\text{and } \mathscr{V}_{\mathsf{inv}}\{\langle \mathcal{L}_3, {}^\star \rangle\} \subset \mathscr{V}_{\mathsf{inv}}\{\langle \mathcal{L}_5, {}^\star \rangle\} \subset \mathscr{V}_{\mathsf{inv}}\{\langle \mathcal{L}_7, {}^\star \rangle\} \subset \cdots$$

constitute two incomparable infinite chains in the lattice $\mathfrak{L}_{\mathsf{inv}}$. *Consequently,*

$$\mathscr{V}_{\mathsf{sem}}\{\mathcal{L}_\ell\} \subset \mathscr{V}_{\mathsf{sem}}\{\mathcal{L}_{\ell+1}\} \text{ and } \mathscr{V}_{\mathsf{inv}}\{\langle \mathcal{L}_\ell, {}^\star \rangle\} \not\subseteq \mathscr{V}_{\mathsf{inv}}\{\langle \mathcal{L}_{\ell+1}, {}^\star \rangle\}$$

for all $\ell \geq 2$.

Details on the existence and uniqueness of the involution * on the semigroup $\mathcal{L}_\ell$ and the proof of Theorem 1.38 are given in Sect. 9.2.

Recent examples have shown that even if both inclusions in (1.6) hold for some finite involution semigroups $\langle S, {}^\star \rangle$ and $\langle T, {}^\star \rangle$, the intervals

$$[\mathscr{V}_{\mathsf{sem}}\{S\}, \mathscr{V}_{\mathsf{sem}}\{T\}] \text{ and } [\mathscr{V}_{\mathsf{inv}}\{\langle S, {}^\star \rangle\}, \mathscr{V}_{\mathsf{inv}}\{\langle T, {}^\star \rangle\}]$$

need not resemble one another: one can be a chain of order two, while the other can be as large as having a chain with cardinality of the continuum (Lee 2022).

1.5.3 Relationship Between an Involution Semigroup and Its Semigroup Reduct

In view of the examples of (E1) and (E2) in Sect. 1.5.1—which demonstrate that an involution semigroup and its semigroup reduct can behave independently with respect to the finite basis property—it is natural to consider the *relationship problem*.

Problem 1.39 With respect to the finite basis property, identify conditions under which there is some relationship between an involution semigroup and its semigroup reduct.

A relationship within the class of finite inverse semigroups can be deduced from results of Kleĭman (1977) and Volkov (2019).

Theorem 1.40 (Kleĭman 1977; Volkov 2019) *A finite inverse semigroup is finitely based if its semigroup reduct is finitely based.*

Even though inverse semigroups are quite specialized due to their characteristics of being regular and having commuting idempotents, no progress on Problem 1.39 has been made on more general involution semigroups. The situation for involution semigroups that are non-inverse turns out to be more tantalizing than expected.

Recall that a *semilattice* is a semigroup that is both commutative and idempotent. An involution semilattice is *twisted* if its involution is nontrivial. The smallest nontrivial semilattice is $\mathcal{S}\ell_2 = \{\mathtt{0}, \mathtt{1}\}$, and the smallest twisted involution semilattice is $\langle \mathcal{S}\ell_3, {}^\star \rangle$, where $\mathcal{S}\ell_3 = \{\mathtt{0}, \mathsf{e}, \mathsf{f}\}$ is the semilattice such that $\mathsf{ef} = \mathsf{fe} = \mathtt{0}$ and $\mathsf{e}^\star = \mathsf{f}$. An involution semigroup $\langle S, {}^\star \rangle$ is *twisted* if the variety $\mathscr{V}_{\mathsf{inv}}\{\langle S, {}^\star \rangle\}$ contains some twisted involution semilattice. Since $\langle \mathcal{S}\ell_3, {}^\star \rangle$ embeds into any twisted involution semilattice, $\langle S, {}^\star \rangle$ is twisted if and only if $\langle \mathcal{S}\ell_3, {}^\star \rangle \in \mathscr{V}_{\mathsf{inv}}\{\langle S, {}^\star \rangle\}$. Some examples of twisted involution semigroups are $\langle \mathcal{A}_2^1, {}^\star \rangle$, $\langle \mathcal{B}_2^1, {}^\star \rangle$, $\langle \mathcal{A}_2^{\mathsf{g}}, {}^\star \rangle$, and $\langle \mathcal{L}_{2n}, {}^\star \rangle$ for all $n \geq 1$. On the other hand, any semigroup with idempotent-fixing involution is non-twisted; these include any involution group and $\langle \mathcal{L}_{2n+1}, {}^\star \rangle$ for all $n \geq 1$. Inverse semigroups, such as $\langle \mathcal{B}_2^1, {}^{-1} \rangle$, are also non-twisted since the regularity axiom $xx^\star x \approx x$ excludes $\langle \mathcal{S}\ell_3, {}^\star \rangle$ from the variety of inverse semigroups.

In general, any involution semigroup $\langle S, {}^\star \rangle$ that is not already twisted is embeddable in the twisted involution semigroup $\langle S, {}^\star \rangle \times \langle \mathcal{S}\ell_3, {}^\star \rangle$. In fact, a much smaller twisted involution semigroup containing $\langle S, {}^\star \rangle$ is available: the disjoint union $\langle S, {}^\star \rangle \cup \langle \mathcal{S}\ell_3, {}^\star \rangle$, where for any $x \in S$ and $y \in \mathcal{S}\ell_3$, the products xy and yx are equal to the zero element of $\mathcal{S}\ell_3$. Consequently, twisted involution semigroups constitute a large class of involution semigroups.

In Chap. 10, twistedness is shown to be a condition sought by Problem 1.39.

Theorem 1.41 *A twisted involution semigroup is non-finitely based if its semigroup reduct is non-finitely based.*

This result can only be viewed as a very loose converse of Theorem 1.40 because the class of twisted involution semigroups is disjoint from the class of inverse semigroups.

Now since the variety $\mathscr{V}_{\mathsf{inv}}\{\langle \mathcal{S}\ell_3, {}^\star \rangle\}$ is an atom in the lattice $\mathfrak{L}_{\mathsf{inv}}$ (Fajtlowicz 1971), the twistedness assumption in Theorem 1.41 is a weakest assumption—the only way to weaken it is to omit it altogether. But twistedness cannot be omitted because by Theorem 1.35, the non-twisted involution semigroup $\langle \mathcal{L}_3, {}^\star \rangle$ is finitely based while its reduct $\mathcal{L}_3$ is not. Further, the converse of the theorem does not hold since the twisted involution semigroup $\langle \mathcal{A}_0^1, {}^\star \rangle$ is non-finitely based but its reduct $\mathcal{A}_0^1$ is finitely based (Gao et al. 2020a). Consequently, Theorem 1.41 has reached its full potential.

As discussed in Sect. 1.5.1, over the years, equational properties of involution semigroups have received much less attention than equational properties of semigroups. As a result, there is a disparity of discoveries made between the two areas, as one can readily observe from the results surveyed by Shevrin and Volkov (1985) and Volkov (2001). Theorem 1.41 narrows this disparity by converting many results on equational properties of semigroups to results applicable to involution semigroups. It is infeasible to provide full details in all such instances, hence only a few representative or unexpected examples are provided in the remainder of this section, while other results will be surveyed elsewhere.

1.5.3.1 Non-Twisted Involution Semigroups

As demonstrated by the involution semigroup $\langle \mathcal{L}_3, {}^\star \rangle$, a non-twisted involution semigroup $\langle S, {}^\star \rangle$ with non-finitely based reduct S need not be non-finitely based. However, since the union $S \cup \mathcal{S}\ell_3$ is non-finitely based if and only if S is non-finitely based (Mel'nik 1970), the twisted involution semigroup $\langle S, {}^\star \rangle \cup \langle \mathcal{S}\ell_3, {}^\star \rangle$ with non-finitely based reduct S is non-finitely based by Theorem 1.41. Therefore, for any involution semigroup $\langle S, {}^\star \rangle$ with non-finitely based reduct S, either $\langle S, {}^\star \rangle$ or $\langle S, {}^\star \rangle \cup \langle \mathcal{S}\ell_3, {}^\star \rangle$ is non-finitely based.

Theorem 1.42 *Any involution semigroup with non-finitely based semigroup reduct is embeddable in some non-finitely based involution semigroup with at most three more elements.*

Recall from Sect. 1.4.2.3 that the semigroup $\mathcal{L}_\ell$ is finitely based if and only if $\ell = 2$. But the finite basis problem for the involution semigroup $\langle \mathcal{L}_\ell, {}^\star \rangle$ lacks a complete solution. If $\ell \geq 2$ is even, then $\langle \mathcal{L}_\ell, {}^\star \rangle$ is twisted, so that Theorem 1.41 is applicable whenever the reduct $\mathcal{L}_\ell$ is non-finitely based. Specifically, $\langle \mathcal{L}_{2n}, {}^\star \rangle$ is non-finitely based for all $n \geq 2$, but it is unknown if $\langle \mathcal{L}_2, {}^\star \rangle$ is finitely based. For odd $\ell \geq 3$, however, the situation for $\langle \mathcal{L}_\ell, {}^\star \rangle$ is not straightforward since it is non-twisted. The involution semigroup $\langle \mathcal{L}_3, {}^\star \rangle$ is finitely based by Theorem 1.35, while the finite basis property of $\langle \mathcal{L}_{2n+1}, {}^\star \rangle$ is unknown for all $n \geq 2$.

In contrast, the finite basis problem for the involution monoid $\langle \mathcal{L}_\ell^1, {}^\star \rangle$ and its reduct $\mathcal{L}_\ell^1$ is completely solved: $\langle \mathcal{L}_\ell^1, {}^\star \rangle$ is non-finitely based for all $\ell \geq 2$ (see Gao et al. 2021) and $\mathcal{L}_\ell^1$ is non-finitely based if and only if $\ell \geq 3$ (see Mikhailova and Sapir 2018; O.B. Sapir 2018).

A summary of the aforementioned results is given in Table 1.6.

Problem 1.43

(i) Is the involution semigroup $\langle \mathcal{L}_2, {}^\star \rangle$ finitely based?
(ii) For which odd $\ell \geq 5$ is the involution semigroup $\langle \mathcal{L}_\ell, {}^\star \rangle$ finitely based?

Presently, the inverse Brandt monoid $\langle \mathcal{B}_2^1, {}^{-1} \rangle$ is the smallest known example of a non-twisted involution semigroup that is non-finitely based (Kleĭman 1979).

1.5.3.2 Inherent Non-Finite Basis Property

The inherent non-finite basis property is an equational property that has some relevance to Problem 1.39: if an involution semigroup is inherently non-finitely based, then its semigroup reduct is also inherently non-finitely based (Auinger et al. 2014). However, the converse of this result does not hold in general since the Brandt monoid $\mathcal{B}_2^1$ is inherently non-finitely based, while the inverse Brandt monoid $\langle \mathcal{B}_2^1, {}^{-1} \rangle$ is weakly finitely based (M.V. Sapir 1987b, 1993). Nevertheless, using Theorem 1.32, Auinger et al. (2014)

Table 1.6 Finite basis property for $\langle \mathcal{L}_\ell, {}^\star \rangle$, $\langle \mathcal{L}_\ell^1, {}^\star \rangle$, and their semigroup reducts

	$\ell = 2$	$\ell = 4, 6, 8, \ldots$	$\ell = 3$	$\ell = 5, 7, 9, \ldots$
$\langle \mathcal{L}_\ell, {}^\star \rangle$ is finitely based	Unknown	No	Yes	Unknown
$\mathcal{L}_\ell$ is finitely based	Yes	No	No	No
$\langle \mathcal{L}_\ell^1, {}^\star \rangle$ is finitely based	No	No	No	No
$\mathcal{L}_\ell^1$ is finitely based	Yes	No	No	No

established a result on inherently non-finitely based involution semigroups that parallels Theorem 1.41.

Theorem 1.44 (Auinger et al. 2014, Theorem 3.1) *A twisted involution semigroup is inherently non-finitely based if its semigroup reduct is inherently non-finitely based.*

This theorem turns out to be an easy consequence of Theorem 1.41. Indeed, a twisted weakly finitely based involution semigroup $\langle S, {}^\star\rangle$ belongs to some finitely based locally finite variety $\mathscr{V}_{\mathsf{inv}}\{\langle U, {}^\star\rangle\}$, whence the semigroup reduct S belongs to the locally finite variety $\mathscr{V}_{\mathsf{sem}}\{U\}$; since the variety $\mathscr{V}_{\mathsf{sem}}\{U\}$ is finitely based by Theorem 1.41, the semigroup S is weakly finitely based.

As observed after Theorem 1.32, the involution semigroups $\langle \mathcal{A}_2^1, {}^\star\rangle$ and $\langle \mathcal{B}_2^1, {}^\star\rangle$ are inherently non-finitely based (Auinger et al. 2012b). This result also follows from Theorem 1.44 since $\langle \mathcal{A}_2^1, {}^\star\rangle$ and $\langle \mathcal{B}_2^1, {}^\star\rangle$ are twisted involution semigroups with inherently non-finitely based semigroup reducts. As for the inverse semigroup $\langle \mathcal{B}_2^1, {}^{-1}\rangle$, even though it is weakly finitely based (M.V. Sapir 1993), the direct product $\langle \mathcal{B}_2^1, {}^{-1}\rangle \times \langle \mathcal{S}\ell_3, {}^\star\rangle$ is twisted and so is inherently non-finitely based by Theorem 1.44. Since $\langle \mathcal{S}\ell_3, {}^\star\rangle$ is finitely based, the class $\mathfrak{WFB}_{\mathsf{inv}}$ of weakly finitely based finite involution semigroups is not closed under the formation of direct products.

Proposition 1.45 *The class $\mathfrak{WFB}_{\mathsf{inv}}$ is not a pseudovariety.*

In contrast, it follows from Theorem 1.21 that the class $\mathfrak{WFB}_{\mathsf{sem}}$ of weakly finitely based finite semigroups is a pseudovariety. In fact, the pseudovariety $\mathfrak{WFB}_{\mathsf{sem}}$ has a finite pseudoidentity basis (Volkov 2000, Corollary 4.1).

Recently, a result similar to Theorem 1.41 has been established in Lee (2018b): for any twisted involution semigroup $\langle S, {}^\star\rangle$, if the variety $\mathscr{V}_{\mathsf{sem}}\{S\}$ contains continuum many subvarieties that are closed under anti-isomorphism, then the variety $\mathscr{V}_{\mathsf{inv}}\{\langle S, {}^\star\rangle\}$ contains continuum many subvarieties. The following result on inherently non-finitely based finite involution semigroups is then a consequence.

Theorem 1.46 (Lee 2018b) *Every twisted inherently non-finitely based finite involution semigroup generates a variety with continuum many subvarieties. In particular, each of the varieties $\mathscr{V}_{\mathsf{inv}}\{\langle \mathcal{A}_2^1, {}^\star\rangle\}$, $\mathscr{V}_{\mathsf{inv}}\{\langle \mathcal{B}_2^1, {}^\star\rangle\}$, and $\mathscr{V}_{\mathsf{inv}}\{\langle \mathcal{B}_2^1, {}^{-1}\rangle \times \langle \mathcal{S}\ell_3, {}^\star\rangle\}$ contains continuum many subvarieties.*

It is of interest to note that the varietal join

$$\mathscr{V}_{\mathsf{inv}}\{\langle \mathcal{B}_2^1, {}^{-1}\rangle\} \vee \mathscr{V}_{\mathsf{inv}}\{\langle \mathcal{S}\ell_3, {}^\star\rangle\} = \mathscr{V}_{\mathsf{inv}}\{\langle \mathcal{B}_2^1, {}^{-1}\rangle \times \langle \mathcal{S}\ell_3, {}^\star\rangle\}$$

contains continuum many subvarieties even though the varieties $\mathscr{V}_{\mathsf{inv}}\{\langle \mathcal{B}_2^1, {}^{-1}\rangle\}$ and $\mathscr{V}_{\mathsf{inv}}\{\langle \mathcal{S}\ell_3, {}^\star\rangle\}$ are small (Fajtlowicz 1971; Kleĭman 1977). For varieties of involution monoids, it is even possible for the join of two Cross varieties to contain continuum many subvarieties (Lee 2019b).

1.5.3.3 Sufficient Conditions for the Non-Finite Basis Property

As mentioned in Sect. 1.4.2, many sufficient conditions have been established under which a semigroup is non-finitely based. By Theorem 1.41, each of these sufficient conditions is also applicable to any twisted involution semigroup $\langle S, {}^\star \rangle$ since any result that guarantees the non-finite basis property of the semigroup reduct S also guarantees the same property for $\langle S, {}^\star \rangle$. For instance, the following is a consequence of Theorems 1.26 and 1.41.

Theorem 1.47 *Any twisted involution semigroup that satisfies the identities*

$$x^{2n} \approx x^n, \quad x^n y_1^n y_2^n \cdots y_m^n x^n \approx x^n y_m^n y_{m-1}^n \cdots y_1^n x^n, \quad m \in \{2, 3, 4, \ldots\}$$

for some fixed $n \geq 2$ but violates the identity $(x^n y^n x^n)^n \approx x^n y^n x^n$ is non-finitely based.

Recall that Theorem 1.30 is an involution version of Volkov's $\mathcal{A}_2$-theorem (Theorem 1.20). Using Theorem 1.41, a version that more closely resembles Theorem 1.20 can be obtained.

Theorem 1.48 *Let $\langle S, {}^\star \rangle$ be any twisted involution semigroup and S_E be the subsemigroup of S generated by its idempotents. Suppose that $\mathcal{A}_2 \in \mathscr{V}_{\mathsf{sem}}\{S\}$ and that some group exists in $\mathscr{V}_{\mathsf{sem}}\{S\} \backslash \mathscr{V}_{\mathsf{sem}}\{S_E\}$. Then $\langle S, {}^\star \rangle$ is non-finitely based.*

It follows that the direct product of $\langle \mathcal{A}_2, {}^\star \rangle$ with any involution group, such as the cyclic group $\langle \mathcal{Z}_p, {}^{-1} \rangle$ of prime order p, is non-finitely based. However, $\langle \mathcal{A}_2, {}^\star \rangle \times \langle \mathcal{Z}_p, {}^{-1} \rangle$ does not generate a limit variety (Lee 2020b) even though by Theorem 1.8, its reduct $\mathcal{A}_2 \times \mathcal{Z}_p$ generates such a variety.

Recall from Theorem 1.17 that up to isomorphism, the semigroups $\mathcal{A}_2^1$, $\mathcal{B}_2^1$, $\mathcal{A}_2^{\mathsf{g}}$, and $\mathcal{L}_3$ of order six are the only minimal non-finitely based semigroups. The involution semigroups having these four semigroups as reducts are $\langle \mathcal{A}_2^1, {}^\star \rangle$, $\langle \mathcal{B}_2^1, {}^\star \rangle$, $\langle \mathcal{B}_2^1, {}^{-1} \rangle$, $\langle \mathcal{A}_2^{\mathsf{g}}, {}^\star \rangle$, and $\langle \mathcal{L}_3, {}^\star \rangle$. The non-finite basis property of $\langle \mathcal{A}_2^1, {}^\star \rangle$, $\langle \mathcal{B}_2^1, {}^\star \rangle$, and $\langle \mathcal{B}_2^1, {}^{-1} \rangle$ has already been addressed, and $\langle \mathcal{A}_2^{\mathsf{g}}, {}^\star \rangle$ is also non-finitely based by Theorem 1.48. Since $\langle \mathcal{L}_3, {}^\star \rangle$ is finitely based by Theorem 1.35, it is the only finitely based involution semigroup whose semigroup reduct is minimal non-finitely based. In other words, $\langle \mathcal{L}_3, {}^\star \rangle$ is the unique smallest example of (E1) in Sect. 1.5.1.

Proposition 1.49 *Up to isomorphism, $\langle \mathcal{L}_3, {}^\star \rangle$ is the unique smallest finitely based involution semigroup with a non-finitely based semigroup reduct.*

Regarding examples of (E2)—non-finitely based involution semigroups with a finitely based semigroup reduct—recall that $\langle \mathcal{A}_0^1, {}^\star \rangle$ is one of order five (Gao et al. 2020a). Presently, it is unknown if a smaller example exists.

1.6 Varieties of Monoids

In the semigroup signature, monoids have played a conspicuous role in the study of varieties with extreme properties. For instance, there exist small monoids, such as $\mathcal{A}_2^1$ and $\mathcal{B}_2^1$, that generate a variety of semigroups with continuum many subvarieties (Edmunds et al. 2010; Jackson 2000; Trahtman 1988). As evidenced from Sect. 1.4, monoids also constitute a rich source of non-finitely based finite semigroups—$\mathcal{B}_2^1$ and $\mathcal{P}_{25}$ being the first published examples (Perkins 1969). Further, every inherently non-finitely based finite semigroup contains some submonoid that is inherently non-finitely based (M.V. Sapir 1987a,b).

Given how close monoids and semigroups are, it is not surprising that a number of equational properties are satisfied by a finite monoid independent of the signature type.

Theorem 1.50 (See Volkov 2001, Sect. 1) *A finite monoid is finitely based in the monoid signature if and only if it is finitely based in the semigroup signature.*

Theorem 1.51 (Jackson and Lee 2018, Theorem 4.3) *A finite monoid is inherently non-finitely based in the monoid signature if and only if it is inherently non-finitely based in the semigroup signature.*

1.6.1 Rees Quotients of Free Monoids

Another similarity between monoids and semigroups is that the class of monoids—even its subclass of Rees quotients of the free monoid $\mathscr{A}_\varnothing^+$—behaves as erratically as the class of semigroups. For instance, the class of finitely based semigroups and the class of non-finitely based semigroups are not closed under the formation of direct products (M.V. Sapir 1991) while the same is also true for the class of Rees quotients of $\mathscr{A}_\varnothing^+$.

Example 1.52 (Jackson and Sapir 2000) There exist finitely based finite monoids M_1 and M_2 such that the direct product $M_1 \times M_2$ is non-finitely based. For example,

(i) $M_1 = \mathcal{R}_\mathcal{Q}\{x^2y^2x, xyxyx, xy^2x^2\}$ and $M_2 = \mathcal{R}_\mathcal{Q}\{x^2y^2, xyxy, xy^2x, yx^3y\}$;
(ii) $M_1 = \mathcal{R}_\mathcal{Q}\{xyzxy, xyzyx\}$ and $M_2 = \mathcal{R}_\mathcal{Q}\{x^n y^n\}$ for any $n \geq 2$.

Example 1.53 (Jackson and Sapir 2000) There exist non-finitely based finite monoids M_1 and M_2 such that the direct product $M_1 \times M_2$ is finitely based. For example,

(i) $M_1 = \mathcal{R}_{\mathrm{Q}}\{xyxy\}$ and $M_2 = \mathcal{R}_{\mathrm{Q}}\{x^2y^2, xy^2x\}$;
(ii) $M_1 = \mathcal{R}_{\mathrm{Q}}\{xyzxy\}$ and $M_2 = \mathcal{R}_{\mathrm{Q}}\{xyzyx\}$.

In fact, Jackson and Sapir (2000) presented methods for exhibiting as many Rees quotients of $\mathscr{A}_{\varnothing}^{+}$ for Examples 1.52 and 1.53 as desired.

However, depending on the signature type, results at the varietal level can differ drastically. For a monoid M in general, there is an order-embedding of $\mathfrak{L}(\mathscr{V}_{\mathsf{mon}}\{M\})$ into $\mathfrak{L}(\mathscr{V}_{\mathsf{sem}}\{M\})$. Therefore, the lattice $\mathfrak{L}(\mathscr{V}_{\mathsf{sem}}\{M\})$ is as complicated as the lattice $\mathfrak{L}(\mathscr{V}_{\mathsf{mon}}\{M\})$; but the latter is often much smaller than the former, due to the strength that the nullary operation 1 of a monoid has in equational deduction. A classical example is the monoid $\mathcal{R}_{\mathrm{Q}}\{x\}$ of order three: the lattice $\mathfrak{L}(\mathscr{V}_{\mathsf{sem}}\{\mathcal{R}_{\mathrm{Q}}\{x\}\})$ is countably infinite but the lattice $\mathfrak{L}(\mathscr{V}_{\mathsf{mon}}\{\mathcal{R}_{\mathrm{Q}}\{x\}\})$ is only a chain of order three (Evans 1971, Fig. 5(b)). In fact, there exists a monoid M with as few as six elements such that the lattice $\mathfrak{L}(\mathscr{V}_{\mathsf{sem}}\{M\})$ has cardinality of the continuum but the lattice $\mathfrak{L}(\mathscr{V}_{\mathsf{mon}}\{M\})$ is finite (Lee 2023); another example with this property is the monoid $\mathcal{R}_{\mathrm{Q}}\{xyx\}$ of order seven (Jackson 2000, 2005b).

Despite the enormous difference, varieties of monoids have received much less attention than varieties of semigroups. For many years, the most notable results were concerned with the variety $\mathbb{C}\mathrm{OM}$ of commutative monoids and the variety $\mathbb{I}\mathrm{DEM}$ of idempotent monoids; for descriptions of the lattices $\mathfrak{L}(\mathbb{C}\mathrm{OM})$ and $\mathfrak{L}(\mathbb{I}\mathrm{DEM})$, see Head (1968) and Wismath (1986), respectively. The situation changed after the mid-1990s when Rees quotients of $\mathscr{A}_{\varnothing}^{+}$ became popular in the study of the finite basis problem; see Sect. 1.4.2.3. Besides the examples and results surveyed so far, Rees quotients of $\mathscr{A}_{\varnothing}^{+}$ also play an important role in the study of subvarieties of the class $\mathbb{A}^{\mathsf{cen}}$ of aperiodic monoids with central idempotents. For any set $\mathscr{W}$ of words, let $\mathbb{R}_{\mathrm{Q}}\mathscr{W}$ denote the variety of monoids generated by the Rees quotient $\mathcal{R}_{\mathrm{Q}}\mathscr{W}$ of $\mathscr{A}_{\varnothing}^{+}$.

The monoid $\mathcal{A}_2^1$ of order six is the first finite example found to generate a variety of semigroups with continuum many subvarieties (Trahtman 1988). Finite monoids that generate a variety of monoids with continuum many subvarieties were not discovered until after the turn of the millennium (Jackson and McKenzie 2006); these monoids, the smallest of which is of order 20, are obtained from a graph-encoding technique and thus have nontrivial descriptions. Recently, simpler examples were more systematically found.

Example 1.54 (Jackson and Lee 2018) Finitely generated varieties of monoids with continuum many subvarieties include

(i) the variety $\mathbb{R}_{\mathrm{Q}}\{xyxy\}$;
(ii) the variety $\mathbb{M}_4$ generated by all monoids of order four;
(iii) the varietal join $\mathbb{R}_{\mathrm{Q}}\{x^2\} \vee \mathscr{V}_{\mathsf{mon}}\{\mathcal{S}\mathrm{YM}_3\}$;
(iv) the variety generated by any inherently non-finitely based finite monoid.

Remark 1.55

(i) As noted after Theorem 1.24, the monoid $\mathcal{R}_{\mathcal{Q}}\{xyxy\}$ of order nine that generates the variety $\mathbb{R}_{\mathbb{Q}}\{xyxy\}$ is the smallest non-finitely based Rees quotient of $\mathscr{A}_{\varnothing}^{+}$ (Jackson and Sapir 2000). It follows that the non-finitely based variety

$$\mathbb{P}_{25} = \mathscr{V}_{\mathsf{mon}}\{\mathcal{P}_{25}\} = \mathbb{R}_{\mathbb{Q}}\{xyxy, xyzyx, xzyxy, x^2z\}$$

also contains continuum many subvarieties.

(ii) The variety $\mathbb{M}_4$, on the other hand, is finitely based (Lee and Li 2015).

(iii) Since the varieties $\mathbb{R}_{\mathbb{Q}}\{x^2\}$ and $\mathscr{V}_{\mathsf{mon}}\{\mathcal{S}\mathcal{Y}\mathcal{M}_3\}$ are Cross (Head 1968; Oates and Powell 1964), Example 1.54(iii) demonstrates that the varietal join of two Cross varieties of monoids can contain continuum many subvarieties; see Gusev (2019) for another example. In contrast, it is unknown if the join of two small varieties of semigroups can contain continuum many subvarieties (Jackson 2000, Question 3.15).

(iv) The varieties $\mathbb{A}_2^1 = \mathscr{V}_{\mathsf{mon}}\{\mathcal{A}_2^1\}$ and $\mathbb{B}_2^1 = \mathscr{V}_{\mathsf{mon}}\{\mathcal{B}_2^1\}$ contain continuum many subvarieties because $\mathcal{A}_2^1$ and $\mathcal{B}_2^1$ are, by Theorems 1.21 and 1.51, inherently non-finitely based as monoids. In fact, even the interval $[\mathbb{B}_2^1, \mathbb{A}_2^1]$ contains continuum many varieties (Jackson and Zhang 2021).

1.6.2 Limit Varieties and Hereditarily Finitely Based Varieties

The first examples of limit varieties of monoids are generated by Rees quotients of $\mathscr{A}_{\varnothing}^{+}$. These limit varieties, due to Jackson (2005b), have very simple lattices of subvarieties.

Theorem 1.56 (Jackson 2005b) *The varieties*

$$\mathbb{J}_1 = \mathbb{R}_{\mathbb{Q}}\{xhxyty\} \text{ and } \mathbb{J}_2 = \mathbb{R}_{\mathbb{Q}}\{xhytxy, xyhxty\}$$

are limit varieties, and their lattices of subvarieties are given in Fig. 1.4.

Jackson (2005b, page 180) commented that he was unable to find other examples of finitely generated limit subvarieties of $\mathbb{A}^{\mathsf{cen}}$ and thus posted a few questions.

Problem 1.57 (Jackson 2005b, Question 1)

(i) Are $\mathbb{J}_1$ and $\mathbb{J}_2$ the only limit subvarieties of $\mathbb{A}^{\mathsf{cen}}$?

(ii) Is there a finitely generated non-finitely based variety of monoids that does not contain $\mathbb{J}_1$ and $\mathbb{J}_2$?

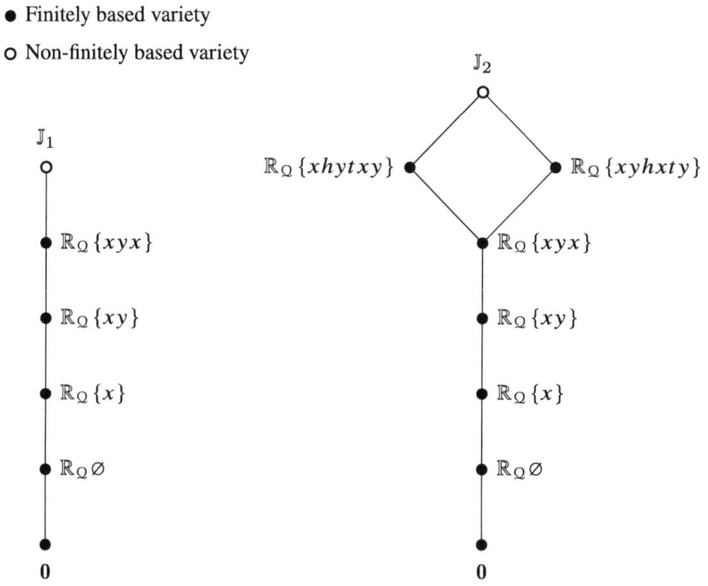

Fig. 1.4 The lattices $\mathfrak{L}(\mathbb{J}_1)$ and $\mathfrak{L}(\mathbb{J}_2)$

Jackson (2005b) observed that any limit subvariety of $\mathbb{A}^{\mathsf{cen}}$ different from $\mathbb{J}_1$ and $\mathbb{J}_2$ has to be contained in the variety $\mathbb{O}$ of monoids defined by the identity system

$$xhxyty \approx xhyxty, \quad xhytxy \approx xhytyx \tag{1.7}$$

or its dual variety $\mathbb{O}^{\lhd}$; see Lemmas 2.7 and 2.8. Investigation of the finite basis problem for subvarieties of $\mathbb{A}^{\mathsf{cen}} \cap \mathbb{O}$ is thus unavoidable in the pursuit of a complete solution to Problem 1.57(i). If Problem 1.57(i) were to be negatively solved by the discovery of a new finitely generated limit subvariety of $\mathbb{A}^{\mathsf{cen}}$, then this limit variety would be an example that positively solves Problem 1.57(ii).

In Chap. 11, the finite basis problem for subvarieties of not only the variety $\mathbb{A}^{\mathsf{cen}} \cap \mathbb{O}$, but also the larger variety $\mathbb{O}$, is addressed. It is first shown that each noncommutative subvariety of $\mathbb{O}$ can be defined within $\mathbb{O}$ by some set of identities of the following two types:

$$x^{e_0} \prod_{i=1}^{r} (h_i x^{e_i}) \approx x^{f_0} \prod_{i=1}^{r} (h_i x^{f_i}),$$

$$x^{e_0} y^{f_0} \prod_{i=1}^{r} (h_i x^{e_i} y^{f_i}) \approx y^{f_0} x^{e_0} \prod_{i=1}^{r} (h_i x^{e_i} y^{f_i}),$$

where $e_0, f_0, e_1, f_1, \ldots, e_r, f_r$ are nonnegative integers that satisfy certain conditions. By a general result of Volkov (1990), any set of identities of the first type defines a finitely based variety. On the other hand, Higman's lemma (1952) implies that identities of the second type are well-quasi-ordered under deduction within the equational theory of $\mathbb{O}$, whence any set of these identities defines a finitely based subvariety of $\mathbb{O}$. Consequently, any noncommutative subvariety of $\mathbb{O}$ is finitely based. But since every commutative variety of monoids is finitely based (Head 1968), the variety $\mathbb{O}$ is hereditarily finitely based. This result facilitates the characterization of several large classes of hereditarily finitely based varieties of monoids.

Theorem 1.58 *Let $\mathbb{V}$ be any variety of monoids such that one of the following conditions holds*: $\mathbb{V} \subseteq \mathbb{A}^{\mathsf{cen}}$, $\mathbb{V} \subseteq \mathbb{B}_2^1$, $\mathcal{R}_Q\{xyx\} \in \mathbb{V}$, *and* $\mathbb{C}\text{OM} \subseteq \mathbb{V}$. *Then the following statements on $\mathbb{V}$ are equivalent*:

(a) $\mathbb{V}$ *is hereditarily finitely based*;
(b) $\mathbb{J}_1, \mathbb{J}_2 \nsubseteq \mathbb{V}$;
(c) $\mathbb{V} \subseteq \mathbb{O}$ *or* $\mathbb{V} \subseteq \mathbb{O}^{\triangleleft}$.

It follows that $\mathbb{J}_1$ and $\mathbb{J}_2$ are the only limit subvarieties of $\mathbb{A}^{\mathsf{cen}}$, $\mathbb{B}_2^1$, and $\mathbb{P}_{25}$; in particular, Problem 1.57(i) is positively solved.

Recently, Problem 1.57(ii) was positively solved by several new examples of finitely generated limit varieties of monoids (Gusev 2020a; Gusev and Sapir 2022; Zhang and Luo 2016). Some of these varieties, together with $\mathbb{J}_1$ and $\mathbb{J}_2$, exhaust all limit varieties of $\mathscr{J}$-trivial monoids (Gusev and Sapir 2022).

Recall that a variety of monoids is *overcommutative* if it contains the variety $\mathbb{C}\text{OM}$ of commutative monoids. Overcommutative varieties of monoids constitute the interval $[\mathbb{C}\text{OM}, \mathbb{M}\text{ON}]$, where $\mathbb{M}\text{ON}$ is the variety of monoids. Unlike subvarieties of $\mathbb{A}^{\mathsf{cen}}$, $\mathbb{B}_2^1$, and $\mathbb{P}_{25}$, which contain limit varieties, the situation is very different for the interval $[\mathbb{C}\text{OM}, \mathbb{M}\text{ON}]$.

Theorem 1.59 *Overcommutative limit varieties of monoids do not exist.*

Indeed, any non-finitely based overcommutative variety of monoids cannot be a limit variety because by Theorem 1.58, it properly contains one of the non-finitely based varieties $\mathbb{J}_1$ and $\mathbb{J}_2$.

Since the hereditarily finitely based varieties $\mathbb{O}$ and $\mathbb{O}^{\triangleleft}$ are overcommutative, Theorem 1.58 can easily be used to establish their maximality and uniqueness.

Theorem 1.60 *The varieties $\mathbb{O}$ and $\mathbb{O}^{\triangleleft}$ are maximal hereditarily finitely based varieties of monoids. Further, among overcommutative varieties, $\mathbb{O}$ and $\mathbb{O}^{\triangleleft}$ are the only maximal hereditarily finitely based varieties.*

Consequently, hereditarily finitely based overcommutative varieties of monoids constitute the intervals $[\mathbb{C}\text{OM}, \mathbb{O}]$ and $[\mathbb{C}\text{OM}, \mathbb{O}^{\lhd}]$. These intervals have a very complex lattice structure since even their subinterval $[\mathbb{C}\text{OM}, \mathbb{O} \cap \mathbb{O}^{\lhd}]$ contains an isomorphic copy of every finite lattice (Gusev and Lee 2020).

Theorems 1.59 and 1.60 are both counterintuitive because the first published example of a limit variety of semigroups is overcommutative (Volkov 1982), while maximal hereditarily finitely based varieties of semigroups do not exist (Shevrin and Volkov 1985, Proposition 15.2).

It is currently unknown if there is a maximal hereditarily finitely based variety of monoids that is different from $\mathbb{O}$ and $\mathbb{O}^{\lhd}$. By Theorem 1.60, if such a variety exists, then it is periodic.

Problem 1.61 Is there a maximal hereditarily finitely based variety of monoids that is periodic?

Recall that Problem 1.23, the finite basis problem for finite Rees quotients of $\mathscr{A}_{\varnothing}^{+}$, is presently open. But since these monoids belong to $\mathbb{A}^{\mathsf{cen}}$, by Theorem 1.58, their hereditary finite basis problem has an easy solution in the monoid signature.

Theorem 1.62 *For any finite set $\mathscr{W}$ of words, the Rees quotient $\mathcal{R}_{\mathbb{Q}}\mathscr{W}$ is hereditarily finitely based in the monoid signature if and only if it satisfies the identity system* (1.7) *or its dual system.*

Specifically, by referring to Theorems 1.24 and 1.58, it is routinely checked that for any word $\mathbf{w} \in \{x, y\}^+$, the Rees quotient $\mathcal{R}_{\mathbb{Q}}\{\mathbf{w}\}$ is hereditarily finitely based in the monoid signature if and only if $\mathbf{w}$ is a factor of either $x^m yx^n$ or $y^m xy^n$ for some $m, n \geq 1$.

Revisiting Theorem 1.56—the result that motivated the study of the varieties $\mathbb{O}$ and $\mathbb{O}^{\lhd}$—it is easily seen that the monoids $\mathcal{R}_{\mathbb{Q}}\{xhytxy\}$ and $\mathcal{R}_{\mathbb{Q}}\{xyhxty\}$ are hereditarily finitely based in the monoid signature but their direct product is not. The following easy consequences of Theorem 1.58 provide several ways to obtain similar examples.

Proposition 1.63

(i) *For any monoid M that belongs to neither $\mathbb{O}$ nor $\mathbb{O}^{\lhd}$, the direct product $\mathcal{R}_{\mathbb{Q}}\{xyx\} \times M$ is not hereditarily finitely based in the monoid signature.*
(ii) *For any cancellative monoid M, the direct product $\mathcal{R}_{\mathbb{Q}}\{xyx\} \times M$ is hereditarily finitely based in the monoid signature if and only if M is commutative.*

For instance, in the monoid signature, $\mathcal{R}_{\mathbb{Q}}\{xyx\}$ and $\mathcal{A}_0^1$ are hereditarily finitely based by Theorems 1.9 and 1.56, but since $\mathcal{A}_0^1 \notin \mathbb{O} \cup \mathbb{O}^{\lhd}$, the direct product $\mathcal{R}_{\mathbb{Q}}\{xyx\} \times \mathcal{A}_0^1$ is not hereditarily finitely based by Proposition 1.63(i). Examples can also be constructed from finite groups because they are hereditarily finitely based (Oates and Powell 1964) and cancellative; specifically, for any noncommutative group G, the direct product $\mathcal{R}_Q\{xyx\} \times G$ is not hereditarily finitely based by Proposition 1.63(ii).

1.6.3 Cross Varieties and Inherently Non-Finitely Generated Varieties

Since every Cross variety is hereditarily finitely based, it is instinctive to consider refining Theorem 1.58 into a characterization that involves Cross varieties. Fortunately, this task is given a head start by Theorem 1.56 since the limit varieties $\mathbb{J}_1$ and $\mathbb{J}_2$ are also almost Cross. Another relevant result is due to Straubing (1982): each finite monoid in $\mathbb{A}^{\mathsf{cen}}$ belongs to the variety $\mathbb{S}_n$ of monoids defined by the single identity

$$x\prod_{i=1}^{n-1}(h_i x) \approx x^n \prod_{i=1}^{n-1} h_i$$

for all sufficiently large $n \geq 2$.

Recall that the *index* of an aperiodic monoid M is the least integer $n \geq 1$ such that M satisfies the identity $x^{n+1} \approx x^n$. Let $\mathbb{A}_n^{\mathsf{cen}}$ denote the subvariety of $\mathbb{A}^{\mathsf{cen}}$ consisting of monoids of index at most n. Then the variety $\mathbb{A}_n^{\mathsf{cen}}$ is defined by the identity system $\{x^{n+1} \approx x^n, x^n y \approx yx^n\}$, and the inclusions

$$\mathbb{A}_1^{\mathsf{cen}} \subset \mathbb{A}_2^{\mathsf{cen}} \subset \mathbb{A}_3^{\mathsf{cen}} \subset \cdots \subset \mathbb{A}^{\mathsf{cen}}$$

hold and are proper. The class $\mathbb{A}^{\mathsf{cen}}$ is not a variety, but each of its subvarieties is contained in $\mathbb{A}_n^{\mathsf{cen}}$ for all sufficiently large n. For instance, the almost Cross varieties $\mathbb{J}_1$ and $\mathbb{J}_2$ are subvarieties of $\mathbb{A}_2^{\mathsf{cen}}$, and $\mathbb{P}_{25}$ is a subvariety of $\mathbb{A}_3^{\mathsf{cen}}$ but not of $\mathbb{A}_2^{\mathsf{cen}}$.

In Chap. 12, a description of Cross subvarieties of $\mathbb{A}^{\mathsf{cen}}$ is established. The subvariety

$$\mathbb{K} = \mathbb{A}_2^{\mathsf{cen}} \cap \mathbb{O} \cap \mathbb{O}^{\triangleleft}$$

of $\mathbb{A}_2^{\mathsf{cen}}$ is first shown in Sect. 12.2 to be almost Cross. Unlike the almost Cross subvarieties $\mathbb{J}_1$ and $\mathbb{J}_2$ of $\mathbb{A}_2^{\mathsf{cen}}$—which are non-finitely based, finitely generated, and small—the variety $\mathbb{K}$ is finitely based and non-finitely generated, and its subvarieties constitute the countably infinite chain

$$\begin{aligned}\mathbf{0} &\subset \mathbb{R}_Q\varnothing \subset \mathbb{R}_Q\{x\} \subset \mathbb{R}_Q\{xy\} \\ &\subset \mathbb{R}_Q\{xy_1x\} \subset \mathbb{R}_Q\{xy_1xy_2x\} \subset \mathbb{R}_Q\{xy_1xy_2xy_3x\} \subset \cdots \subset \mathbb{K}.\end{aligned}$$

(In contrast, the variety $\mathbb{A}_n^{\mathsf{cen}} \cap \mathbb{O} \cap \mathbb{O}^{\triangleleft}$ is finitely universal for each $n \geq 3$ (Gusev and Lee 2020).) Based on this description of the lattice $\mathfrak{L}(\mathbb{K})$, the variety $\mathbb{K}$ is also shown to be a subvariety of $\mathbb{B}_2^1$.

It turns out that by the aforementioned result of Straubing (1982), each subvariety of $\mathbb{A}^{\mathsf{cen}} \cap \mathbb{O}$ that excludes $\mathbb{K}$ is contained in $\mathbb{S}_n$ for some $n \geq 2$. This result, together with the characterization of hereditarily finitely based subvarieties of $\mathbb{A}^{\mathsf{cen}}$, is essential in Sect. 12.3 in establishing the following description of Cross subvarieties of $\mathbb{A}^{\mathsf{cen}}$.

Theorem 1.64 *The following statements on any subvariety* $\mathbb{V}$ *of* $\mathbb{A}^{\mathsf{cen}}$ *are equivalent*:

(a) $\mathbb{V}$ *is Cross*;
(b) $\mathbb{J}_1, \mathbb{J}_2, \mathbb{K} \nsubseteq \mathbb{V}$;
(c) $\mathbb{V} \subseteq \mathbb{A}_n^{\mathsf{cen}} \cap \mathbb{O} \cap \mathbb{S}_n$ *or* $\mathbb{V} \subseteq \mathbb{A}_n^{\mathsf{cen}} \cap \mathbb{O}^{\triangleleft} \cap \mathbb{S}_n$ *for some* $n \geq 2$.

Therefore, the subvarieties of $\mathbb{A}_n^{\mathsf{cen}} \cap \mathbb{O} \cap \mathbb{S}_n$ and $\mathbb{A}_n^{\mathsf{cen}} \cap \mathbb{O}^{\triangleleft} \cap \mathbb{S}_n$, where $n \geq 2$, are precisely all Cross subvarieties of $\mathbb{A}^{\mathsf{cen}}$, and $\mathbb{J}_1$, $\mathbb{J}_2$, and $\mathbb{K}$ are the only almost Cross subvarieties of $\mathbb{A}^{\mathsf{cen}}$. In particular, $\mathbb{K}$ is the unique non-finitely generated almost Cross subvariety of $\mathbb{A}^{\mathsf{cen}}$. Other examples of non-finitely generated almost Cross varieties of monoids include the classical examples $\mathbb{C}\text{OM}$ and $\mathbb{I}\text{DEM}$ (Head 1968; Wismath 1986), the maximal subvariety of the variety generated by a certain monoid of order six (Jackson and Lee 2018, Proposition 5.11), and the subvariety of $\mathscr{V}_{\mathsf{mon}}\{\mathcal{M}_5\}$ consisting of monoids with commuting idempotents (Gusev and Vernikov 2018, proof of Corollary 7.6). Up to isomorphism and anti-isomorphism, the monoid $\mathcal{M}_5$ of order five is the unique smallest monoid that generates a non-Cross variety of monoids (Lee and Zhang 2014).

Now it is easily shown that every variety $\mathbb{S}_n$ excludes the variety $\mathbb{K}$. But since each finite monoid in $\mathbb{A}^{\mathsf{cen}}$ belongs to some $\mathbb{S}_n$ (Straubing 1982), any subvariety of $\mathbb{A}^{\mathsf{cen}}$ that contains $\mathbb{K}$ is non-finitely generated, so that $\mathbb{K}$ is *inherently non-finitely generated within* $\mathbb{A}^{\mathsf{cen}}$. In fact, $\mathbb{K}$ is shown in Sect. 12.4 to be contained in every variety that is inherently non-finitely generated within $\mathbb{A}^{\mathsf{cen}}$.

Theorem 1.65 *The following statements on any subvariety* $\mathbb{V}$ *of* $\mathbb{A}^{\mathsf{cen}}$ *are equivalent*:

(a) $\mathbb{V}$ *is inherently non-finitely generated within* $\mathbb{A}^{\mathsf{cen}}$;
(b) $\mathbb{K} \subseteq \mathbb{V}$;
(c) $\mathbb{V} \nsubseteq \mathbb{S}_n$ *for all* $n \geq 2$.

Consequently, $\mathbb{K}$ is the variety that is smallest with respect to being inherently non-finitely generated within $\mathbb{A}^{\mathsf{cen}}$, and the subvarieties of $\mathbb{A}^{\mathsf{cen}}$ containing $\mathbb{K}$ are precisely varieties that are inherently non-finitely generated within $\mathbb{A}^{\mathsf{cen}}$. However, $\mathbb{K}$ is not inherently non-finitely generated within the variety $\mathbb{M}\text{ON}$ of all monoids since it is contained in the finitely generated variety $\mathbb{B}_2^1$. The variety $\mathbb{C}\text{OM}$ is clearly inherently non-finitely generated within $\mathbb{M}\text{ON}$; a less trivial example is the variety $\mathbb{I}\text{DEM}$ (O.B. Sapir 2005).

A variety that is inherently non-finitely generated within $\mathbb{A}^{\mathsf{cen}}$ is vacuously non-finitely generated. Intuitively, the existence of a non-finitely generated subvariety of $\mathbb{A}^{\mathsf{cen}}$ that is not inherently non-finitely generated within $\mathbb{A}^{\mathsf{cen}}$ seems normal, but no such examples have been found. More specifically, it has been unknown if the following problem has a positive solution.

Problem 1.66 (Jackson, Email Communication, 27 Jan 2008) Is there a finite set $\mathscr{W}$ of words such that the variety $\mathbb{R}_{\mathrm{Q}}\mathscr{W}$ of monoids contains a non-finitely generated subvariety?

In Sect. 12.5, the variety $\mathbb{R}_{\mathrm{Q}}\{x^2y^2\}$, which is finitely based by Theorem 1.24, is shown to contain a non-finitely generated subvariety. Therefore, Problem 1.66 is positively answered by the simple set $\mathscr{W} = \{x^2y^2\}$.

A next step in the investigation is to generalize results from Theorems 1.64 and 1.65 to larger classes of monoids containing $\mathbb{A}^{\mathsf{cen}}$. One natural class that properly contains $\mathbb{A}^{\mathsf{cen}}$ is the class $\mathbb{A}^{\mathsf{com}}$ of all aperiodic monoids with commuting idempotents. Let $\mathbb{A}_n^{\mathsf{com}}$ denote the subvariety of $\mathbb{A}^{\mathsf{com}}$ consisting of monoids of index at most n. Similar to its subclass $\mathbb{A}^{\mathsf{cen}}$, the class $\mathbb{A}^{\mathsf{com}}$ is not a variety, but each of its subvarieties is contained in $\mathbb{A}_n^{\mathsf{com}}$ for all sufficiently large n.

As observed earlier, the variety $\mathbb{K}$ is contained in the finitely generated subvariety $\mathbb{B}_2^1$ of $\mathbb{A}^{\mathsf{com}}$. Therefore, $\mathbb{K}$ is not inherently non-finitely generated within $\mathbb{A}^{\mathsf{com}}$, whence Theorem 1.65 does not generalize directly to subvarieties of $\mathbb{A}^{\mathsf{com}}$. But the almost Cross subvarieties $\mathbb{J}_1$, $\mathbb{J}_2$, and $\mathbb{K}$ of $\mathbb{A}^{\mathsf{cen}}$ do play a vital role in the generalization of Theorem 1.64 to some subclass of $\mathbb{A}^{\mathsf{com}}$.

For any class $\mathbb{V}$ of monoids, let $\widetilde{\mathbb{V}}$ denote the class of monoids from $\mathbb{V}$ that satisfy the identity $x^2yx \approx xyx^2$. In Chap. 13, the following characterization of Cross subvarieties of $\widetilde{\mathbb{A}}^{\mathsf{com}}$ is established.

Theorem 1.67 *A subvariety of $\widetilde{\mathbb{A}}^{\mathsf{com}}$ is Cross if and only if it excludes $\mathbb{J}_1$, $\mathbb{J}_2$, and $\mathbb{K}$.*

Consequently, $\mathbb{J}_1$, $\mathbb{J}_2$, and $\mathbb{K}$ are the only almost Cross subvarieties of $\widetilde{\mathbb{A}}^{\mathsf{com}}$.

Although satisfaction of the identity $x^2yx \approx xyx^2$ is a substantial restriction, the class $\widetilde{\mathbb{A}}^{\mathsf{com}}$ is still highly nontrivial since it contains the infinite chain

$$\widetilde{\mathbb{A}}_1^{\mathsf{com}} \subset \widetilde{\mathbb{A}}_2^{\mathsf{com}} \subset \widetilde{\mathbb{A}}_3^{\mathsf{com}} \subset \cdots$$

of subvarieties, where even $\widetilde{\mathbb{A}}_2^{\mathsf{com}}$ contains continuum many subvarieties due to its subvariety $\mathbb{B}_2^1$ having the same property (Jackson and Lee 2018). Incidentally, since the inclusions $\mathbb{J}_1, \mathbb{J}_2, \mathbb{K} \subseteq \mathbb{B}_2^1 \subset \widetilde{\mathbb{A}}^{\mathsf{com}}$ hold, Theorem 1.67 also applies to subvarieties of $\mathbb{B}_2^1$.

Corollary 1.68 *The varieties $\mathbb{J}_1$, $\mathbb{J}_2$, and $\mathbb{K}$ are the only almost Cross subvarieties of $\mathbb{B}_2^1$.*

In view of Theorem 1.67, it is logical to ask if $\mathbb{J}_1$, $\mathbb{J}_2$, and $\mathbb{K}$ are the only almost Cross subvarieties of $\mathbb{A}^{\mathsf{com}}$. The answer to this question is negative, as Gusev (2020a,b) recently proved that there are precisely nine almost Cross subvarieties of $\mathbb{A}^{\mathsf{com}}$, four of which are also limit varieties.

1.6.4 Further Examples Involving Rees Quotients of Free Monoids

In Chap. 14, the last chapter of the monograph, several finite monoids with interesting equational properties are constructed from Rees quotients of $\mathscr{A}_{\varnothing}^{+}$.

As shown in Example 1.52 and Theorem 1.56, Rees quotients of $\mathscr{A}_{\varnothing}^{+}$ provide pairs of finitely based finite monoids M_1 and M_2 whose direct product $M_1 \times M_2$ is non-finitely based. However, the monoids involved are quite large:

- $|\mathcal{R}_Q\{x^2y^2x, xyxyx, xy^2x^2\}| = 22$ and $|\mathcal{R}_Q\{x^2y^2, xyxy, xy^2x, yx^3y\}| = 21$;
- $|\mathcal{R}_Q\{xyzxy, xyzyx\}| = 21$ and $|\mathcal{R}_Q\{x^ny^n\}| = (n+1)^2 + 1$ for any $n \geq 2$;
- $|\mathcal{R}_Q\{xyhxty\}| = |\mathcal{R}_Q\{xhytxy\}| = 21$.

If one looks beyond the class of Rees quotients of $\mathscr{A}_{\varnothing}^{+}$ or the larger class $\mathbb{A}^{\mathsf{cen}}$, then much smaller examples are available. For instance, the monoid $\mathcal{R}_Q\{xyx\}$ of order seven and the monoid $\mathcal{A}_0^1$ of order five are finitely based but their direct product $\mathcal{R}_Q\{xyx\} \times \mathcal{A}_0^1$ is not (O.B. Sapir 2015b). Recall that this direct product was earlier shown by Proposition 1.63(i) to satisfy the weaker property of being not hereditarily finitely based.

Since the aforementioned monoids are all aperiodic, it is of interest to locate examples with nontrivial subgroups. In view of Proposition 1.63(ii) and since every finite group is finitely based (Oates and Powell 1964), a natural candidate for consideration is the direct product $\mathcal{R}_Q\{xyx\} \times G$, where G is a finite noncommutative group. In Sect. 14.1, such a direct product is shown by the syntactic method to be non-finitely based if G is a noncommutative group of finite exponent; in particular, the monoid $\mathcal{R}_Q\{xyx\} \times \mathcal{S}ym_n$ that first appeared in Fig. 1.2 is non-finitely based for all $n \geq 3$. On the other hand, by Proposition 1.63(ii), the monoid $\mathcal{R}_Q\{xyx\} \times G$ is finitely based if G is a commutative group. Consequently, the finite basis problem for the direct product of $\mathcal{R}_Q\{xyx\}$ with any group of finite exponent has a straightforward solution.

Theorem 1.69 *For any group G of finite exponent, the direct product $\mathcal{R}_Q\{xyx\} \times G$ is finitely based if and only if G is commutative.*

Besides the formation of direct products, other constructions under which the class of finitely based semigroups is not closed include the formation of homomorphic images and of subsemigroups; see Volkov (2001, Table 1) for more information.

The class of finitely based semigroups is also not closed under the operator $S \mapsto S^1$. For instance, the semigroups $\mathcal{A}_2$ and $\mathcal{B}_2$ are finitely based (Reilly 2008a; Trahtman 1994) but the monoids $\mathcal{A}_2^1$ and $\mathcal{B}_2^1$ are not (M.V. Sapir 1987b). Rees quotients of $\mathscr{A}_{\varnothing}^{+}$ also provide many examples: for any finite $\mathcal{R}_Q\mathscr{W}$ that is non-finitely based, the semigroup $S = \mathcal{R}_Q\mathscr{W}\backslash\{\mathtt{1}\}$ is nilpotent and so is finitely based (Perkins 1969), but the monoid $S^1 = \mathcal{R}_Q\mathscr{W}$ is non-finitely based. These examples led Shneerson (1989) to question the existence of semigroups with the "opposite" property: is there a semigroup S that is non-finitely based such that the monoid S^1 is finitely based? It is convenient to say that such a semigroup S is *conformable*, as suggested by Jackson (email communication, 6 Nov 2013). Shneerson (1989) provided a positive answer by exhibiting the first example of a conformable semigroup:

$$\mathcal{T} = \langle \mathsf{a}, \mathsf{b} \mid \mathsf{aba} = \mathsf{ba} \rangle.$$

But unlike the finite examples that motivated his investigation, the semigroup $\mathcal{T}$ is infinite. Further, no other conformable semigroup has since been found. Therefore, the restriction of Shneerson's question to finite semigroups is of fundamental interest.

Problem 1.70 Do finite conformable semigroups exist?

It so happens that by a result of Volkov (1984b) that preceded the publication of Shneerson (1989), an abundance of finite conformable semigroups can be constructed.

Theorem 1.71 *Suppose that S and N are any semigroups such that S^1 is non-finitely based, N is nilpotent, and $S^1 \times N^1$ is finitely based. Then the direct product $P = S^1 \times N$ is conformable.*

It is easily seen that the nilpotent semigroups $S = M_1 \backslash \{1\}$ and $N = M_2 \backslash \{1\}$ from Example 1.53 can be used to form a conformable semigroup $P = S^1 \times N$. Specifically, if

$$S = \mathcal{R}_{\mathcal{Q}}\{xyxy\} \backslash \{1\} \text{ and } N = \mathcal{R}_{\mathcal{Q}}\{x^2y^2, xy^2x\} \backslash \{1\},$$

then $P = S^1 \times N$ is conformable by Theorem 1.71. By Jackson and Sapir (2000, Corollary 5.2), finite semigroups S and N that satisfy the hypothesis of Theorem 1.71 can be systematically and easily selected from the class of Rees quotients of $\mathscr{A}^+_\varnothing$.

Corollary 1.72 *There exist finite semigroups $P_0 \subset P_1 \subset P_2 \subset \cdots$ such that for each $k \geq 0$,*

(a) *P_{2k} is finitely based but P^1_{2k} is non-finitely based;*
(b) *P_{2k+1} is non-finitely based but P^1_{2k+1} is finitely based.*

In Sect. 14.2, Theorem 1.71 and Corollary 1.72 are established and a conformable semigroup of order 20 is constructed. Although it is unknown if there is a smaller example, it is shown that any conformable semigroup is of order at least seven. Given that $\mathcal{A}_2$ and $\mathcal{B}_2$ are the smallest examples of finitely based semigroups S with non-finitely based monoids S^1, it is relevant to pose the following problem.

Problem 1.73 What is the smallest possible order of a conformable semigroup?

2 Preliminaries

Acquaintance with the rudiments of semigroup theory and universal algebra is assumed. Refer to the monographs of Howie (1995) and Burris and Sankappanavar (1981) for more information.

2.1 Identities and Deducibility

Let $\mathscr{A}$ be a countably infinite alphabet. For any subset $\mathscr{X}$ of $\mathscr{A}$, let $\mathscr{X}^+$ and $\mathscr{X}^+_\varnothing = \mathscr{X}^+ \cup \{\varnothing\}$ denote the free semigroup and free monoid over $\mathscr{X}$, respectively. Elements of $\mathscr{A}$ are called *variables* and elements of $\mathscr{A}^+_\varnothing$ are called *words*. For any word $\mathbf{w} \in \mathscr{A}^+_\varnothing$:

- the *content* of $\mathbf{w}$, denoted by $\mathsf{con}(\mathbf{w})$, is the set of variables occurring in $\mathbf{w}$;
- the *head* of $\mathbf{w}$, denoted by $\mathsf{h}(\mathbf{w})$, is the first variable occurring in $\mathbf{w}$;
- the *tail* of $\mathbf{w}$, denoted by $\mathsf{t}(\mathbf{w})$, is the last variable occurring in $\mathbf{w}$;
- the *reversal* of $\mathbf{w}$, denoted by $\mathbf{w}^{\triangleleft}$, is the word obtained by writing $\mathbf{w}$ in reverse;
- the number of times a variable $x \in \mathscr{A}$ occurs in $\mathbf{w}$ is denoted by $\mathsf{occ}(x, \mathbf{w})$;
- a variable $x \in \mathscr{A}$ in $\mathbf{w}$ is *simple* if $\mathsf{occ}(x, \mathbf{w}) = 1$ and *non-simple* if $\mathsf{occ}(x, \mathbf{w}) \geq 2$;
- the set of simple variables of $\mathbf{w}$ is denoted by $\mathsf{sim}(\mathbf{w})$;
- the set of non-simple variables of $\mathbf{w}$ is denoted by $\mathsf{non}(\mathbf{w})$.

Note that $\mathsf{con}(\mathbf{w}) = \mathsf{sim}(\mathbf{w}) \cup \mathsf{non}(\mathbf{w})$ and $\mathsf{sim}(\mathbf{w}) \cap \mathsf{non}(\mathbf{w}) = \varnothing$. A word is *simple* if all its variables are simple; otherwise, it is *non-simple*. A simple word that contains only one variable is called a *singleton*.

Let $\stackrel{\circ}{=}$ denote the relation on $\mathscr{A}^+_\varnothing$ given by $\mathbf{w} \stackrel{\circ}{=} \mathbf{w}'$ if $\mathsf{occ}(x, \mathbf{w}) = \mathsf{occ}(x, \mathbf{w}')$ for all $x \in \mathscr{A}$. Equivalently, two words $\mathbf{w}, \mathbf{w}' \in \mathscr{A}^+_\varnothing$ are $\stackrel{\circ}{=}$-related if they can be obtained from one another by rearrangement of the variables.

E. W. H. Lee, *Advances in the Theory of Varieties of Semigroups*, Frontiers in Mathematics, https://doi.org/10.1007/978-3-031-16497-2_2

An *identity* is an expression of the form $\mathbf{w} \approx \mathbf{w}'$, where $\mathbf{w}, \mathbf{w}' \in \mathscr{A}^+$. An identity $\mathbf{w} \approx \mathbf{w}'$ is *balanced* if $\mathbf{w} \stackrel{\circ}{=} \mathbf{w}'$; it is *trivial* if $\mathbf{w} = \mathbf{w}'$. A *permutation identity* is a balanced identity formed by a pair of simple words. In other words, a permutation identity is of the form

$$x_1 x_2 \cdots x_k \approx x_{1\pi} x_{2\pi} \cdots x_{k\pi},$$

where π is some permutation of $\{1, 2, \ldots, k\}$.

A semigroup S *satisfies* an identity $\mathbf{w} \approx \mathbf{w}'$, indicated by $S \models \mathbf{w} \approx \mathbf{w}'$, if under every substitution $\varphi : \mathscr{A} \to S$, the elements $\mathbf{w}\varphi$ and $\mathbf{w}'\varphi$ of S are equal. More generally, a class $\mathfrak{C}$ of semigroups *satisfies* an identity $\mathbf{w} \approx \mathbf{w}'$, indicated by $\mathfrak{C} \models \mathbf{w} \approx \mathbf{w}'$, if every semigroup in $\mathfrak{C}$ satisfies $\mathbf{w} \approx \mathbf{w}'$. A semigroup S *violates* an identity $\mathbf{w} \approx \mathbf{w}'$ if $S \not\models \mathbf{w} \approx \mathbf{w}'$. For any class $\mathfrak{C}$ of semigroups, let $\mathsf{id}\,\mathfrak{C}$ denote the set of all identities satisfied by every semigroup in $\mathfrak{C}$, called the *equational theory* of $\mathfrak{C}$, and let $\mathsf{id}_n\mathfrak{C}$ denote the set of identities in $\mathsf{id}\,\mathfrak{C}$ that involve at most n distinct variables.

An identity $\mathbf{w} \approx \mathbf{w}'$ is *directly deducible* from an identity $\mathbf{u} \approx \mathbf{u}'$ if there exist words $\mathbf{e}, \mathbf{f} \in \mathscr{A}^+_\varnothing$ and a substitution $\varphi : \mathscr{A} \to \mathscr{A}^+$ such that $\mathbf{w} = \mathbf{e}(\mathbf{u}\varphi)\mathbf{f}$ and $\mathbf{w}' = \mathbf{e}(\mathbf{u}'\varphi)\mathbf{f}$. An identity $\mathbf{w} \approx \mathbf{w}'$ is *deducible* from a set Σ of identities, indicated by $\Sigma \vdash \mathbf{w} \approx \mathbf{w}'$ or $\mathbf{w} \stackrel{\Sigma}{\approx} \mathbf{w}'$, if there exists a finite sequence

$$\mathbf{w} = \mathbf{w}_1, \mathbf{w}_2, \ldots, \mathbf{w}_m = \mathbf{w}'$$

of words, where each identity $\mathbf{w}_i \approx \mathbf{w}_{i+1}$ is directly deducible from some identity in Σ.

Deducibility can be defined in the obvious manner for algebras of any general type; see, for example, Sect. 2.5 for the definition of deducibility of identities in the involution semigroup signature. In general, two sets of identities Σ_1 and Σ_2 are *equivalent*, indicated by $\Sigma_1 \sim \Sigma_2$, if the deductions $\Sigma_1 \vdash \Sigma_2$ and $\Sigma_2 \vdash \Sigma_1$ hold.

2.2 Varieties and Identity Bases

The variety *generated* by a class $\mathfrak{C}$ of algebras is the smallest variety containing $\mathfrak{C}$; such a smallest variety exists because the intersection of varieties is a variety. A variety generated by a single finite algebra is *finitely generated*. A variety is *locally finite* if every finitely generated algebra in it is finite. It is well known that every finitely generated variety is locally finite; see, for example, Burris and Sankappanavar (1981, Theorem II.10.16).

The variety *defined* by a set Σ of identities is the class of all algebras that satisfy all identities in Σ; in this case, Σ is an *identity basis* for the variety. More generally, an identity basis for a variety $\mathbf{V}$ is any subset Σ of $\mathsf{id}\,\mathbf{V}$ such that $\Sigma \vdash \mathsf{id}\,\mathbf{V}$. For an algebra A, a subset Σ of $\mathsf{id}\{A\}$ is an *identity basis* for A if Σ is an identity basis for the variety generated by A. A variety or algebra is *finitely based* if it possesses some finite identity basis. For any variety $\mathbf{V}$ and set Σ of identities, the subvariety of $\mathbf{V}$ defined by Σ is denoted by $\mathbf{V}\Sigma$. For any varieties $\mathbf{U}$ and $\mathbf{V}$ such that $\mathbf{U} \subseteq \mathbf{V}$, the *interval* $[\mathbf{U}, \mathbf{V}]$ is the set of subvarieties of $\mathbf{V}$ containing $\mathbf{U}$. The lattice of subvarieties of a variety $\mathbf{V}$ is denoted by $\mathfrak{L}(\mathbf{V})$; equivalently, $\mathfrak{L}(\mathbf{V}) = [\mathbf{0}, \mathbf{V}]$, where $\mathbf{0}$ is the trivial variety.

The *dual* of a semigroup S with binary operation $\cdot$ is the semigroup $S^{\triangleleft} = \{a^{\triangleleft} \mid a \in S\}$ with binary operation $\circ$ given by $a^{\triangleleft} \circ b^{\triangleleft} = (b \cdot a)^{\triangleleft}$ for all $a^{\triangleleft}, b^{\triangleleft} \in S^{\triangleleft}$. The *dual* of a variety $\mathbf{V}$ is the variety

$$\mathbf{V}^{\triangleleft} = \{S^{\triangleleft} \mid S \in \mathbf{V}\}.$$

Equivalently, the dual of a variety with identity basis Σ is the variety defined by

$$\Sigma^{\triangleleft} = \{\mathbf{u}^{\triangleleft} \approx \mathbf{v}^{\triangleleft} \mid \mathbf{u} \approx \mathbf{v} \in \Sigma\}.$$

The identity $x^m \approx x^n$, where $m > n \geq 1$, is called a *periodicity identity*. A variety is *periodic* if it satisfies some periodicity identity. In particular, a variety that satisfies a periodicity identity of the form $x^{n+1} \approx x^n$ excludes nontrivial groups and so is aperiodic. Let $\mathbf{P}_n$ denote the variety of semigroups defined by the periodicity identity

$$x^{2n} \approx x^n. \tag{$\wp_n$}$$

It is easily shown that a variety of semigroups is periodic if and only if it is contained in $\mathbf{P}_n$ for some $n \geq 1$. A semigroup is *uniformly periodic* if it satisfies the identity $\wp_n$ for some $n \geq 1$.

2.3 Connected Words and Identities

In this subsection, it is shown that certain varieties containing the semigroup

$$\mathcal{A}_0 = \langle \mathsf{e}, \mathsf{f} \mid \mathsf{e}^2 = \mathsf{e},\ \mathsf{f}^2 = \mathsf{f},\ \mathsf{ef} = 0 \rangle$$

can be defined by identities of a very specific form. Results of the subsection will be useful in Chaps. 3, 5, and 6.

Two words $\mathbf{w}_1, \mathbf{w}_2 \in \mathscr{A}_{\varnothing}^{+}$ are *disjoint* if $\mathsf{con}(\mathbf{w}_1) \cap \mathsf{con}(\mathbf{w}_2) = \varnothing$. A non-simple word $\mathbf{w}$ is *connected* if it cannot be decomposed into a product of two disjoint nonempty words; in other words, if $\mathbf{w} = \mathbf{w}_1\mathbf{w}_2$ for some $\mathbf{w}_1, \mathbf{w}_2 \in \mathscr{A}_{\varnothing}^{+}$, then $\mathsf{con}(\mathbf{w}_1) \cap \mathsf{con}(\mathbf{w}_2) \neq \varnothing$. An identity is *connected* if it is formed by a pair of connected words.

Lemma 2.1 (Edmunds 1980, Proof of Part 4 of the First Proposition) *Suppose that* $\mathbf{w}, \mathbf{w}' \in \mathscr{A}^{+}$ *are any connected words such that* $\mathsf{con}(\mathbf{w}) = \mathsf{con}(\mathbf{w}')$. *Then* $\mathcal{A}_0 \models \mathbf{w} \approx \mathbf{w}'$.

Proof For any substitution $\varphi : \mathscr{A} \to \mathcal{A}_0$, one of the following conditions holds:

(A) $x\varphi \in \{0, \mathsf{fe}\}$ for some $x \in \mathsf{con}(\mathbf{w}) = \mathsf{con}(\mathbf{w}')$;
(B) $x\varphi = \mathsf{e}$ and $y\varphi = \mathsf{f}$ for some $x, y \in \mathsf{con}(\mathbf{w}) = \mathsf{con}(\mathbf{w}')$;
(C) $x\varphi = \mathsf{e}$ for all $x \in \mathsf{con}(\mathbf{w}) = \mathsf{con}(\mathbf{w}')$;
(D) $x\varphi = \mathsf{f}$ for all $x \in \mathsf{con}(\mathbf{w}) = \mathsf{con}(\mathbf{w}')$.

Each connected word either begins and ends with the same variable or is of the overlapping form

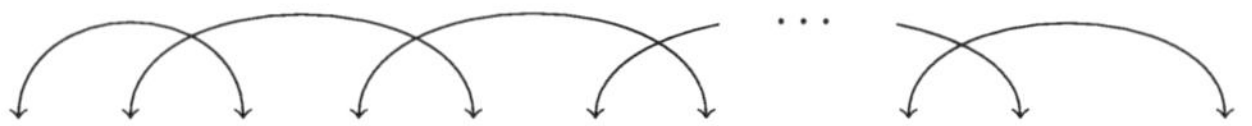

$$x_1\, \mathbf{a}_1\, x_2\, \mathbf{a}_2\, x_1\, \mathbf{b}_1\, x_3\, \mathbf{a}_3\, x_2\, \mathbf{b}_2\, x_4\, \mathbf{a}_4\, x_3\, \mathbf{b}_3 \cdots x_r\, \mathbf{a}_r\, x_{r-1}\, \mathbf{b}_{r-1}\, x_r,$$

where $x_1, x_2, \ldots, x_r \in \mathscr{A}$ are distinct and $\mathbf{a}_i, \mathbf{b}_i \in \mathscr{A}^+_\varnothing$. (Note that $\mathbf{a}_i$ follows the first x_i and $\mathbf{b}_i$ follows the last x_i.) Hence, it is routinely checked that $\mathbf{w}\varphi = 0 = \mathbf{w}'\varphi$ in (A) and (B), $\mathbf{w}\varphi = \mathsf{e} = \mathbf{w}'\varphi$ in (C), and $\mathbf{w}\varphi = \mathsf{f} = \mathbf{w}'\varphi$ in (D). □

A semigroup S is *left idempotent-separable* if for any distinct elements $a, b \in S$, there exists an idempotent $e \in S$ such that $ea \neq eb$; a *right idempotent-separable* semigroup is dually defined. A semigroup is *idempotent-separable* if it is both left idempotent-separable and right idempotent-separable. The following result is routinely verified.

Lemma 2.2 *The following semigroups are idempotent-separable*:

(i) $\mathcal{A}_0$, $\mathcal{A}_2$, $\mathcal{B}_0$, $\mathcal{B}_2$, *and* $\mathcal{L}_3$;
(ii) *monoids*;
(iii) *direct products of idempotent-separable semigroups.*

Lemma 2.3 (Lee and Volkov 2007, Proposition 3.2(i)) *Let S be any idempotent-separable semigroup such that $\mathcal{A}_0 \in \mathscr{V}_{\mathsf{sem}}\{S\}$ and $\mathbf{w} \approx \mathbf{w}'$ be any identity satisfied by S.*

(i) *If either $\mathbf{w}$ or $\mathbf{w}'$ is simple, then $\mathbf{w} = \mathbf{w}'$.*
(ii) *If $\mathbf{w}$ and $\mathbf{w}'$ are non-simple, then*

$$\mathbf{w} = \mathbf{s}_0\mathbf{w}_1\mathbf{s}_1\mathbf{w}_2\mathbf{s}_2 \cdots \mathbf{w}_m\mathbf{s}_m \quad \textit{and} \quad \mathbf{w}' = \mathbf{s}_0\mathbf{w}'_1\mathbf{s}_1\mathbf{w}'_2\mathbf{s}_2 \cdots \mathbf{w}'_m\mathbf{s}_m,$$

where

- $\mathbf{s}_0, \mathbf{s}_1, \ldots, \mathbf{s}_m \in \mathscr{A}^+_\varnothing$ *are simple words*;
- $\mathbf{w}_1, \mathbf{w}_2, \ldots, \mathbf{w}_m, \mathbf{w}'_1, \mathbf{w}'_2, \ldots, \mathbf{w}'_m \in \mathscr{A}^+$ *are connected words*;
- $\mathbf{s}_0, \mathbf{w}_1, \mathbf{s}_1, \mathbf{w}_2, \mathbf{s}_2, \ldots, \mathbf{w}_m, \mathbf{s}_m$ *are pairwise disjoint*;

- $\mathbf{s}_0, \mathbf{w}'_1, \mathbf{s}_1, \mathbf{w}'_2, \mathbf{s}_2, \ldots, \mathbf{w}'_m, \mathbf{s}_m$ *are pairwise disjoint;*
- $\mathsf{con}(\mathbf{w}_i) = \mathsf{con}(\mathbf{w}'_i)$ *and* $S \models \mathbf{w}_i \approx \mathbf{w}'_i$ *for all* i.

Proof It is well known and easily checked that identities satisfied by any nontrivial semilattice are homotypical. Since the subsemigroup $\{0, \mathsf{e}\}$ of $\mathcal{A}_0$ is isomorphic to $\mathcal{S}\ell_2$ and $\mathcal{A}_0 \models \mathbf{w} \approx \mathbf{w}'$, it follows that $\mathsf{con}(\mathbf{w}) = \mathsf{con}(\mathbf{w}')$.

(i) Suppose that $\mathbf{w} = x_1 x_2 \cdots x_m$ is simple. Then $\mathsf{con}(\mathbf{w}') = \{x_1, x_2, \ldots, x_m\}$. If $\mathbf{w}'$ is non-simple, say $\mathsf{occ}(x_r, \mathbf{w}') \geq 2$ for some $r \in \{1, 2, \ldots, m\}$, then under the substitution $\varphi : \mathscr{A} \to \mathcal{A}_0$ given by

$$x \mapsto \begin{cases} \mathsf{f} & \text{if } x \in \{x_1, x_2, \ldots, x_{r-1}\}, \\ \mathsf{fe} & \text{if } x = x_r, \\ \mathsf{e} & \text{otherwise,} \end{cases}$$

the contradiction $\mathbf{w}\varphi = \mathsf{fe} \neq 0 = \mathbf{w}'\varphi$ is obtained. Hence, $\mathbf{w}'$ is simple. Further, if x_r precedes x_{r-1} in $\mathbf{w}'$ for some $r \in \{2, 3, \ldots, m\}$, then $\mathbf{w}\varphi = \mathsf{fe} \neq 0 = \mathbf{w}'\varphi$, which is a contradiction. Therefore, x_{i-1} precedes x_i in $\mathbf{w}'$ for all $i \in \{2, 3, \ldots, m\}$, whence $\mathbf{w} = \mathbf{w}'$.

(ii) It suffices to assume that $\mathbf{w} = \mathbf{uv}$ is a decomposition into disjoint words $\mathbf{u}, \mathbf{v} \in \mathscr{A}^+$, and then show that $\mathbf{w}'$ can also be decomposed into $\mathbf{w}' = \mathbf{u}'\mathbf{v}'$, where $\mathbf{u}', \mathbf{v}' \in \mathscr{A}^+$ are such that $\mathsf{con}(\mathbf{u}) = \mathsf{con}(\mathbf{u}')$, $\mathsf{con}(\mathbf{v}) = \mathsf{con}(\mathbf{v}')$, and $S \models \{\mathbf{u} \approx \mathbf{u}', \mathbf{v} \approx \mathbf{v}'\}$. Let $\varphi : \mathscr{A} \to \mathcal{A}_0$ denote the substitution

$$x \mapsto \begin{cases} \mathsf{f} & \text{if } x \in \mathsf{con}(\mathbf{u}), \\ \mathsf{e} & \text{otherwise.} \end{cases}$$

Then $\mathbf{w}'\varphi = \mathbf{w}\varphi = \mathsf{fe}$. If within $\mathbf{w}'$, some $y \in \mathsf{con}(\mathbf{v})$ precedes some $x \in \mathsf{con}(\mathbf{u})$, then the contradiction $\mathbf{w}'\varphi = 0$ is obtained. Therefore, within $\mathbf{w}'$, every $x \in \mathsf{con}(\mathbf{u})$ precedes every $y \in \mathsf{con}(\mathbf{v})$, whence the aforementioned decomposition $\mathbf{w}' = \mathbf{u}'\mathbf{v}'$ follows.

It remains to verify that $S \models \mathbf{u} \approx \mathbf{u}'$ and $S \models \mathbf{v} \approx \mathbf{v}'$. Seeking a contradiction, suppose that $S \not\models \mathbf{u} \approx \mathbf{u}'$. Then there exists some substitution $\chi_1 : \mathsf{con}(\mathbf{u}) \to S$ such that $\mathbf{u}\chi_1 \neq \mathbf{u}'\chi_1$. Since S is idempotent-separable, there exists some idempotent $e \in S$ such that $(\mathbf{u}\chi_1)e \neq (\mathbf{u}'\chi_1)e$. But under the substitution $\chi_2 : \mathscr{A} \to S$ given by

$$x \mapsto \begin{cases} x\chi_1 & \text{if } x \in \mathsf{con}(\mathbf{u}), \\ e & \text{otherwise,} \end{cases}$$

the contradiction $\mathbf{u}\chi_2 \neq \mathbf{u}'\chi_2$ is obtained. Therefore, $S \models \mathbf{u} \approx \mathbf{u}'$. A similar contradiction can be deduced from $S \not\models \mathbf{v} \approx \mathbf{v}'$. □

A semigroup S in the variety $\mathbf{P}_n$ is said to be *absorbing* if for any connected word $\mathbf{w}$, there exist some nontrivial prefix $\mathbf{e}$ of $\mathbf{w}$ and some nontrivial suffix $\mathbf{f}$ of $\mathbf{w}$ such that S satisfies the identity $\mathbf{e}^n\mathbf{w}\mathbf{f}^n \approx \mathbf{w}$. A variety is said to be *absorbing* if it is generated by some absorbing semigroup.

Example 2.4 Any semigroup in $\mathbf{P}_n$ that satisfies $x^{n+1}yx^{n+1} \approx xyx$ or $(xy)^{n+1}x \approx xyx$ is absorbing. For instance, $\mathcal{A}_0, \mathcal{A}_2, \mathcal{B}_0, \mathcal{B}_2, \mathcal{L}_3 \in \mathbf{P}_2$ and any group with a finite exponent are absorbing.

Let $\widehat{\mathbb{N}}$ denote the countably infinite chain

$$0 < 1 < 2 < \cdots < \omega < \widehat{\omega}.$$

For any $\ell, r \in \widehat{\mathbb{N}}$, define the words

$$\mathbf{a}_{(\ell)} = \begin{cases} \varnothing & \text{if } \ell = 0, \\ a_1a_2\cdots a_\ell & \text{if } 1 \leq \ell < \omega, \\ a_1^n a_2 & \text{if } \ell = \omega, \\ a_1^n & \text{if } \ell = \widehat{\omega} \end{cases} \quad \text{and } \mathbf{b}_{(r)} = \begin{cases} \varnothing & \text{if } r = 0, \\ b_1b_2\cdots b_r & \text{if } 1 \leq r < \omega, \\ b_1b_2^n & \text{if } r = \omega, \\ b_1^n & \text{if } r = \widehat{\omega}. \end{cases}$$

Proposition 2.5 *Let $\mathbf{V}$ be any absorbing subvariety of $\mathbf{P}_n$. Suppose that $\mathcal{A}_0 \in \mathbf{V}$. Then each variety in the interval $[\mathbf{A_0}, \mathbf{V}]$ is defined within $\mathbf{V}$ by identities of the form*

$$\mathbf{a}_{(\ell)}\mathbf{w}\mathbf{b}_{(r)} \approx \mathbf{a}_{(\ell)}\mathbf{w}'\mathbf{b}_{(r)} \tag{2.1}$$

satisfying all of the following properties:

(a) $\mathbf{w}, \mathbf{w}' \in \mathscr{A}^+$ *are connected words such that* $\mathsf{con}(\mathbf{w}) = \mathsf{con}(\mathbf{w}')$ *and* $\mathcal{A}_0 \models \mathbf{w} \approx \mathbf{w}'$;
(b) $\mathbf{a}_{(\ell)}, \mathbf{w}, \mathbf{b}_{(r)}$ *are pairwise disjoint and* $\mathbf{a}_{(\ell)}, \mathbf{w}', \mathbf{b}_{(r)}$ *are pairwise disjoint.*

Proof It suffices to consider a variety in $[\mathbf{A_0}, \mathbf{V}]$ of the form $\mathbf{V}\{\sigma\}$, where

$$\sigma : \mathbf{z} \approx \mathbf{z}'$$

is any nontrivial identity satisfied by $\mathcal{A}_0$, and show that

$$\mathbf{V}\{\sigma\} = \mathbf{V}\{\sigma_1, \sigma_2, \ldots, \sigma_m\} \tag{2.2}$$

for some identities $\sigma_1, \sigma_2, \ldots, \sigma_m$ of the form (2.1) that satisfy (a) and (b). If either $\mathbf{z}$ or $\mathbf{z}'$ is simple, then the identity σ is contradictorily trivial by Lemma 2.3(i). Therefore, the words $\mathbf{z}$ and $\mathbf{z}'$ are non-simple, so that by Lemma 2.3(ii),

$$\mathbf{z} = \mathbf{s}_0\mathbf{w}_1\mathbf{s}_1\mathbf{w}_2\mathbf{s}_2 \cdots \mathbf{w}_m\mathbf{s}_m \text{ and } \mathbf{z}' = \mathbf{s}_0\mathbf{w}'_1\mathbf{s}_1\mathbf{w}'_2\mathbf{s}_2 \cdots \mathbf{w}'_m\mathbf{s}_m,$$

where

- $\mathbf{s}_0, \mathbf{s}_1, \ldots, \mathbf{s}_m \in \mathscr{A}^+_\varnothing$ are simple;
- $\mathbf{w}_1, \mathbf{w}_2, \ldots, \mathbf{w}_m, \mathbf{w}'_1, \mathbf{w}'_2, \ldots, \mathbf{w}'_m \in \mathscr{A}^+$ are connected;
- $\mathbf{s}_0, \mathbf{w}_1, \mathbf{s}_1, \mathbf{w}_2, \mathbf{s}_2, \ldots, \mathbf{w}_m, \mathbf{s}_m$ are pairwise disjoint;
- $\mathbf{s}_0, \mathbf{w}'_1, \mathbf{s}_1, \mathbf{w}'_2, \mathbf{s}_2, \ldots, \mathbf{w}'_m, \mathbf{s}_m$ are pairwise disjoint;
- $\mathsf{con}(\mathbf{w}_i) = \mathsf{con}(\mathbf{w}'_i)$ and $\mathcal{A}_0 \models \mathbf{w}_i \approx \mathbf{w}'_i$ for all i.

Generality is not lost by assuming that $\mathsf{con}(\mathbf{zz}') \cap \mathsf{con}(\mathbf{a}_{(\ell)}\mathbf{b}_{(r)}) = \varnothing$ for all $\ell, r \in \widehat{\mathbb{N}}$. Further, since $\mathbf{V}$ is an absorbing subvariety of $\mathbf{P}_n$, it satisfies the identities $\wp_n$ and

$$\mathbf{e}_i^n\mathbf{w}_i \approx \mathbf{w}_i, \quad \mathbf{w}_i\mathbf{f}_i^n \approx \mathbf{w}_i, \quad (\mathbf{e}'_i)^n\mathbf{w}'_i \approx \mathbf{w}'_i, \quad \mathbf{w}'_i(\mathbf{f}'_i)^n \approx \mathbf{w}'_i \tag{2.3}$$

for some $\mathbf{e}_i, \mathbf{f}_i, \mathbf{e}'_i, \mathbf{f}'_i \in \mathscr{A}^+$. If $m = 1$, then the identity σ is $\mathbf{s}_0\mathbf{w}_1\mathbf{s}_1 \approx \mathbf{s}_0\mathbf{w}'_1\mathbf{s}_1$, whence (2.2) holds with σ_1 being $\mathbf{a}_{(\ell)}\mathbf{w}_1\mathbf{b}_{(r)} \approx \mathbf{a}_{(\ell)}\mathbf{w}'_1\mathbf{b}_{(r)}$, where $\ell = |\mathbf{s}_0|$ and $r = |\mathbf{s}_1|$.

Therefore, it remains to assume that $m \geq 2$. Let φ_1 denote the substitution

$$x \mapsto \begin{cases} b_1 & \text{if } x = \mathsf{h}(\mathbf{s}_1), \\ b_2^n & \text{if } x \text{ occurs after } \mathsf{h}(\mathbf{s}_1) \text{ in } \mathbf{z}. \end{cases}$$

Then

$$\begin{aligned} \{\wp_n, (2.3), \sigma\} &\vdash \mathbf{s}_0\mathbf{w}_1 b_1 b_2^n \overset{\wp_n}{\approx} \mathbf{z}\varphi_1 \overset{\sigma}{\approx} \mathbf{z}'\varphi_1 \overset{\wp_n}{\approx} \mathbf{s}_0\mathbf{w}'_1 b_1 b_2^n \\ &\vdash \sigma_1 : \mathbf{s}_0\mathbf{w}_1 b_1 b_2^n \approx \mathbf{s}_0\mathbf{w}'_1 b_1 b_2^n, \end{aligned}$$

where $b_1 = \varnothing$ if and only if $\mathbf{s}_1 = \varnothing$, whence $\sigma_1 \sim \mathbf{a}_{(\ell)}\mathbf{w}_1\mathbf{b}_{(r)} \approx \mathbf{a}_{(\ell)}\mathbf{w}'_1\mathbf{b}_{(r)}$ with $\ell = |\mathbf{s}_0|$ and $r \in \{\omega, \widehat{\omega}\}$. For each i such that $1 < i < m$, let φ_i denote the substitution

$$x \mapsto \begin{cases} a_1^n & \text{if } x \text{ occurs before } \mathsf{t}(\mathbf{s}_{i-1}) \text{ in } \mathbf{z}, \\ a_2 & \text{if } x = \mathsf{t}(\mathbf{s}_{i-1}), \\ b_1 & \text{if } x = \mathsf{h}(\mathbf{s}_i), \\ b_2^n & \text{if } x \text{ occurs after } \mathsf{h}(\mathbf{s}_i) \text{ in } \mathbf{z}. \end{cases}$$

Then

$$\begin{aligned} \{\wp_n, (2.3), \sigma\} &\vdash a_1^n a_2 \mathbf{w}_i b_1 b_2^n \stackrel{\wp_n}{\approx} \mathbf{z}\varphi_i \stackrel{\sigma}{\approx} \mathbf{z}'\varphi_i \stackrel{\wp_n}{\approx} a_1^n a_2 \mathbf{w}_i' b_1 b_2^n \\ &\vdash \sigma_i : a_1^n a_2 \mathbf{w}_i b_1 b_2^n \approx a_1^n a_2 \mathbf{w}_i' b_1 b_2^n, \end{aligned}$$

where $a_2 = \varnothing$ if and only if $\mathbf{s}_{i-1} = \varnothing$, and $b_1 = \varnothing$ if and only if $\mathbf{s}_i = \varnothing$, whence $\sigma_i \sim \mathbf{a}_{(\ell)}\mathbf{w}_1\mathbf{b}_{(r)} \approx \mathbf{a}_{(\ell)}\mathbf{w}_1'\mathbf{b}_{(r)}$ with $\ell, r \in \{\omega, \widehat{\omega}\}$. Finally, let φ_m denote the substitution

$$x \mapsto \begin{cases} a_1^n & \text{if } x \text{ occurs before } \mathsf{t}(\mathbf{s}_{m-1}) \text{ in } \mathbf{z}, \\ a_2 & \text{if } x = \mathsf{t}(\mathbf{s}_{m-1}). \end{cases}$$

Then

$$\begin{aligned} \{\wp_n, (2.3), \sigma\} &\vdash a_1^n a_2 \mathbf{w}_m \mathbf{s}_m \stackrel{\wp_n}{\approx} \mathbf{z}\varphi_m \stackrel{\sigma}{\approx} \mathbf{z}'\varphi_m \stackrel{\wp_n}{\approx} a_1^n a_2 \mathbf{w}_m' \mathbf{s}_m \\ &\vdash \sigma_m : a_1^n a_2 \mathbf{w}_m \mathbf{s}_m \approx a_1^n a_2 \mathbf{w}_m' \mathbf{s}_m, \end{aligned}$$

where $a_2 = \varnothing$ if and only if $\mathbf{s}_{m-1} = \varnothing$, whence $\sigma_m \sim \mathbf{a}_{(\ell)}\mathbf{w}_m\mathbf{b}_{(r)} \approx \mathbf{a}_{(\ell)}\mathbf{w}_m'\mathbf{b}_{(r)}$ with $\ell \in \{\omega, \widehat{\omega}\}$ and $r = |\mathbf{s}_m|$. Therefore, the deduction $\{\wp_n, (2.3), \sigma\} \vdash \{\sigma_1, \sigma_2, \ldots, \sigma_m\}$ holds, and so the inclusion $\mathbf{V}\{\sigma\} \subseteq \mathbf{V}\{\sigma_1, \sigma_2, \ldots, \sigma_m\}$ follows. Conversely, since

$$\begin{aligned} \{(2.3), \sigma_1, \sigma_2, \ldots, \sigma_m\} \vdash \mathbf{z} &= \mathbf{s}_0\mathbf{w}_1\mathbf{s}_1\mathbf{w}_2\mathbf{s}_2\mathbf{w}_3\mathbf{s}_3 \cdots \mathbf{w}_m\mathbf{s}_m \\ &\stackrel{(2.3)}{\approx} (\mathbf{s}_0\mathbf{w}_1\mathbf{s}_1\mathbf{e}_2^n)\mathbf{w}_2\mathbf{s}_2\mathbf{w}_3\mathbf{s}_3 \cdots \mathbf{w}_m\mathbf{s}_m \\ &\stackrel{\sigma_1}{\approx} \mathbf{s}_0\mathbf{w}_1'\mathbf{s}_1\mathbf{e}_2^n\mathbf{w}_2\mathbf{s}_2\mathbf{w}_3\mathbf{s}_3 \cdots \mathbf{w}_m\mathbf{s}_m \\ &\stackrel{(2.3)}{\approx} \mathbf{s}_0\mathbf{w}_1'((\mathbf{f}_1')^n\mathbf{s}_1\mathbf{w}_2\mathbf{s}_2\mathbf{e}_3^n)\mathbf{w}_3\mathbf{s}_3 \cdots \mathbf{w}_m\mathbf{s}_m \\ &\stackrel{\sigma_2}{\approx} \mathbf{s}_0\mathbf{w}_1'(\mathbf{f}_1')^n\mathbf{s}_1\mathbf{w}_2'\mathbf{s}_2\mathbf{e}_3^n\mathbf{w}_3\mathbf{s}_3 \cdots \mathbf{w}_m\mathbf{s}_m \\ &\;\;\vdots \\ &\stackrel{(2.3)}{\approx} \mathbf{s}_0\mathbf{w}_1'\mathbf{s}_1\mathbf{w}_2'\mathbf{s}_2 \cdots \mathbf{w}_{m-1}'((\mathbf{f}_{m-1}')^n\mathbf{s}_{m-1}\mathbf{w}_m\mathbf{s}_m) \\ &\stackrel{\sigma_m}{\approx} \mathbf{s}_0\mathbf{w}_1'\mathbf{s}_1\mathbf{w}_2'\mathbf{s}_2 \cdots \mathbf{w}_{m-1}'(\mathbf{f}_{m-1}')^n\mathbf{s}_{m-1}\mathbf{w}_m'\mathbf{s}_m \end{aligned}$$

$$\overset{(2.3)}{\approx} \mathbf{s}_0\mathbf{w}_1'\mathbf{s}_1\mathbf{w}_2'\mathbf{s}_2\cdots\mathbf{w}_{m-1}'\mathbf{s}_{m-1}\mathbf{w}_m'\mathbf{s}_m$$
$$= \mathbf{z}' \vdash \sigma,$$

the inclusion $\mathbf{V}\{\sigma_1, \sigma_2, \ldots, \sigma_m\} \subseteq \mathbf{V}\{\sigma\}$ holds. □

2.4 Rees Quotients of Free Monoids

Recall that a *monoid* is a semigroup with a unit element $\mathtt{1}$ that is regarded as a nullary operation. For any word $\mathbf{w} \in \mathscr{A}^+$ and any set $\mathscr{X}$ of variables, let $\mathbf{w}[\mathscr{X}]$ denote the word obtained from $\mathbf{w}$ by retaining the variables in $\mathscr{X}$; if $\mathscr{X} = \{x_1, x_2, \ldots, x_r\}$, then simply write $\mathbf{w}[\mathscr{X}] = \mathbf{w}[x_1, x_2, \ldots, x_r]$. A monoid that satisfies an identity $\mathbf{w} \approx \mathbf{w}'$ clearly also satisfies the identity $\mathbf{w}[\mathscr{X}] \approx \mathbf{w}'[\mathscr{X}]$ for all $\mathscr{X} \subseteq \mathscr{A}$. In other words, the identity $\mathbf{w}[\mathscr{X}] \approx \mathbf{w}'[\mathscr{X}]$ is deducible from $\mathbf{w} \approx \mathbf{w}'$ within the equational theory of monoids.

For any set $\mathscr{W} \subseteq \mathscr{A}_\varnothing^+$ of words, let $\mathcal{R}_{\mathrm{Q}}\mathscr{W}$ denote the Rees quotient of $\mathscr{A}_\varnothing^+$ over the ideal of all words that are not factors of any word in $\mathscr{W}$. Equivalently, $\mathcal{R}_{\mathrm{Q}}\mathscr{W}$ is the monoid that consists of every factor of every word in $\mathscr{W}$, together with a zero element $\mathtt{0}$, with binary operation $\cdot$ given by

$$\mathbf{u} \cdot \mathbf{v} = \begin{cases} \mathbf{uv} & \text{if } \mathbf{uv} \text{ is a factor of some word in } \mathscr{W}, \\ \mathtt{0} & \text{otherwise.} \end{cases}$$

The empty factor, more conveniently written as $\mathtt{1}$, is the unit element of $\mathcal{R}_{\mathrm{Q}}\mathscr{W}$. The variety of monoids generated by $\mathcal{R}_{\mathrm{Q}}\mathscr{W}$ is denoted by $\mathbb{R}_{\mathrm{Q}}\mathscr{W}$.

A word $\mathbf{w}$ is an *isoterm* for a class $\mathfrak{C}$ of monoids if $\mathfrak{C}$ violates any nontrivial identity of the form $\mathbf{w} \approx \mathbf{w}'$. The set of all isoterms for $\mathfrak{C}$ is denoted by $\mathsf{iso}\,\mathfrak{C}$. The concept of an isoterm provides a convenient method to determine if a Rees quotient of $\mathscr{A}_\varnothing^+$ belongs to a variety.

Lemma 2.6 (Jackson 2005b, Lemma 3.3) *Let $\mathbb{V}$ be any variety of monoids and $\mathscr{W}$ be any set of words. Then $\mathcal{R}_{\mathrm{Q}}\mathscr{W} \in \mathbb{V}$ if and only if $\mathscr{W} \subseteq \mathsf{iso}\,\mathbb{V}$.*

Proof It is clear that every word in $\mathscr{W}$ is an isoterm for $\mathcal{R}_{\mathrm{Q}}\mathscr{W}$. Hence, $\mathcal{R}_{\mathrm{Q}}\mathscr{W} \in \mathbb{V}$ implies that $\mathscr{W} \subseteq \mathsf{iso}\{\mathcal{R}_{\mathrm{Q}}\mathscr{W}\} \subseteq \mathsf{iso}\,\mathbb{V}$. Conversely, suppose that $\mathscr{W} \subseteq \mathsf{iso}\,\mathbb{V}$. Let $\mathbf{u} \approx \mathbf{v}$ be any identity such that $\mathcal{R}_{\mathrm{Q}}\mathscr{W} \nvDash \mathbf{u} \approx \mathbf{v}$. Then there exists some substitution $\varphi : \mathscr{A} \to \mathcal{R}_{\mathrm{Q}}\mathscr{W}$ such that $\mathbf{u}\varphi \neq \mathbf{v}\varphi$ in $\mathcal{R}_{\mathrm{Q}}\mathscr{W}$. Generality is not lost by assuming that $\mathbf{u}\varphi \neq \mathtt{0}$, so that $\mathbf{u}\varphi$ is a factor of some word $\mathbf{w} \in \mathscr{W} \subseteq \mathsf{iso}\,\mathbb{V}$, say $\mathbf{w} = \mathbf{p}(\mathbf{u}\varphi)\mathbf{q}$ for some $\mathbf{p}, \mathbf{q} \in \mathscr{A}_\varnothing^+$. Then $\mathbf{p}(\mathbf{u}\varphi)\mathbf{q} \approx \mathbf{p}(\mathbf{v}\varphi)\mathbf{q}$ is a nontrivial identity that is directly deducible from $\mathbf{u} \approx \mathbf{v}$. Now if $\mathbb{V}$ satisfies $\mathbf{u} \approx \mathbf{v}$, then it also satisfies $\mathbf{p}(\mathbf{u}\varphi)\mathbf{q} \approx \mathbf{p}(\mathbf{v}\varphi)\mathbf{q}$, whence the contradiction $\mathbf{w} = \mathbf{p}(\mathbf{u}\varphi)\mathbf{q} \notin \mathsf{iso}\,\mathbb{V}$ is obtained. Consequently, $\mathbb{V} \nvDash \mathbf{u} \approx \mathbf{v}$. □

Lemma 2.7 (O.B. Sapir 2015b, Fact 3.1) *Let $\mathbb{V}$ be any variety of monoids such that $\mathcal{R}_{\mathcal{Q}}\{xyx\} \in \mathbb{V}$. Then the following equivalences hold*:

(i) $\mathcal{R}_{\mathcal{Q}}\{xhxyty\} \notin \mathbb{V}$ *if and only if* $\mathbb{V} \vDash xhxyty \approx xhyxty$;
(ii) $\mathcal{R}_{\mathcal{Q}}\{xhytxy\} \notin \mathbb{V}$ *if and only if* $\mathbb{V} \vDash xhytxy \approx xhytyx$;
(iii) $\mathcal{R}_{\mathcal{Q}}\{xyhxty\} \notin \mathbb{V}$ *if and only if* $\mathbb{V} \vDash xyhxty \approx yxhxty$.

Proof It suffices to establish part (i) since parts (ii) and (iii) are similar. Suppose that $\mathcal{R}_{\mathcal{Q}}\{xhxyty\} \notin \mathbb{V}$. Then by Lemma 2.6, the variety $\mathbb{V}$ satisfies $xhxyty \approx \mathbf{w}$ for some $\mathbf{w} \neq xhxyty$, whence it also satisfies the identities

$$xhx \approx \mathbf{w}[x,h], \quad yty \approx \mathbf{w}[y,t], \quad ht \approx \mathbf{w}[h,t], \quad x^2t \approx \mathbf{w}[x,t], \quad hy^2 \approx \mathbf{w}[y,h].$$

It follows from Lemma 2.6 that xyx—and so also its factor xy—are isoterms for $\mathbb{V}$. Therefore,

$$\mathbf{w}[x,h] = xhx, \quad \mathbf{w}[y,t] = yty, \quad \mathbf{w}[h,t] = ht, \quad \mathbf{w}[x,t] \neq xtx, \quad \mathbf{w}[y,h] \neq yhy.$$

It is then routinely checked that $\mathbf{w} = xhyxty$ is the only possibility.

Conversely, it is obvious that if $\mathbb{V} \vDash xhxyty \approx xhyxty$, then $\mathcal{R}_{\mathcal{Q}}\{xhxyty\} \notin \mathbb{V}$. □

Lemma 2.8 (Jackson 2005b, Lemma 4.1) *Suppose that $\mathbb{V}$ is any subvariety of $\mathbb{A}^{\mathsf{cen}}$ such that $\mathcal{R}_{\mathcal{Q}}\{xyx\} \notin \mathbb{V}$. Then $\mathbb{V}$ satisfies either $xyx \approx x^2y$ or $xyx \approx yx^2$.*

Proof By Lemma 2.6, the variety $\mathbb{V}$ satisfies $xyx \approx \mathbf{w}$ for some $\mathbf{w} \neq xyx$.

Case 1: $\mathsf{occ}(x,\mathbf{w}) \leq 1$ or $\mathsf{occ}(y,\mathbf{w}) \neq 1$. Since $\mathbb{V} \subseteq \mathbb{A}^{\mathsf{cen}}$ is aperiodic, it is routinely checked that $\mathbb{V}$ is idempotent and so commutative. Hence, $\mathbb{V} \vDash xyx \approx x^2y$.

Case 2: $\mathsf{occ}(x,\mathbf{w}) \geq 2$ and $\mathsf{occ}(y,\mathbf{w}) = 1$. If $\mathsf{occ}(x,\mathbf{w}) = 2$, then $\mathbf{w} \in \{x^2y, yx^2\}$ and the result holds. Hence, assume that $\mathsf{occ}(x,\mathbf{w}) \geq 3$, so that $\mathbf{w} = x^pyx^q$ for some $p,q \geq 0$ such that $p+q \geq 3$. Since $\mathbb{V} \vDash x^2 \approx x^{p+q}$ and $\mathbb{V} \subseteq \mathbb{A}^{\mathsf{cen}}$, it follows that $\mathbb{V} \vDash \{x^3 \approx x^2, x^2y \approx yx^2\}$. Now $p+q \geq 3$ implies that either $p \geq 2$ or $q \geq 2$. Then it is straightforwardly verified that $\mathbb{V} \vDash x^pyx^q \approx yx^2$ if $p \geq 2$ and $\mathbb{V} \vDash x^pyx^q \approx x^2y$ if $q \geq 2$. Consequently, either $\mathbb{V} \vDash xyx \approx x^2y$ or $\mathbb{V} \vDash xyx \approx yx^2$. □

Lemma 2.9 *The varieties* $\mathbb{J}_1 = \mathbb{R}_{\mathcal{Q}}\{xhxyty\}$ *and* $\mathbb{J}_2 = \mathbb{R}_{\mathcal{Q}}\{xhytxy, xyhxty\}$ *are subvarieties of* $\mathbb{B}_2^1$ *and* $\mathbb{P}_{25}$.

Proof It is easy to check that $xyx \in \mathsf{iso}\{\mathcal{B}_2^1\}$, so that $\mathcal{R}_{\mathcal{Q}}\{xyx\} \in \mathbb{B}_2^1$ by Lemma 2.6. Further, the semigroup $\mathcal{B}_2^1$ violates the identities in Lemma 2.7 under the substitution $(x, y, h, t) \mapsto (\mathsf{a}, \mathsf{b}, \mathtt{1}, \mathtt{1})$. Therefore, the variety $\mathbb{B}_2^1$ contains $\mathcal{R}_{\mathcal{Q}}\{xhxyty\}$, $\mathcal{R}_{\mathcal{Q}}\{xhytxy\}$, and $\mathcal{R}_{\mathcal{Q}}\{xyhxty\}$, whence $\mathbb{J}_1, \mathbb{J}_2 \subseteq \mathbb{B}_2^1$ by Theorem 1.56.

The inclusions $\mathbb{J}_1 \subseteq \mathbb{P}_{25}$ and $\mathbb{J}_2 \subseteq \mathbb{P}_{25}$ are similarly established. □

2.5 Involution Semigroups

For any semigroup S, let S^0 denote the semigroup obtained by adjoining a new zero element $\mathtt{0}$ to S and let S^1 denote the monoid obtained by adjoining a new unit element $\mathtt{1}$ to S.

Recall that an *involution semigroup* is a unary semigroup $\langle S, {}^\star \rangle$ that satisfies the following identities:

$$(x^\star)^\star \approx x, \quad (xy)^\star \approx y^\star x^\star. \tag{inv}$$

Any involution on a semigroup maps an idempotent to an idempotent; in particular, it fixes any zero and unit elements. It follows that any involution semigroup $\langle S, {}^\star \rangle$ can be extended to the involution semigroups $\langle S^0, {}^\star \rangle$ and $\langle S^1, {}^\star \rangle$ by defining $\mathtt{0}^\star = \mathtt{0}$ and $\mathtt{1}^\star = \mathtt{1}$, respectively.

2.5.1 Terms, Words, and Plain Words

Let $\mathscr{A}^\star = \{x^\star \mid x \in \mathscr{A}\}$ be a disjoint copy of the alphabet $\mathscr{A}$. Elements of $\mathscr{A} \cup \mathscr{A}^\star$ are called *variables*. The set of *terms* over $\mathscr{A}$ is the smallest set $\mathsf{T}(\mathscr{A})$ containing $\mathscr{A} \cup \{\varnothing\}$ that is closed under concatenation and *. The *subterms* of a term $\mathbf{t} \in \mathsf{T}(\mathscr{A})$ are thus defined as follows:

- $\varnothing$ and $\mathbf{t}$ are subterms of $\mathbf{t}$;
- if $\mathbf{t}_1\mathbf{t}_2$ is a subterm of $\mathbf{t}$ with $\mathbf{t}_1, \mathbf{t}_2 \in \mathsf{T}(\mathscr{A})$, then $\mathbf{t}_1$ and $\mathbf{t}_2$ are subterms of $\mathbf{t}$;
- if $\mathbf{s}^\star$ is a subterm of $\mathbf{t}$ with $\mathbf{s} \in \mathsf{T}(\mathscr{A})$, then $\mathbf{s}$ is a subterm of $\mathbf{t}$.

The *free involution semigroup* over the alphabet $\mathscr{A}$ is the free semigroup $(\mathscr{A} \cup \mathscr{A}^\star)^+$ over the disjoint union $\mathscr{A} \cup \mathscr{A}^\star$ with unary operation * defined by $(x^\star)^\star = x$ for all $x \in \mathscr{A}$ and

$$(x_1x_2\cdots x_m)^\star = x_m^\star x_{m-1}^\star \cdots x_1^\star$$

for all $x_1, x_2, \ldots, x_m \in \mathscr{A} \cup \mathscr{A}^\star$. Elements of $(\mathscr{A} \cup \mathscr{A}^\star)^+_\varnothing = (\mathscr{A} \cup \mathscr{A}^\star)^+ \cup \{\varnothing\}$ are called *words* while elements of $\mathscr{A}^+_\varnothing = \mathscr{A}^+ \cup \{\varnothing\}$ are called *plain words*.

Remark 2.10

(i) The proper inclusion $(\mathscr{A} \cup \mathscr{A}^\star)^+_\varnothing \subset \mathsf{T}(\mathscr{A})$ holds.
(ii) The identities (inv) can be used to convert each term $\mathbf{t}$ in $\mathsf{T}(\mathscr{A})$ into some unique word $\lfloor \mathbf{t} \rfloor$ in $(\mathscr{A} \cup \mathscr{A}^\star)^+_\varnothing$. For instance, $\lfloor (x^3(yx^\star)^\star)^\star x(z^\star)^\star z^4 \rfloor = y(x^\star)^4 xz^5$.
(iii) For any subterm $\mathbf{s}$ of a term $\mathbf{t}$, either $\lfloor \mathbf{s} \rfloor$ or $\lfloor \mathbf{s}^\star \rfloor$ is a factor of the word $\lfloor \mathbf{t} \rfloor$.

2.5.2 Identities and Deducibility

In the involution semigroup signature, an identity—or more specifically, a *term identity*—is an expression of the form $\mathbf{t} \approx \mathbf{t}'$, where $\mathbf{t}, \mathbf{t}' \in \mathsf{T}(\mathscr{A})$. An identity $\mathbf{w} \approx \mathbf{w}'$ is a *word identity* if $\mathbf{w}, \mathbf{w}' \in (\mathscr{A} \cup \mathscr{A}^\star)^+$; it is a *plain identity* if $\mathbf{w}, \mathbf{w}' \in \mathscr{A}^+$. For any involution semigroup $\langle S, {}^\star \rangle$, let $\mathsf{id}\{\langle S, {}^\star \rangle\}$ denote the set of identities satisfied by $\langle S, {}^\star \rangle$. Let $\mathsf{id}_\mathsf{W}\{\langle S, {}^\star \rangle\}$ and $\mathsf{id}_\mathsf{P}\{\langle S, {}^\star \rangle\}$ denote the subsets of $\mathsf{id}\{\langle S, {}^\star \rangle\}$ that consist of word identities and plain identities, respectively. Clearly, the inclusions

$$\mathsf{id}\{S\} = \mathsf{id}_\mathsf{P}\{\langle S, {}^\star \rangle\} \subset \mathsf{id}_\mathsf{W}\{\langle S, {}^\star \rangle\} \subset \mathsf{id}\{\langle S, {}^\star \rangle\}$$

hold, where $\mathsf{id}\{S\}$ is the set of identities satisfied by the reduct S.

A term identity $\mathbf{t} \approx \mathbf{t}'$ is *directly deducible* from a term identity $\mathbf{s} \approx \mathbf{s}'$ if there exists some substitution $\varphi : \mathscr{A} \to \mathsf{T}(\mathscr{A})$ such that $\mathbf{s}\varphi$ is a subterm of $\mathbf{t}$, and replacing this particular subterm $\mathbf{s}\varphi$ of $\mathbf{t}$ with $\mathbf{s}'\varphi$ results in the term $\mathbf{t}'$. A term identity $\mathbf{t} \approx \mathbf{t}'$ is *deducible* from a set Σ of term identities, indicated by $\Sigma \vdash \mathbf{t} \approx \mathbf{t}'$ or $\mathbf{t} \overset{\Sigma}{\approx} \mathbf{t}'$, if there exists a finite sequence

$$\mathbf{t} = \mathbf{t}_1, \mathbf{t}_2, \ldots, \mathbf{t}_m = \mathbf{t}'$$

of terms, where each term identity $\mathbf{t}_i \approx \mathbf{t}_{i+1}$ is directly deducible from some term identity in Σ.

Lemma 2.11 *Suppose that* $\mathbf{t} \approx \mathbf{t}'$ *is any identity deducible from* (inv). *Then* $\lfloor \mathbf{t} \rfloor = \lfloor \mathbf{t}' \rfloor$.

Proof This result holds because the semigroup $\langle \mathsf{M}_2(\mathbb{R}), ^T \rangle$ of 2×2 real matrices with transposition is an involution semigroup that violates every nontrivial word identity (Auinger et al. 2012b, Theorem 3.7). □

Let $\mathbf{w}$ be any word in $(\mathscr{A} \cup \mathscr{A}^\star)^+$, so that

$$\mathbf{w} = x_1^{\circledast_1} x_2^{\circledast_2} \cdots x_m^{\circledast_m}$$

for some $x_1, x_2, \ldots, x_m \in \mathscr{A}$ and $\circledast_1, \circledast_2, \ldots, \circledast_m \in \{1, \star\}$. Then the *plain projection* of $\mathbf{w}$ is the plain word

$$\overline{\mathbf{w}} = x_1 x_2 \cdots x_m$$

obtained by removing all occurrences of the symbol * from $\mathbf{w}$. Similar to the case of plains words, the *reversal* of $\mathbf{w}$ is the word

$$\mathbf{w}^{\triangleleft} = x_m^{\circledast_m} x_{m-1}^{\circledast_{m-1}} \cdots x_1^{\circledast_1}$$

obtained by writing $\mathbf{w}$ in reverse. A set Σ of word identities is *closed under reversal* if $\mathbf{w}^{\triangleleft} \approx (\mathbf{w}')^{\triangleleft} \in \Sigma$ for all $\mathbf{w} \approx \mathbf{w}' \in \Sigma$.

Lemma 2.12 *For any involution semigroup $\langle S, ^\star \rangle$, the set* $\mathsf{id}_\mathsf{W}\{\langle S, ^\star \rangle\}$ *is closed under reversal.*

Proof Let φ denote the substitution given by $x \mapsto x^\star$ for all $x \in \mathscr{A}$. Then it is routinely checked that $\lfloor (\mathbf{w}\varphi)^\star \rfloor = \mathbf{w}^{\triangleleft}$ for all $\mathbf{w} \in (\mathscr{A} \cup \mathscr{A}^\star)^+_\varnothing$. Therefore, if $\langle S, ^\star \rangle$ satisfies a word identity $\mathbf{w} \approx \mathbf{w}'$, then since

$$\begin{aligned} \mathbf{w} \approx \mathbf{w}' &\vdash \lfloor (\mathbf{w}\varphi)^\star \rfloor \approx \lfloor (\mathbf{w}'\varphi)^\star \rfloor \\ &\vdash \mathbf{w}^{\triangleleft} \approx (\mathbf{w}')^{\triangleleft}, \end{aligned}$$

it also satisfies $\mathbf{w}^{\triangleleft} \approx (\mathbf{w}')^{\triangleleft}$. □

Part I

Semigroups

3 Aperiodic Rees–Suschkewitsch Varieties

As described in Sect. 1.2.3, the present chapter is a detailed investigation of subvarieties of $\mathbf{A_2}$, which are precisely all aperiodic Rees–Suschkewitsch varieties. In Sect. 3.1, some preliminary results on the lattice $\mathfrak{L}(\mathbf{A_2})$ and some of its varieties are established; most crucially, the proper subvarieties of $\mathbf{A_2}$ are shown to constitute five pairwise disjoint intervals. In Sect. 3.2, all subvarieties of $\mathbf{A_2}$ are shown to be finitely based, thus proving Theorem 1.6. In Sect. 3.3, the subinterval $[\mathbf{B_0}, \mathbf{A_2}]$ of $\mathfrak{L}(\mathbf{A_2})$ is completely described and shown to be a distributive lattice. In Sect. 3.4, the characterization in Theorem 1.7 of Cross subvarieties of $\mathbf{A_2}$ is established. Other results concerning subvarieties of $\mathbf{A_2}$ that are Cross, finitely generated, or non-finitely generated are also deduced.

3.1 Background Information on $\mathfrak{L}(\mathbf{A_2})$

In Sect. 3.1.1, identity bases for the varieties $\mathbf{A_2}$, $\mathbf{A_0}$, $\mathbf{B_2}$, and $\mathbf{B_0}$ are given. In Sect. 3.1.2, varieties in the interval $[\mathbf{A_0}, \mathbf{A_2}]$ are shown to be definable within $\mathbf{A_2}$ by identities of a very specific form. In Sect. 3.1.3, a decomposition of the lattice $\mathfrak{L}(\overline{\mathbf{A_2}})$ into five pairwise disjoint intervals is given.

3.1.1 Identity Bases for Some Subvarieties of $\mathbf{A_2}$

For any $\mathbf{w} \in \mathscr{A}_{\varnothing}^{+}$, let $\mathsf{fac}_2(\mathbf{w})$ denote the set of factors of $\mathbf{w}$ of length two, that is,

$$\mathsf{fac}_2(\mathbf{w}) = \{xy \mid x, y \in \mathscr{A} \text{ and } \mathbf{w} = \mathbf{u}xy\mathbf{v} \text{ for some } \mathbf{u}, \mathbf{v} \in \mathscr{A}_{\varnothing}^{+}\}.$$

E. W. H. Lee, *Advances in the Theory of Varieties of Semigroups*, Frontiers in Mathematics, https://doi.org/10.1007/978-3-031-16497-2_3

Lemma 3.1 (Trahtman 1981b, 1994) *The identities*

$$x^3 \approx x^2, \quad xyxyx \approx xyx, \quad xyxzx \approx xzxyx \tag{3.1}$$

constitute an identity basis for the variety $\mathbf{A_2}$. *More generally,*

$$\mathsf{id}\,\mathbf{A_2} = \left\{ \mathbf{w} \approx \mathbf{w}' \;\middle|\; \begin{array}{l} \mathbf{w}, \mathbf{w}' \in \mathscr{A}^+, \mathsf{fac}_2(\mathbf{w}) = \mathsf{fac}_2(\mathbf{w}'), \\ \mathsf{h}(\mathbf{w}) = \mathsf{h}(\mathbf{w}'), \textit{and } \mathsf{t}(\mathbf{w}) = \mathsf{t}(\mathbf{w}') \end{array} \right\}. \tag{3.2}$$

It is routinely checked that the semigroups $\mathcal{A}_0$, $\mathcal{B}_0$, and $\mathcal{B}_2$ satisfy the identities (3.1), so that the varieties $\mathbf{A_0}$, $\mathbf{B_0}$, and $\mathbf{B_2}$ are subvarieties of $\mathbf{A_2}$.

Lemma 3.2 *The varieties* $\mathbf{A_0}$, $\mathbf{B_2}$, *and* $\mathbf{B_0}$ *are finitely based. Specifically,*

(i) $\mathbf{A_0} = \mathbf{A_2}\{xyxy \approx yxyx\}$;
(ii) $\mathbf{B_2} = \mathbf{A_2}\{x^2y^2 \approx y^2x^2\}$;
(iii) $\mathbf{B_0} = \mathbf{A_0} \cap \mathbf{B_2} = \mathbf{A_2}\{xyxy \approx yxyx, x^2y^2 \approx y^2x^2\}$;
(iv) $\mathbf{A_0} \vee \mathbf{B_2} = \mathbf{A_2}\{x^2y^2x^2 \approx y^2x^2y^2\}$.

Consequently, these varieties constitute the sublattice of $\mathfrak{L}(\mathbf{A_2})$ *in Fig. 3.1.*

Proof

(i) This follows from Lee and Volkov (2007, Theorem 4.1).
(ii) See Lee and Volkov (2007, Proposition 3.5) or Reilly (2008a, Theorem 5.4).
(iii) This follows from Lee (2004, Sect. 4) and parts (i) and (ii).
(iv) See Lee and Volkov (2007, Theorem 4.3(i)). □

3.1.2 Identities Defining Varieties in $[\mathbf{A_0}, \mathbf{A_2}]$

As observed in Example 2.4, the subvariety $\mathbf{A_2}$ of $\mathbf{P_2}$ is absorbing. Therefore, by Proposition 2.5 with $n = 2$, each variety in the interval $[\mathbf{A_0}, \mathbf{A_2}]$ is defined within $\mathbf{A_2}$

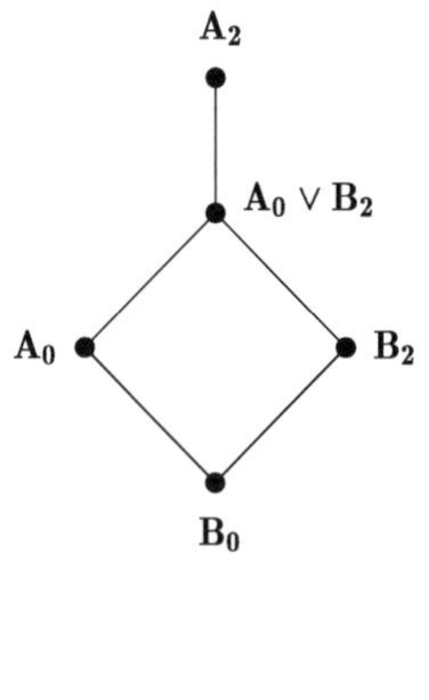

Fig. 3.1 A sublattice of $\mathfrak{L}(\mathbf{A_2})$

by identities of the form

$$\xi : \mathbf{a}_{(\ell)}\mathbf{w}\mathbf{b}_{(r)} \approx \mathbf{a}_{(\ell)}\mathbf{w}'\mathbf{b}_{(r)}$$

that satisfy all of the following properties:

(a) $\mathbf{w}, \mathbf{w}' \in \mathscr{A}^+$ are connected words such that $\mathsf{con}(\mathbf{w}) = \mathsf{con}(\mathbf{w}')$ and $\mathcal{A}_0 \models \mathbf{w} \approx \mathbf{w}'$;
(b) $\mathbf{a}_{(\ell)}, \mathbf{w}, \mathbf{b}_{(r)}$ are pairwise disjoint and $\mathbf{a}_{(\ell)}, \mathbf{w}', \mathbf{b}_{(r)}$ are pairwise disjoint.

Note that since $n = 2$,

$$\mathbf{a}_{(\omega)} = a_1^2 a_2, \quad \mathbf{a}_{(\widehat{\omega})} = a_1^2, \quad \mathbf{b}_{(\omega)} = b_1 b_2^2, \quad \text{and} \quad \mathbf{b}_{(\widehat{\omega})} = b_1^2.$$

The following result shows that if either $\mathsf{h}(\mathbf{w}) = \mathsf{h}(\mathbf{w}')$ or $\mathsf{t}(\mathbf{w}) = \mathsf{t}(\mathbf{w}')$, then the prefix $\mathbf{a}_{(\ell)}$ or the suffix $\mathbf{b}_{(r)}$ in the identity ξ can be chosen to be empty.

Lemma 3.3

$$\mathbf{A_2}\{\xi\} = \begin{cases} \mathbf{A_2}\{\mathbf{a}_{(\ell)}\mathbf{w} \approx \mathbf{a}_{(\ell)}\mathbf{w}'\} & \textit{if } \mathsf{h}(\mathbf{w}) \neq \mathsf{h}(\mathbf{w}') \textit{ and } \mathsf{t}(\mathbf{w}) = \mathsf{t}(\mathbf{w}'), \\ \mathbf{A_2}\{\mathbf{w}\mathbf{b}_{(r)} \approx \mathbf{w}'\mathbf{b}_{(r)}\} & \textit{if } \mathsf{h}(\mathbf{w}) = \mathsf{h}(\mathbf{w}') \textit{ and } \mathsf{t}(\mathbf{w}) \neq \mathsf{t}(\mathbf{w}'), \\ \mathbf{A_2}\{\mathbf{w} \approx \mathbf{w}'\} & \textit{if } \mathsf{h}(\mathbf{w}) = \mathsf{h}(\mathbf{w}') \textit{ and } \mathsf{t}(\mathbf{w}) = \mathsf{t}(\mathbf{w}'). \end{cases}$$

Proof It suffices to assume that $\mathsf{h}(\mathbf{w}) \neq \mathsf{h}(\mathbf{w}')$ and $\mathsf{t}(\mathbf{w}) = \mathsf{t}(\mathbf{w}') = t$ since the other cases are similar. Since the words $\mathbf{w}$ and $\mathbf{w}'$ are connected, $\mathbf{w} = \mathbf{u}t\mathbf{v}t$ and $\mathbf{w}' = \mathbf{u}'t\mathbf{v}'t$ for some $\mathbf{u}, \mathbf{v}, \mathbf{u}', \mathbf{v}' \in \mathscr{A}_\varnothing^+$. Hence, the deduction (3.1) $\vdash \{\mathbf{w}\mathbf{v}t \approx \mathbf{w}, \mathbf{w}'\mathbf{v}'t \approx \mathbf{w}'\}$ holds. The deduction $\{(3.1), \xi\} \vdash \mathbf{a}_{(\ell)}\mathbf{w} \approx \mathbf{a}_{(\ell)}\mathbf{w}'$ also holds because

$$\begin{aligned} \mathbf{a}_{(\ell)}\mathbf{w} &\stackrel{(3.1)}{\approx} \mathbf{a}_{(\ell)}\mathbf{w}(\mathbf{v}t)^{|\mathbf{b}_{(r)}|} \\ &\stackrel{\xi}{\approx} \mathbf{a}_{(\ell)}\mathbf{w}'(\mathbf{v}t)^{|\mathbf{b}_{(r)}|} \\ &\stackrel{(3.1)}{\approx} \mathbf{a}_{(\ell)}\mathbf{w}'(\mathbf{v}'t)^{|\mathbf{b}_{(r)}|}(\mathbf{v}t)^{|\mathbf{b}_{(r)}|} \\ &\stackrel{\xi}{\approx} \mathbf{a}_{(\ell)}\mathbf{w}(\mathbf{v}'t)^{|\mathbf{b}_{(r)}|}(\mathbf{v}t)^{|\mathbf{b}_{(r)}|} \\ &\stackrel{(3.1)}{\approx} \mathbf{a}_{(\ell)}\mathbf{w}(\mathbf{v}t)^{|\mathbf{b}_{(r)}|}(\mathbf{v}'t)^{|\mathbf{b}_{(r)}|} \\ &\stackrel{(3.1)}{\approx} \mathbf{a}_{(\ell)}\mathbf{w}(\mathbf{v}'t)^{|\mathbf{b}_{(r)}|} \\ &\stackrel{\xi}{\approx} \mathbf{a}_{(\ell)}\mathbf{w}'(\mathbf{v}'t)^{|\mathbf{b}_{(r)}|} \\ &\stackrel{(3.1)}{\approx} \mathbf{a}_{(\ell)}\mathbf{w}'. \end{aligned}$$

It follows that the inclusion $\mathbf{A_2}\{\xi\} \subseteq \mathbf{A_2}\{\mathbf{a}_{(\ell)}\mathbf{w} \approx \mathbf{a}_{(\ell)}\mathbf{w}'\}$ holds. The reverse inclusion $\mathbf{A_2}\{\mathbf{a}_{(\ell)}\mathbf{w} \approx \mathbf{a}_{(\ell)}\mathbf{w}'\} \subseteq \mathbf{A_2}\{\xi\}$ holds vacuously. □

3.1.3 A Decomposition of $\mathfrak{L}(\overline{\mathbf{A}_2})$

For any subvariety $\mathbf{V}$ of $\mathbf{A_2}$, the unique largest subvariety of $\mathbf{A_2}$ that excludes $\mathbf{V}$, if it exists, is denoted by $\overline{\mathbf{V}}$. In the present subsection, the existence of $\overline{\mathbf{A}_2}$, $\overline{\mathbf{B}_2}$, $\overline{\mathbf{A}_0}$, and $\overline{\mathbf{B}_0}$ is addressed. These subvarieties are then used to decompose the lattice $\mathfrak{L}(\overline{\mathbf{A}_2})$ into five pairwise disjoint intervals.

Lemma 3.4 *Let S be any semigroup in $\mathbf{P}_n$ for some $n \geq 2$. Then $\mathcal{A}_2 \notin \mathscr{V}_{\mathsf{sem}}\{S\}$ if and only if S satisfies the identity*

$$((x^n y)^n (yx^n)^n)^n \approx (x^n yx^n)^n. \tag{3.3}$$

Proof The semigroup $\mathcal{A}_2$ violates the identity (3.3) under the substitution $(x, y) \mapsto (\mathsf{e}, \mathsf{a})$. Hence, if S satisfies the identity (3.3), then $\mathcal{A}_2 \notin \mathscr{V}_{\mathsf{sem}}\{S\}$. Conversely, suppose that $\mathcal{A}_2 \notin \mathscr{V}_{\mathsf{sem}}\{S\}$. Then it follows from Almeida and Escada (2002, Lemma 5.5) and Escada (2000, Proposition 5.3) that S satisfies the identity $\mathbf{w}^{n+1} \approx \mathbf{w}$, where $\mathbf{w} = x^n(y^n zx^n)^n y^n$. Let φ denote the substitution $(x, y, z) \mapsto (yx^n, x^n y, x^n)$. Then since

$$\begin{aligned} \mathbf{w}\varphi &= (yx^n)^n \big((x^n y)^n x^n (yx^n)^n\big)^n (x^n y)^n \\ &\overset{\wp_n}{\approx} (yx^n)^n (x^n y)^n, \end{aligned}$$

it is routinely verified that $x^n(\mathbf{w}\varphi)x^n \overset{\wp_n}{\approx} (x^n yx^n)^n$ and

$$\begin{aligned} x^n(\mathbf{w}^{n+1}\varphi)x^n &= x^n(\mathbf{w}\varphi)^{n+1}x^n \\ &\overset{\wp_n}{\approx} x^n(yx^n)^n\big((x^n y)^n(yx^n)^n\big)^n(x^n y)^n x^n \\ &= (x^n y)^n x^n\big((x^n y)^n(yx^n)^n\big)^n x^n(yx^n)^n \\ &\overset{\wp_n}{\approx} ((x^n y)^n(yx^n)^n)^n. \end{aligned}$$

It follows that S satisfies the identity (3.3). □

Proposition 3.5 *The varieties* $\overline{\mathbf{A_2}}$, $\overline{\mathbf{B_2}}$, $\overline{\mathbf{A_0}}$, *and* $\overline{\mathbf{B_0}}$ *exist and are finitely based. Specifically,*

(i) $\overline{\mathbf{A_2}} = \mathbf{A_2}\{x^2y^2x^2 \approx x^2yx^2\}$;
(ii) $\overline{\mathbf{B_2}} = \mathbf{A_2}\{xy^2x \approx xyx\}$;
(iii) $\overline{\mathbf{A_0}} = \mathbf{A_2}\{x^2y^2x^2y^2 \approx x^2y^2\}$;
(iv) $\overline{\mathbf{B_0}} = \mathbf{A_2}\{x^2y^2z^2 \approx x^2yz^2\}$.

Proof

(i) The semigroup $\mathcal{A}_2$ violates the identity $x^2y^2x^2 \approx x^2yx^2$ under the substitution $(x, y) \mapsto (\mathsf{e}, \mathsf{a})$. Therefore, the variety $\mathbf{A_2}\{x^2y^2x^2 \approx x^2yx^2\}$ excludes $\mathbf{A_2}$. Suppose that $\mathbf{V}$ is any subvariety of $\mathbf{A_2}$ such that $\mathbf{A_2} \nsubseteq \mathbf{V}$. Then by Lemma 3.4, the variety $\mathbf{V}$ satisfies the identity (3.3) with $n = 2$; it also satisfies the identity $x^2y^2x^2 \approx x^2yx^2$ because

$$\begin{aligned} x^2y^2x^2 &\overset{(3.2)}{\approx} ((x^2y)^2(yx^2)^2)^2 \\ &\overset{(3.3)}{\approx} (x^2yx^2)^2 \\ &\overset{(3.1)}{\approx} x^2yx^2. \end{aligned}$$

Therefore, the inclusion $\mathbf{V} \subseteq \mathbf{A_2}\{x^2y^2x^2 \approx x^2yx^2\}$ holds.
(ii) See Lee (2004, Theorem 3.6) or Volkov (2005, Proposition 2(iv)).
(iii) This follows from Lee (2017b, Proposition 2.3).
(iv) See Lee (2007d, Lemma 6, Proposition 10, and Theorem 14). □

Lemma 3.6 *The inclusions* $\overline{\mathbf{B_0}} \subseteq \overline{\mathbf{A_0}} \subseteq \overline{\mathbf{A_2}}$ *and* $\overline{\mathbf{B_0}} \subseteq \overline{\mathbf{B_2}} \subseteq \overline{\mathbf{A_2}}$ *hold.*

Proof For any $\mathbf{U}, \mathbf{V} \in \mathfrak{L}(\mathbf{A_2})$ such that $\mathbf{U} \subseteq \mathbf{V}$, it is easily shown that if $\overline{\mathbf{U}}$ and $\overline{\mathbf{V}}$ exist, then the inclusion $\overline{\mathbf{U}} \subseteq \overline{\mathbf{V}}$ holds. The result then follows from the inclusions in Lemma 3.2. □

Proposition 3.7 *The lattice* $\mathfrak{L}(\overline{\mathbf{A_2}})$ *is the disjoint union of the following intervals*:

$$\mathfrak{I}_1 = [\mathbf{A_0} \vee \mathbf{B_2}, \overline{\mathbf{A_2}}], \qquad \mathfrak{I}_2 = [\mathbf{A_0}, \overline{\mathbf{B_2}}], \qquad \mathfrak{I}_3 = [\mathbf{B_2}, \overline{\mathbf{A_0}}],$$

$$\mathfrak{I}_4 = [\mathbf{B_0}, \overline{\mathbf{A_0}} \cap \overline{\mathbf{B_2}}], \qquad \mathfrak{I}_5 = \mathfrak{L}(\overline{\mathbf{B_0}}).$$

Proof The inclusion $\bigcup_{i=1}^{5} \mathfrak{I}_i \subseteq \mathfrak{L}(\overline{\mathbf{A_2}})$ is obvious since $\overline{\mathbf{A_2}}$ is the unique maximal subvariety of $\mathbf{A_2}$. Conversely, if $\mathbf{U} \in \mathfrak{L}(\overline{\mathbf{A_2}})$, then

- $\mathcal{A}_0, \mathcal{B}_2 \in \mathbf{U}$ implies that $\mathbf{U} \in \mathfrak{I}_1$;
- $\mathcal{A}_0 \in \mathbf{U}$ and $\mathcal{B}_2 \notin \mathbf{U}$ imply that $\mathbf{U} \in \mathfrak{I}_2$;
- $\mathcal{A}_0 \notin \mathbf{U}$ and $\mathcal{B}_2 \in \mathbf{U}$ imply that $\mathbf{U} \in \mathfrak{I}_3$;
- $\mathcal{A}_0, \mathcal{B}_2 \notin \mathbf{U}$ and $\mathcal{B}_0 \in \mathbf{U}$ imply that $\mathbf{U} \in \mathfrak{I}_4$;
- $\mathcal{B}_0 \notin \mathbf{U}$ implies that $\mathbf{U} \in \mathfrak{I}_5$.

Therefore, $\mathfrak{L}(\overline{\mathbf{A}_2}) = \bigcup_{i=1}^{5} \mathfrak{I}_i$. It remains to show that the five intervals are pairwise disjoint. For this purpose, observe that for any $\mathbf{V} \in \mathfrak{L}(\mathbf{A}_2)$,

- if $\mathbf{A_0} \subseteq \mathbf{V}$, then $\mathbf{V} \nsubseteq \overline{\mathbf{A_0}}$;
- if $\mathbf{B_2} \subseteq \mathbf{V}$, then $\mathbf{V} \nsubseteq \overline{\mathbf{B_2}}$;
- if either $\mathbf{A_0} \subseteq \mathbf{V}$ or $\mathbf{B_2} \subseteq \mathbf{V}$, then $\mathbf{B_0} \subseteq \mathbf{V}$ by Lemma 3.2, so that $\mathbf{V} \nsubseteq \overline{\mathbf{B_0}}$.

Now suppose that $\mathbf{V} \in \mathfrak{I}_i \cap \mathfrak{I}_j$ for some distinct $i, j \in \{1, 2, 3, 4, 5\}$. Then there are four cases.

CASE 1: $\mathbf{V} \in \mathfrak{I}_1$. Then $\mathbf{A_0}, \mathbf{B_2} \subseteq \mathbf{V}$, so that $\mathbf{V} \nsubseteq \overline{\mathbf{A_0}}, \overline{\mathbf{B_2}}, \overline{\mathbf{B_0}}$. Hence, $\mathbf{V} \notin \mathfrak{I}_i$ for all $i \in \{2, 3, 4, 5\}$.
CASE 2: $\mathbf{V} \in \mathfrak{I}_2$. Then $\mathbf{A_0} \subseteq \mathbf{V}$, so that $\mathbf{V} \nsubseteq \overline{\mathbf{A_0}}, \overline{\mathbf{B_0}}$. Hence, $\mathbf{V} \notin \mathfrak{I}_i$ for all $i \in \{3, 4, 5\}$.
CASE 3: $\mathbf{V} \in \mathfrak{I}_3$. Then $\mathbf{B_2} \subseteq \mathbf{V}$, so that $\mathbf{V} \nsubseteq \overline{\mathbf{B_2}}, \overline{\mathbf{B_0}}$. Hence, $\mathbf{V} \notin \mathfrak{I}_4$ and $\mathbf{V} \notin \mathfrak{I}_5$.
CASE 4: $\mathbf{V} \in \mathfrak{I}_4$. Then $\mathbf{B_0} \subseteq \mathbf{V}$, so that $\mathbf{V} \nsubseteq \overline{\mathbf{B_0}}$. Hence, $\mathbf{V} \notin \mathfrak{I}_5$.

Since every case is impossible, the variety $\mathbf{V}$ does not exist. □

3.2 Finite Basis Property for Subvarieties of $\mathbf{A}_2$

Recall from Proposition 3.7 that the proper subvarieties of $\overline{\mathbf{A_2}}$ constitute the pairwise disjoint intervals $\mathfrak{I}_1$, $\mathfrak{I}_2$, $\mathfrak{I}_3$, $\mathfrak{I}_4$, and $\mathfrak{I}_5$. The varieties in these five intervals are shown to be finitely based in Sects. 3.2.1–3.2.4. Therefore, all proper subvarieties of $\mathbf{A_2}$ are finitely based. Since the variety $\mathbf{A_2}$ itself is finitely based by Lemma 3.1, the proof of Theorem 1.6 is complete.

3.2.1 Varieties in $\mathfrak{I}_2 = [\mathbf{A}_0, \overline{\mathbf{B}_2}]$

In the present subsection, all varieties in $\mathfrak{I}_2 = [\mathbf{A_0}, \overline{\mathbf{B_2}}]$ are shown to be finitely based.

Lemma 3.8

(i) *A connected identity* $\mathbf{w} \approx \mathbf{w}'$ *is satisfied by* $\overline{\mathbf{B}}_2$ *if and only if* $\mathsf{con}(\mathbf{w}) = \mathsf{con}(\mathbf{w}')$, $\mathsf{h}(\mathbf{w}) = \mathsf{h}(\mathbf{w}')$, *and* $\mathsf{t}(\mathbf{w}) = \mathsf{t}(\mathbf{w}')$.

(ii) *The identities*

$$\left\{ \mathbf{w} \approx \mathbf{w}' \,\middle|\, \begin{array}{l} \mathbf{w}, \mathbf{w}' \in \mathscr{A}^+ \textit{ are connected words such that} \\ \mathsf{con}(\mathbf{w}) = \mathsf{con}(\mathbf{w}'), \mathsf{h}(\mathbf{w}) = \mathsf{h}(\mathbf{w}'), \textit{and } \mathsf{t}(\mathbf{w}) = \mathsf{t}(\mathbf{w}') \end{array} \right\} \tag{3.4}$$

constitute an identity basis for the variety $\overline{\mathbf{B}}_2$.

Proof

(i) Let $\mathbf{w} \approx \mathbf{w}'$ be any connected identity. Suppose that $\overline{\mathbf{B}}_2$ satisfies $\mathbf{w} \approx \mathbf{w}'$. Then $\mathbf{w} \approx \mathbf{w}'$ is deducible from the identity basis $\{(3.1), xy^2x \approx xyx\}$ for $\overline{\mathbf{B}}_2$ given in Proposition 3.5(ii). Since each identity in this basis is formed by a pair of words that share the same head, the same tail, and the same content, it is easily shown that the identity $\mathbf{w} \approx \mathbf{w}'$ satisfies the same properties.

Conversely, suppose $\mathsf{con}(\mathbf{w}) = \mathsf{con}(\mathbf{w}') = \{x_1, x_2, \ldots, x_m\}$, $\mathsf{h}(\mathbf{w}) = \mathsf{h}(\mathbf{w}') = x_i$, and $\mathsf{t}(\mathbf{w}) = \mathsf{t}(\mathbf{w}') = x_j$. Let F_m be the free object of $\overline{\mathbf{B}}_2$ over $\{x_1, x_2, \ldots, x_m\}$. Then the words $\mathbf{w}$ and $\mathbf{w}'$ are regular elements of F_m (Lee and Volkov 2007, Proposition 2.2). By Theorem 8.1.7 in Almeida (1994) and the remark following its proof, these regular elements are $\mathscr{D}$-related in F_m. Since $\mathsf{fac}_2(\mathbf{w}\mathbf{w}') = \mathsf{fac}_2(\mathbf{w}'\mathbf{w})$, $\mathsf{h}(\mathbf{w}\mathbf{w}') = \mathsf{h}(\mathbf{w}'\mathbf{w}) = x_i$, and $\mathsf{t}(\mathbf{w}\mathbf{w}') = \mathsf{t}(\mathbf{w}'\mathbf{w}) = x_j$, it follows from Lemma 3.1 that the variety $\mathbf{A_2}$ satisfies the identity $\mathbf{w}\mathbf{w}' \approx \mathbf{w}'\mathbf{w}$. Therefore, the elements $\mathbf{w}$ and $\mathbf{w}'$ of F_m commute and so must coincide in F_m.

(ii) This follows from part (i) and Proposition 3.5(ii). □

A word $x_1x_2 \cdots x_m$ is *ordered* if the variables $x_1, x_2, \ldots, x_m$ are in alphabetical order. In this chapter, a word $\mathbf{w}$ is said to be in *canonical form* if one of the following conditions holds:

(CF1) $\mathbf{w} = x\mathbf{s}x$ for some $x \in \mathscr{A}$ and simple ordered $\mathbf{s} \in \mathscr{A}^+_\varnothing$ such that $x \notin \mathsf{con}(\mathbf{s})$;

(CF2) $\mathbf{w} = xy\mathbf{s}xy$ for some distinct $x, y \in \mathscr{A}$ and simple ordered $\mathbf{s} \in \mathscr{A}^+_\varnothing$ such that $x, y \notin \mathsf{con}(\mathbf{s})$.

Every word in canonical form is vacuously connected.

Lemma 3.9 *Let* $\mathbf{w}$ *be any connected word. Then the identities* (3.4) *can be used to convert* $\mathbf{w}$ *into some unique word* $\widetilde{\mathbf{w}}$ *in canonical form such that* $\mathsf{con}(\mathbf{w}) = \mathsf{con}(\widetilde{\mathbf{w}})$, $\mathsf{h}(\mathbf{w}) = \mathsf{h}(\widetilde{\mathbf{w}})$, *and* $\mathsf{t}(\mathbf{w}) = \mathsf{t}(\widetilde{\mathbf{w}})$.

Proof This is an easy consequence of Lemma 3.8. □

The following identities are crucial to the description of subvarieties of $\mathbf{A_2}$:

$$\mathbf{a}_{(\ell)}x^2y^2x^2y^2 \approx \mathbf{a}_{(\ell)}y^2x^2y^2, \qquad \langle\ell]$$

$$x^2y^2x^2y^2\mathbf{b}_{(r)} \approx x^2y^2x^2\mathbf{b}_{(r)}, \qquad [r\rangle$$

where $\ell, r \in \widehat{\mathbb{N}}$. It is routine to establish the following deductions.

Lemma 3.10 *Let* $\ell, \ell', r, r' \in \widehat{\mathbb{N}}$. *Then*

(i) $\ell < \ell'$ *implies that* $\{x^3 \approx x^2, \langle\ell]\} \vdash \langle\ell']$;
(ii) $r < r'$ *implies that* $\{x^3 \approx x^2, [r\rangle\} \vdash [r'\rangle$.

Lemma 3.11 *Each variety in the interval* $\mathfrak{I}_2 = [\mathbf{A_0}, \overline{\mathbf{B}}_\mathbf{2}]$ *is of the form* $\overline{\mathbf{B}}_\mathbf{2}\Sigma$, *where* Σ *is some subset of* $\{\langle\ell], [r\rangle \mid \ell, r \in \widehat{\mathbb{N}}\}$.

Proof As shown earlier in Sect. 3.1.2, each variety $\mathbf{V}$ in the interval $\mathfrak{I}_2 = [\mathbf{A_0}, \overline{\mathbf{B}}_\mathbf{2}]$ is defined within $\overline{\mathbf{B}}_\mathbf{2}$ by identities of the form

$$\xi : \mathbf{a}_{(\ell)}\mathbf{w}\mathbf{b}_{(r)} \approx \mathbf{a}_{(\ell)}\mathbf{w}'\mathbf{b}_{(r)}$$

that satisfy all of the following conditions:

(A) $\mathbf{w}, \mathbf{w}' \in \mathscr{A}^+$ are connected words such that $\mathsf{con}(\mathbf{w}) = \mathsf{con}(\mathbf{w}')$ and $\mathcal{A}_0 \models \mathbf{w} \approx \mathbf{w}'$;
(B) $\mathbf{a}_{(\ell)}, \mathbf{w}, \mathbf{b}_{(r)}$ are pairwise disjoint and $\mathbf{a}_{(\ell)}, \mathbf{w}', \mathbf{b}_{(r)}$ are pairwise disjoint;
(C) if $\mathsf{h}(\mathbf{w}) = \mathsf{h}(\mathbf{w}')$, then $\ell = 0$;
(D) if $\mathsf{t}(\mathbf{w}) = \mathsf{t}(\mathbf{w}')$, then $r = 0$.

Suppose that $\mathsf{h}(\mathbf{w}) = \mathsf{h}(\mathbf{w}')$ and $\mathsf{t}(\mathbf{w}) = \mathsf{t}(\mathbf{w}')$, so that by (C) and (D), the identity ξ is $\mathbf{w} \approx \mathbf{w}'$. Then $\overline{\mathbf{B}}_\mathbf{2} \models \mathbf{w} \approx \mathbf{w}'$ by Lemma 3.8(i), whence the identity ξ is redundant in the definition of $\mathbf{V}$ and so can be omitted from its identity basis. Hence, assume that

(E) $\mathsf{h}(\mathbf{w}) \neq \mathsf{h}(\mathbf{w}')$ or $\mathsf{t}(\mathbf{w}) \neq \mathsf{t}(\mathbf{w}')$.

Further, in view of Lemma 3.9, the words $\mathbf{w}$ and $\mathbf{w}'$ can be assumed to be in canonical form. The proof of the present lemma is thus established by proving that

$$\overline{\mathbf{B}}_2\{\xi\} = \begin{cases} \overline{\mathbf{B}}_2\{\langle\ell],[r\rangle\} & \text{if } \mathsf{h}(\mathbf{w}) \neq \mathsf{h}(\mathbf{w}') \text{ and } \mathsf{t}(\mathbf{w}) \neq \mathsf{t}(\mathbf{w}'), \\ \overline{\mathbf{B}}_2\{\langle\ell]\} & \text{if } \mathsf{h}(\mathbf{w}) \neq \mathsf{h}(\mathbf{w}') \text{ and } \mathsf{t}(\mathbf{w}) = \mathsf{t}(\mathbf{w}'), \\ \overline{\mathbf{B}}_2\{[r\rangle\} & \text{if } \mathsf{h}(\mathbf{w}) = \mathsf{h}(\mathbf{w}') \text{ and } \mathsf{t}(\mathbf{w}) \neq \mathsf{t}(\mathbf{w}'). \end{cases}$$

By symmetry, there are three cases. In each case, $x, x', y, y' \in \mathscr{A}$ represent distinct variables and $\mathbf{s} \in \mathscr{A}_\varnothing^+$ represents a simple word such that $x, x', y, y' \notin \mathsf{con}(\mathbf{s})$.

CASE 1: $\mathbf{w}$ and $\mathbf{w}'$ are both of the form (CF1). Then $\mathsf{h}(\mathbf{w}) = \mathsf{t}(\mathbf{w}) \neq \mathsf{h}(\mathbf{w}') = \mathsf{t}(\mathbf{w}')$ by (E), so that

$$\mathbf{w} \overset{(3.4)}{\approx} xx'\mathbf{s}x \quad \text{and} \quad \mathbf{w}' \overset{(3.4)}{\approx} x'x\mathbf{s}x'.$$

Let φ denote the substitution that maps every variable in $\mathsf{con}(x'\mathbf{s}\mathbf{b}_{(r)})$ to y. Then

$$\begin{aligned} \mathbf{a}_{(\ell)}x^2y^2x^2y^2 &\overset{(3.4)}{\approx} ((\mathbf{a}_{(\ell)}\mathbf{w}\mathbf{b}_{(r)})\varphi)y \\ &\overset{\xi}{\approx} ((\mathbf{a}_{(\ell)}\mathbf{w}'\mathbf{b}_{(r)})\varphi)y \\ &\overset{(3.4)}{\approx} \mathbf{a}_{(\ell)}y^2x^2y^2, \end{aligned}$$

so that the deduction $\{(3.4), \xi\} \vdash \langle\ell]$ holds. The deduction $\{(3.4), \xi\} \vdash [r\rangle$ also holds by symmetry, so that $\overline{\mathbf{B}}_2\{\xi\} \subseteq \overline{\mathbf{B}}_2\{\langle\ell],[r\rangle\}$. Conversely, let χ denote the substitution $y \mapsto x'\mathbf{s}x'$. Then

$$\begin{aligned} \mathbf{a}_{(\ell)}\mathbf{w}\mathbf{b}_{(r)} &\overset{(3.4)}{\approx} (\mathbf{a}_{(\ell)}x^2y^2x^2\mathbf{b}_{(r)})\chi \\ &\overset{[r\rangle}{\approx} (\mathbf{a}_{(\ell)}x^2y^2x^2y^2\mathbf{b}_{(r)})\chi \\ &\overset{\langle\ell]}{\approx} (\mathbf{a}_{(\ell)}y^2x^2y^2\mathbf{b}_{(r)})\chi \\ &\overset{(3.4)}{\approx} \mathbf{a}_{(\ell)}\mathbf{w}'\mathbf{b}_{(r)}, \end{aligned}$$

so that the deduction $\{(3.4), \langle\ell],[r\rangle\} \vdash \xi$ holds. Hence, $\overline{\mathbf{B}}_2\{\xi\} = \overline{\mathbf{B}}_2\{\langle\ell],[r\rangle\}$.

CASE 2: $\mathbf{w}$ is of the form (CF1) and $\mathbf{w}'$ is of the form (CF2).

2.1: $\mathsf{h}(\mathbf{w}) \neq \mathsf{h}(\mathbf{w}')$ and $\mathsf{t}(\mathbf{w}) \neq \mathsf{t}(\mathbf{w}')$. Then

$$\mathbf{w} \overset{(3.4)}{\approx} xx'y'\mathbf{s}x \quad \text{and} \quad \mathbf{w}' \overset{(3.4)}{\approx} x'y'x\mathbf{s}x'y'.$$

Let φ denote the substitution that maps every variable in $\mathsf{con}(x'y'\mathbf{s}\mathbf{b}_{(r)})$ to y. Then

$$\begin{aligned}\mathbf{a}_{(\ell)}x^2y^2x^2y^2 &\overset{(3.4)}{\approx} ((\mathbf{a}_{(\ell)}\mathbf{w}\mathbf{b}_{(r)})\varphi)y\\ &\overset{\xi}{\approx} ((\mathbf{a}_{(\ell)}\mathbf{w}'\mathbf{b}_{(r)})\varphi)y\\ &\overset{(3.4)}{\approx} \mathbf{a}_{(\ell)}y^2x^2y^2,\end{aligned}$$

so that the deduction $\{(3.4), \xi\} \vdash \langle\ell]$ holds. The deduction $\{(3.4), \xi\} \vdash [r\rangle$ also holds by symmetry, so that $\overline{\mathbf{B}}_2\{\xi\} \subseteq \overline{\mathbf{B}}_2\{\langle\ell], [r\rangle\}$. Conversely, let χ denote the substitution $y \mapsto x'\mathbf{s}y'$. Then

$$\begin{aligned}\mathbf{a}_{(\ell)}\mathbf{w}\mathbf{b}_{(r)} &\overset{(3.4)}{\approx} (\mathbf{a}_{(\ell)}x^2y^2x^2\mathbf{b}_{(r)})\chi\\ &\overset{[r\rangle}{\approx} (\mathbf{a}_{(\ell)}x^2y^2x^2y^2\mathbf{b}_{(r)})\chi\\ &\overset{\langle\ell]}{\approx} (\mathbf{a}_{(\ell)}y^2x^2y^2\mathbf{b}_{(r)})\chi\\ &\overset{(3.4)}{\approx} \mathbf{a}_{(\ell)}\mathbf{w}'\mathbf{b}_{(r)},\end{aligned}$$

so that the deduction $\{(3.4), \langle\ell], [r\rangle\} \vdash \xi$ holds. Hence, $\overline{\mathbf{B}}_2\{\xi\} = \overline{\mathbf{B}}_2\{\langle\ell], [r\rangle\}$.

2.2: $\mathsf{h}(\mathbf{w}) = \mathsf{h}(\mathbf{w}')$ and $\mathsf{t}(\mathbf{w}) \neq \mathsf{t}(\mathbf{w}')$. Then

$$\mathbf{w} \overset{(3.4)}{\approx} xx'\mathbf{s}x \ \text{ and } \ \mathbf{w}' \overset{(3.4)}{\approx} xx'\mathbf{s}xx'.$$

Further, $\mathbf{a}_{(\ell)} = \varnothing$ by (C). Let φ denote the substitution that maps every variable in $\mathsf{con}(x'\mathbf{s})$ to y. Since

$$\begin{aligned}x^2y^2x^2\mathbf{b}_{(r)} &\overset{(3.4)}{\approx} (\mathbf{w}\mathbf{b}_{(r)})\varphi\\ &\overset{\xi}{\approx} (\mathbf{w}'\mathbf{b}_{(r)})\varphi\\ &\overset{(3.4)}{\approx} x^2y^2x^2y^2\mathbf{b}_{(r)},\end{aligned}$$

the deduction $\{(3.4), \xi\} \vdash [r\rangle$ holds, so that $\overline{\mathbf{B}}_2\{\xi\} \subseteq \overline{\mathbf{B}}_2\{[r\rangle\}$. Conversely, let χ denote the substitution $y \mapsto \mathbf{s}x'$. Then

$$\mathbf{w}\mathbf{b}_{(r)} \overset{(3.4)}{\approx} (x^2y^2x^2\mathbf{b}_{(r)})\chi$$

$$\overset{[r\rangle}{\approx} (x^2y^2x^2y^2\mathbf{b}_{(r)})\chi$$

$$\overset{(3.4)}{\approx} \mathbf{w}'\mathbf{b}_{(r)},$$

so that the deduction $\{(3.4), [r\rangle\} \vdash \xi$ holds. Therefore, $\overline{\mathbf{B}}_2\{\xi\} = \overline{\mathbf{B}}_2\{[r\rangle\}$.

2.3: $\mathsf{h}(\mathbf{w}) \neq \mathsf{h}(\mathbf{w}')$ and $\mathsf{t}(\mathbf{w}) = \mathsf{t}(\mathbf{w}')$. Then $\overline{\mathbf{B}}_2\{\xi\} = \overline{\mathbf{B}}_2\{\langle\ell]\}$ by an argument symmetrical to Case 2.2.

CASE 3: $\mathbf{w}$ and $\mathbf{w}'$ are both of the form (CF2).

3.1: $\mathsf{h}(\mathbf{w}) \neq \mathsf{h}(\mathbf{w}')$ and $\mathsf{t}(\mathbf{w}) \neq \mathsf{t}(\mathbf{w}')$.

3.1.1: $\mathsf{h}(\mathbf{w}) = \mathsf{t}(\mathbf{w}')$ and $\mathsf{t}(\mathbf{w}) = \mathsf{h}(\mathbf{w}')$. Then

$$\mathbf{w} \overset{(3.4)}{\approx} xx'\mathbf{s}xx' \text{ and } \mathbf{w}' \overset{(3.4)}{\approx} x'x\mathbf{s}x'x.$$

Let φ denote the substitution $(x, x') \mapsto (xx', x'x)$. Then

$$\begin{aligned}\mathbf{a}_{(\ell)}xx'\mathbf{s}x\mathbf{b}_{(r)} &\overset{(3.4)}{\approx} (\mathbf{a}_{(\ell)}\mathbf{w}\mathbf{b}_{(r)})\varphi \\ &\overset{\xi}{\approx} (\mathbf{a}_{(\ell)}\mathbf{w}'\mathbf{b}_{(r)})\varphi \\ &\overset{(3.4)}{\approx} \mathbf{a}_{(\ell)}x'x\mathbf{s}x'\mathbf{b}_{(r)},\end{aligned}$$

so that the deduction $\{(3.4), \xi\} \vdash \mathbf{a}_{(\ell)}xx'\mathbf{s}x\mathbf{b}_{(r)} \approx \mathbf{a}_{(\ell)}x'x\mathbf{s}x'\mathbf{b}_{(r)}$ holds. Therefore,

$$\begin{aligned}\overline{\mathbf{B}}_2\{\xi\} &\subseteq \overline{\mathbf{B}}_2\{\mathbf{a}_{(\ell)}xx'\mathbf{s}x\mathbf{b}_{(r)} \approx \mathbf{a}_{(\ell)}x'x\mathbf{s}x'\mathbf{b}_{(r)}\} \\ &= \overline{\mathbf{B}}_2\{\langle\ell], [r\rangle\}\end{aligned}$$

by Case 1. Conversely, let χ denote the substitution $(x, y) \mapsto (x\mathbf{s}x', x'x)$. Then

$$\begin{aligned}\mathbf{a}_{(\ell)}\mathbf{w}\mathbf{b}_{(r)} &\overset{(3.4)}{\approx} (\mathbf{a}_{(\ell)}x^2y^2x^2\mathbf{b}_{(r)})\chi \\ &\overset{[r\rangle}{\approx} (\mathbf{a}_{(\ell)}x^2y^2x^2y^2\mathbf{b}_{(r)})\chi \\ &\overset{\langle\ell]}{\approx} (\mathbf{a}_{(\ell)}y^2x^2y^2\mathbf{b}_{(r)})\chi \\ &\overset{(3.4)}{\approx} \mathbf{a}_{(\ell)}\mathbf{w}'\mathbf{b}_{(r)},\end{aligned}$$

so that the deduction $\{(3.4), \langle\ell], [r\rangle\} \vdash \xi$ holds. Therefore, $\overline{\mathbf{B}}_2\{\xi\} = \overline{\mathbf{B}}_2\{\langle\ell], [r\rangle\}$.

3.1.2: $\mathsf{h}(\mathbf{w}) = \mathsf{t}(\mathbf{w}')$ and $\mathsf{t}(\mathbf{w}) \neq \mathsf{h}(\mathbf{w}')$. Then

$$\mathbf{w} \overset{(3.4)}{\approx} xyx'\mathbf{s}xy \quad\text{and}\quad \mathbf{w}' \overset{(3.4)}{\approx} x'xy\mathbf{s}x'x.$$

Let φ denote the substitution $(x, y) \mapsto (xx', x'x)$. Then

$$\begin{aligned}\mathbf{a}_{(\ell)}xx'\mathbf{s}x\mathbf{b}_{(r)} &\overset{(3.4)}{\approx} (\mathbf{a}_{(\ell)}\mathbf{w}\mathbf{b}_{(r)})\varphi \\ &\overset{\xi}{\approx} (\mathbf{a}_{(\ell)}\mathbf{w}'\mathbf{b}_{(r)})\varphi \\ &\overset{(3.4)}{\approx} \mathbf{a}_{(\ell)}x'x\mathbf{s}x'\mathbf{b}_{(r)},\end{aligned}$$

so that the deduction $\{(3.4), \xi\} \vdash \mathbf{a}_{(\ell)}xx'\mathbf{s}x\mathbf{b}_{(r)} \approx \mathbf{a}_{(\ell)}x'x\mathbf{s}x'\mathbf{b}_{(r)}$ holds. Therefore, the inclusion $\overline{\mathbf{B}}_2\{\xi\} \subseteq \overline{\mathbf{B}}_2\{\langle\ell], [r\rangle\}$ holds as in Case 3.1.1. Conversely, let χ denote the substitution $(x, y) \mapsto (x\mathbf{s}y, x'x)$. Then

$$\begin{aligned}\mathbf{a}_{(\ell)}\mathbf{w}\mathbf{b}_{(r)} &\overset{(3.4)}{\approx} (\mathbf{a}_{(\ell)}x^2y^2x^2\mathbf{b}_{(r)})\chi \\ &\overset{[r\rangle}{\approx} (\mathbf{a}_{(\ell)}x^2y^2x^2y^2\mathbf{b}_{(r)})\chi \\ &\overset{\langle\ell]}{\approx} (\mathbf{a}_{(\ell)}y^2x^2y^2\mathbf{b}_{(r)})\chi \\ &\overset{(3.4)}{\approx} \mathbf{a}_{(\ell)}\mathbf{w}'\mathbf{b}_{(r)},\end{aligned}$$

so that the deduction $\{(3.4), \langle\ell], [r\rangle\} \vdash \xi$ holds. Therefore, $\overline{\mathbf{B}}_2\{\xi\} = \overline{\mathbf{B}}_2\{\langle\ell], [r\rangle\}$.

3.1.3: $\mathsf{h}(\mathbf{w}) \neq \mathsf{t}(\mathbf{w}')$ and $\mathsf{t}(\mathbf{w}) = \mathsf{h}(\mathbf{w}')$. Then $\overline{\mathbf{B}}_2\{\xi\} = \overline{\mathbf{B}}_2\{\langle\ell], [r\rangle\}$ by an argument symmetrical to Case 3.1.2.

3.1.4: $\mathsf{h}(\mathbf{w}) \neq \mathsf{t}(\mathbf{w}')$ and $\mathsf{t}(\mathbf{w}) \neq \mathsf{h}(\mathbf{w}')$. Then

$$\mathbf{w} \overset{(3.4)}{\approx} xyx'y'\mathbf{s}xy \quad\text{and}\quad \mathbf{w}' \overset{(3.4)}{\approx} x'y'xy\mathbf{s}x'y'.$$

Let φ denote the substitution $(y, y') \mapsto (x, x')$. Then

$$\begin{aligned}\mathbf{a}_{(\ell)}xx'\mathbf{s}x\mathbf{b}_{(r)} &\overset{(3.4)}{\approx} (\mathbf{a}_{(\ell)}\mathbf{w}\mathbf{b}_{(r)})\varphi \\ &\overset{\xi}{\approx} (\mathbf{a}_{(\ell)}\mathbf{w}'\mathbf{b}_{(r)})\varphi \\ &\overset{(3.4)}{\approx} \mathbf{a}_{(\ell)}x'x\mathbf{s}x'\mathbf{b}_{(r)},\end{aligned}$$

so that the deduction $\{(3.4), \xi\} \vdash \mathbf{a}_{(\ell)}xx'\mathbf{s}x\mathbf{b}_{(r)} \approx \mathbf{a}_{(\ell)}x'x\mathbf{s}x'\mathbf{b}_{(r)}$ holds. Therefore, the inclusion $\overline{\mathbf{B}}_2\{\xi\} \subseteq \overline{\mathbf{B}}_2\{\langle\ell], [r\rangle\}$ holds as in Case 3.1.1. Conversely, let χ denote the substitution $(x, y) \mapsto (x\mathbf{s}y, x'y')$. Then

$$\begin{aligned}\mathbf{a}_{(\ell)}\mathbf{w}\mathbf{b}_{(r)} &\overset{(3.4)}{\approx} (\mathbf{a}_{(\ell)}x^2y^2x^2\mathbf{b}_{(r)})\chi \\ &\overset{[r\rangle}{\approx} (\mathbf{a}_{(\ell)}x^2y^2x^2y^2\mathbf{b}_{(r)})\chi \\ &\overset{\langle\ell]}{\approx} (\mathbf{a}_{(\ell)}y^2x^2y^2\mathbf{b}_{(r)})\chi \\ &\overset{(3.4)}{\approx} \mathbf{a}_{(\ell)}\mathbf{w}'\mathbf{b}_{(r)},\end{aligned}$$

so that the deduction $\{(3.4), \langle\ell], [r\rangle\} \vdash \xi$ holds. Therefore, $\overline{\mathbf{B}}_2\{\xi\} = \overline{\mathbf{B}}_2\{\langle\ell], [r\rangle\}$.

3.2: $\mathsf{h}(\mathbf{w}) = \mathsf{h}(\mathbf{w}')$ and $\mathsf{t}(\mathbf{w}) \neq \mathsf{t}(\mathbf{w}')$. Then

$$\mathbf{w} \overset{(3.4)}{\approx} xyx'\mathbf{s}xy \text{ and } \mathbf{w}' \overset{(3.4)}{\approx} xx'y\mathbf{s}xx'.$$

Further, $\mathbf{a}_{(\ell)} = \varnothing$ by (C). Let φ denote the substitution that maps every variable in $\mathsf{con}(x'\mathbf{s})$ to x. Since

$$\begin{aligned}x^2y^2x^2y^2\mathbf{b}_{(r)} &\overset{(3.4)}{\approx} (\mathbf{w}\mathbf{b}_{(r)})\varphi \\ &\overset{\xi}{\approx} (\mathbf{w}'\mathbf{b}_{(r)})\varphi \\ &\overset{(3.4)}{\approx} x^2y^2x^2\mathbf{b}_{(r)},\end{aligned}$$

the deduction $\{(3.4), \xi\} \vdash [r\rangle$ holds, so that $\overline{\mathbf{B}}_2\{\xi\} \subseteq \overline{\mathbf{B}}_2\{[r\rangle\}$. Conversely, let χ denote the substitution $x \mapsto x\mathbf{s}x'$. Then

$$\begin{aligned}\mathbf{w}\mathbf{b}_{(r)} &\overset{(3.4)}{\approx} (x^2y^2x^2y^2\mathbf{b}_{(r)})\chi \\ &\overset{[r\rangle}{\approx} (x^2y^2x^2\mathbf{b}_{(r)})\chi \\ &\overset{(3.4)}{\approx} \mathbf{w}'\mathbf{b}_{(r)},\end{aligned}$$

so that the deduction $\{(3.4), [r\rangle\} \vdash \xi$ holds. Therefore, $\overline{\mathbf{B}}_2\{\xi\} = \overline{\mathbf{B}}_2\{[r\rangle\}$.

3.3: $\mathsf{h}(\mathbf{w}) \neq \mathsf{h}(\mathbf{w}')$ and $\mathsf{t}(\mathbf{w}) = \mathsf{t}(\mathbf{w}')$. Then $\overline{\mathbf{B}}_2\{\xi\} = \overline{\mathbf{B}}_2\{\langle\ell]\}$ by an argument symmetrical to Case 3.2. □

Proposition 3.12 *Every variety in the interval* $\mathfrak{I}_2 = [\mathbf{A_0}, \overline{\mathbf{B}}_2]$ *is finitely based.*

Proof Let $\mathbf{V} \in \mathfrak{I}_2$. Then by Lemma 3.11, there exist some sets $\Sigma_1 \subseteq \{\langle \ell] \mid \ell \in \widehat{\mathbb{N}}\}$ and $\Sigma_2 \subseteq \{[r\rangle \mid r \in \widehat{\mathbb{N}}\}$ such that $\mathbf{V} = \overline{\mathbf{B}}_2(\Sigma_1 \cup \Sigma_2)$. By Lemma 3.10,

$$\overline{\mathbf{B}}_2\Sigma_1 = \begin{cases} \overline{\mathbf{B}}_2\{\langle \ell_0]\} & \text{if } \Sigma_1 \neq \varnothing \text{ and } \ell_0 = \min\{\ell \in \widehat{\mathbb{N}} \mid \langle \ell] \in \Sigma_1\}, \\ \overline{\mathbf{B}}_2 & \text{if } \Sigma_1 = \varnothing; \end{cases}$$

$$\overline{\mathbf{B}}_2\Sigma_2 = \begin{cases} \overline{\mathbf{B}}_2\{[r_0\rangle\} & \text{if } \Sigma_2 \neq \varnothing \text{ and } r_0 = \min\{r \in \widehat{\mathbb{N}} \mid [r\rangle \in \Sigma_2\}, \\ \overline{\mathbf{B}}_2 & \text{if } \Sigma_2 = \varnothing. \end{cases}$$

Since $\overline{\mathbf{B}}_2$ is finitely based by Proposition 3.5, the variety $\mathbf{V} = \overline{\mathbf{B}}_2\Sigma_1 \cap \overline{\mathbf{B}}_2\Sigma_2$ is also finitely based. □

3.2.2 Varieties in $\mathfrak{I}_1 = [\mathbf{A}_0 \vee \mathbf{B}_2, \overline{\mathbf{A}}_2]$

In the present subsection, all varieties in $\mathfrak{I}_1 = [\mathbf{A_0} \vee \mathbf{B}_2, \overline{\mathbf{A}}_2]$ are shown to be finitely based.

Lemma 3.13 *Let* $\mathbf{w} \approx \mathbf{w}'$ *be any identity, where*

(a) $\mathbf{w}$ *and* $\mathbf{w}'$ *are connected words such that either* $\mathsf{h}(\mathbf{w}) \neq \mathsf{h}(\mathbf{w}')$ *or* $\mathsf{t}(\mathbf{w}) \neq \mathsf{t}(\mathbf{w}')$;
(b) $\mathbf{A_0} \models \mathbf{w} \approx \mathbf{w}'$.

Then $\overline{\mathbf{B}}_2\{\mathbf{w} \approx \mathbf{w}'\} = \overline{\mathbf{B}}_2\Sigma$, *where*

$$\Sigma = \begin{cases} \{\langle 0], [0\rangle\} & \textit{if } \mathsf{h}(\mathbf{w}) \neq \mathsf{h}(\mathbf{w}') \textit{ and } \mathsf{t}(\mathbf{w}) \neq \mathsf{t}(\mathbf{w}'), \\ \{\langle 0]\} & \textit{if } \mathsf{h}(\mathbf{w}) \neq \mathsf{h}(\mathbf{w}') \textit{ and } \mathsf{t}(\mathbf{w}) = \mathsf{t}(\mathbf{w}'), \\ \{[0\rangle\} & \textit{if } \mathsf{h}(\mathbf{w}) = \mathsf{h}(\mathbf{w}') \textit{ and } \mathsf{t}(\mathbf{w}) \neq \mathsf{t}(\mathbf{w}'). \end{cases}$$

Proof This is the special case $\ell = r = 0$ in the proof of Lemma 3.11. □

Lemma 3.14 *For any subset* Σ *of* $\{\langle 0], [0\rangle\}$, *the equality* $\overline{\mathbf{A}}_2\Sigma = \overline{\mathbf{B}}_2\Sigma \vee \mathbf{B}_2$ *holds.*

Proof This follows from Lee and Volkov (2007), specifically,

- $\overline{\mathbf{A}}_2 = \overline{\mathbf{B}}_2 \vee \mathbf{B}_2$ (Lee and Volkov 2007, Theorems 4.2(iii) and 4.3(iv));
- $\overline{\mathbf{A}}_2\{\langle 0]\} = \overline{\mathbf{B}}_2\{\langle 0]\} \vee \mathbf{B}_2$ (Lee and Volkov 2007, Theorems 4.2(ii) and 4.3(iii));
- $\overline{\mathbf{A}}_2\{[0\rangle\} = \overline{\mathbf{B}}_2\{[0\rangle\} \vee \mathbf{B}_2$ (Lee and Volkov 2007, Theorems 4.2(i) and 4.3(ii));
- $\overline{\mathbf{A}}_2\{\langle 0], [0\rangle\} = \overline{\mathbf{B}}_2\{\langle 0], [0\rangle\} \vee \mathbf{B}_2$ (Lee and Volkov 2007, Theorems 4.1 and 4.3(i)). □

Lemma 3.15 *Let* $\mathbf{w} \approx \mathbf{w}'$ *be any identity, where*

(a) $\mathbf{w}$ *and* $\mathbf{w}'$ *are connected words such that either* $\mathsf{h}(\mathbf{w}) \neq \mathsf{h}(\mathbf{w}')$ *or* $\mathsf{t}(\mathbf{w}) \neq \mathsf{t}(\mathbf{w}')$;
(b) $\mathbf{A_0} \vee \mathbf{B_2} \models \mathbf{w} \approx \mathbf{w}'$.

Then $\overline{\mathbf{A}}_2\{\mathbf{w} \approx \mathbf{w}'\} = \overline{\mathbf{A}}_2\Sigma$, *where*

$$\Sigma = \begin{cases} \{\langle 0], [0\rangle\} & \text{if } \mathsf{h}(\mathbf{w}) \neq \mathsf{h}(\mathbf{w}') \text{ and } \mathsf{t}(\mathbf{w}) \neq \mathsf{t}(\mathbf{w}'), \\ \{\langle 0]\} & \text{if } \mathsf{h}(\mathbf{w}) \neq \mathsf{h}(\mathbf{w}') \text{ and } \mathsf{t}(\mathbf{w}) = \mathsf{t}(\mathbf{w}'), \\ \{[0\rangle\} & \text{if } \mathsf{h}(\mathbf{w}) = \mathsf{h}(\mathbf{w}') \text{ and } \mathsf{t}(\mathbf{w}) \neq \mathsf{t}(\mathbf{w}'). \end{cases}$$

Proof Suppose that $\mathsf{h}(\mathbf{w}) \neq \mathsf{h}(\mathbf{w}')$. Let φ denote the substitution

$$z \mapsto \begin{cases} x^2 & \text{if } z = \mathsf{h}(\mathbf{w}), \\ y^2 & \text{otherwise.} \end{cases}$$

Then $(\mathbf{w}\varphi)x^2 \overset{(3.1)}{\approx} x^2y^2x^2$ and $(\mathbf{w}'\varphi)x^2 \overset{(3.1)}{\approx} y^2x^2y^2x^2$, so that $\{(3.1), \mathbf{w} \approx \mathbf{w}'\} \vdash \langle 0]$. By symmetry, if $\mathsf{t}(\mathbf{w}) \neq \mathsf{t}(\mathbf{w}')$, then $\{(3.1), \mathbf{w} \approx \mathbf{w}'\} \vdash [0\rangle$. Therefore, $\overline{\mathbf{A}}_2\{\mathbf{w} \approx \mathbf{w}'\} \subseteq \overline{\mathbf{A}}_2\Sigma$. Conversely,

$$\begin{aligned} \overline{\mathbf{A}}_2\Sigma &= \overline{\mathbf{B}}_2\Sigma \vee \mathbf{B}_2 && \text{by Lemma 3.14} \\ &= \overline{\mathbf{B}}_2\{\mathbf{w} \approx \mathbf{w}'\} \vee \mathbf{B}_2 && \text{by Lemma 3.13} \\ &\subseteq \overline{\mathbf{A}}_2\{\mathbf{w} \approx \mathbf{w}'\}, \end{aligned}$$

where the inclusion holds because $\overline{\mathbf{B}}_2\{\mathbf{w} \approx \mathbf{w}'\}$ and $\mathbf{B}_2$ are subvarieties of $\overline{\mathbf{A}}_2$ that satisfy the identity $\mathbf{w} \approx \mathbf{w}'$. □

Lemma 3.16 *The variety* $\overline{\mathbf{A}}_2$ *satisfies the identity*

$$x\mathbf{h}_1y^2z^2\mathbf{h}_2x \approx x\mathbf{h}_1z^2y^2\mathbf{h}_2x, \tag{3.5}$$

where $\mathbf{h}_i \in \{\varnothing, h_i\}$.

Proof The variety $\mathbf{B_2}$ satisfies the identity (3.5) because the idempotents of $\mathcal{B}_2$ commute while the variety $\overline{\mathbf{B_2}}$ satisfies the identity (3.5) due to Lemma 3.8(i). Consequently, the result holds because $\overline{\mathbf{A_2}} = \overline{\mathbf{B_2}} \vee \mathbf{B_2}$ by Lemma 3.14. □

Lemma 3.17 *Let* $\sigma : \mathbf{a}_{(\ell)}\mathbf{w}\mathbf{b}_{(r)} \approx \mathbf{a}_{(\ell)}\mathbf{w}'\mathbf{b}_{(r)}$ *be any identity, where*

(a) $\mathbf{w}$ *and* $\mathbf{w}'$ *are connected words such that* $\mathsf{h}(\mathbf{w}) \neq \mathsf{h}(\mathbf{w}')$ *and* $\mathsf{t}(\mathbf{w}) \neq \mathsf{t}(\mathbf{w}')$;
(b) $\mathbf{A_0} \vee \mathbf{B_2} \models \mathbf{w} \approx \mathbf{w}'$.

Then $\overline{\mathbf{A_2}}\{\sigma\} = \overline{\mathbf{A_2}}\{\langle\ell], [r\rangle\}$.

Proof Let φ denote the substitution

$$z \mapsto \begin{cases} x^2 & \text{if } z \in \mathsf{con}(\mathbf{b}_{(r)}) \cup \{\mathsf{h}(\mathbf{w})\}, \\ y^2 & \text{if } z \in \mathsf{con}(\mathbf{w}) \backslash \{\mathsf{h}(\mathbf{w})\}. \end{cases}$$

Since

$$\begin{aligned} \mathbf{a}_{(\ell)}x^2y^2x^2 &\overset{(3.1)}{\approx} ((\mathbf{a}_{(\ell)}\mathbf{w}\mathbf{b}_{(r)})\varphi)x^2 \\ &\overset{\sigma}{\approx} ((\mathbf{a}_{(\ell)}\mathbf{w}'\mathbf{b}_{(r)})\varphi)x^2 \\ &\overset{(3.1)}{\approx} \mathbf{a}_{(\ell)}y^2x^2y^2x^2, \end{aligned}$$

the deduction $\{(3.1), \sigma\} \vdash \langle\ell]$ holds, so that $\overline{\mathbf{A_2}}\{\sigma\} \subseteq \overline{\mathbf{A_2}}\{\langle\ell]\}$. By symmetry, the inclusion $\overline{\mathbf{A_2}}\{\sigma\} \subseteq \overline{\mathbf{A_2}}\{[r\rangle\}$ also holds. Therefore, $\overline{\mathbf{A_2}}\{\sigma\} \subseteq \overline{\mathbf{A_2}}\{\langle\ell], [r\rangle\}$, and it remains to verify the reverse inclusion.

By Lemma 3.1 and Proposition 3.5(i), the variety $\overline{\mathbf{A_2}}\{\langle 0], [0\rangle\}$ is defined by the identities

$$(3.1), \quad x^2y^2x^2 \approx x^2yx^2, \quad \langle 0], \quad [0\rangle. \tag{3.6}$$

Since $\overline{\mathbf{A_2}}\{\mathbf{w} \approx \mathbf{w}'\} = \overline{\mathbf{A_2}}\{\langle 0], [0\rangle\}$ by Lemma 3.15, there exists a deduction sequence

$$\mathbf{w} = \mathbf{w}_1, \mathbf{w}_2, \ldots, \mathbf{w}_m = \mathbf{w}',$$

where each identity $\mathbf{w}_i \approx \mathbf{w}_{i+1}$ is directly deducible from some identity in (3.6).

CASE 1: $\mathbf{w}_i \approx \mathbf{w}_{i+1}$ is directly deducible from an identity in $\{(3.1), x^2y^2x^2 \approx x^2yx^2\}$. Then the identity $\mathbf{a}_{(\ell)}\mathbf{w}_i\mathbf{b}_{(r)} \approx \mathbf{a}_{(\ell)}\mathbf{w}_{i+1}\mathbf{b}_{(r)}$ is clearly deducible from $\{(3.1), x^2y^2x^2 \approx x^2yx^2\}$.

CASE 2: $\mathbf{w}_i \approx \mathbf{w}_{i+1}$ is directly deducible from $\langle 0]$. Then there exist words $\mathbf{e}, \mathbf{f} \in \mathscr{A}^+_\varnothing$ and a substitution $\gamma : \mathscr{A} \to \mathscr{A}^+$ such that

$$\mathbf{w}_i = \mathbf{e}((x^2y^2x^2y^2)\gamma)\mathbf{f} \text{ and } \mathbf{w}_{i+1} = \mathbf{e}((y^2x^2y^2)\gamma)\mathbf{f}.$$

Hence,

$$\mathbf{a}_{(\ell)}\mathbf{w}_i\mathbf{b}_{(r)} = \mathbf{a}_{(\ell)}\mathbf{e}\mathbf{c}^2\mathbf{d}^2\mathbf{c}^2\mathbf{d}^2\mathbf{f}\,\mathbf{b}_{(r)} \text{ and } \mathbf{a}_{(\ell)}\mathbf{w}_{i+1}\mathbf{b}_{(r)} = \mathbf{a}_{(\ell)}\mathbf{e}\mathbf{d}^2\mathbf{c}^2\mathbf{d}^2\mathbf{f}\,\mathbf{b}_{(r)},$$

where $\mathbf{c} = x\gamma$ and $\mathbf{d} = y\gamma$. If $\mathbf{e} = \varnothing$, then

$$\begin{aligned} \mathbf{a}_{(\ell)}\mathbf{w}_i\mathbf{b}_{(r)} &= \mathbf{a}_{(\ell)}\mathbf{c}^2\mathbf{d}^2\mathbf{c}^2\mathbf{d}^2\mathbf{f}\,\mathbf{b}_{(r)} \\ &\overset{\langle \ell]}{\approx} \mathbf{a}_{(\ell)}\mathbf{d}^2\mathbf{c}^2\mathbf{d}^2\mathbf{f}\,\mathbf{b}_{(r)} \\ &= \mathbf{a}_{(\ell)}\mathbf{w}_{i+1}\mathbf{b}_{(r)}. \end{aligned}$$

If $\mathbf{e} \neq \varnothing$, then $\mathsf{con}(\mathbf{e}) \cap \mathsf{con}(\mathbf{cdf}) \neq \varnothing$ by the connectedness of $\mathbf{w}_i$, whence

$$\begin{aligned} \mathbf{a}_{(\ell)}\mathbf{w}_i\mathbf{b}_{(r)} &= \mathbf{a}_{(\ell)}\mathbf{e}\mathbf{c}^2\mathbf{d}^2\mathbf{c}^2\mathbf{d}^2\mathbf{f}\,\mathbf{b}_{(r)} \\ &\overset{(3.5)}{\approx} \mathbf{a}_{(\ell)}\mathbf{e}\mathbf{d}^2\mathbf{c}^2\mathbf{c}^2\mathbf{d}^2\mathbf{f}\,\mathbf{b}_{(r)} \\ &\overset{(3.1)}{\approx} \mathbf{a}_{(\ell)}\mathbf{e}\mathbf{d}^2\mathbf{c}^2\mathbf{d}^2\mathbf{f}\,\mathbf{b}_{(r)} \\ &= \mathbf{a}_{(\ell)}\mathbf{w}_{i+1}\mathbf{b}_{(r)}. \end{aligned}$$

The identity $\mathbf{a}_{(\ell)}\mathbf{w}_i\mathbf{b}_{(r)} \approx \mathbf{a}_{(\ell)}\mathbf{w}_{i+1}\mathbf{b}_{(r)}$ is thus deducible from $\{(3.1), (3.5), \langle \ell]\}$.

CASE 3: $\mathbf{w}_i \approx \mathbf{w}_{i+1}$ is directly deducible from $[0\rangle$. By an argument symmetrical to Case 2, the identity $\mathbf{a}_{(\ell)}\mathbf{w}_i\mathbf{b}_{(r)} \approx \mathbf{a}_{(\ell)}\mathbf{w}_{i+1}\mathbf{b}_{(r)}$ is deducible from $\{(3.1), (3.5), [r\rangle\}$.

Therefore, in any case, the identity $\mathbf{a}_{(\ell)}\mathbf{w}_i\mathbf{b}_{(r)} \approx \mathbf{a}_{(\ell)}\mathbf{w}_{i+1}\mathbf{b}_{(r)}$ is deducible from

$$(3.1), \quad (3.5), \quad x^2y^2x^2 \approx x^2yx^2, \quad \langle \ell], \quad [r\rangle. \tag{3.7}$$

It follows that the deduction (3.7) $\vdash \sigma$ holds. By Proposition 3.5(i) and Lemma 3.16, the variety $\overline{\mathbf{A}}_2\{\langle \ell], [r\rangle\}$ satisfies (3.7). Consequently, the inclusion $\overline{\mathbf{A}}_2\{\langle \ell], [r\rangle\} \subseteq \overline{\mathbf{A}}_2\{\sigma\}$ holds as required. □

Lemma 3.18 *Let* $\sigma : \mathbf{a}_{(\ell)}\mathbf{w}\mathbf{b}_{(r)} \approx \mathbf{a}_{(\ell)}\mathbf{w}'\mathbf{b}_{(r)}$ *be any identity, where*

(a) $\mathbf{w}$ *and* $\mathbf{w}'$ *are connected words such that either* $\mathsf{h}(\mathbf{w}) \neq \mathsf{h}(\mathbf{w}')$ *or* $\mathsf{t}(\mathbf{w}) \neq \mathsf{t}(\mathbf{w}')$;
(b) $\mathbf{A_0} \vee \mathbf{B_2} \models \mathbf{w} \approx \mathbf{w}'$.

Then $\overline{\mathbf{A}}_2\{\sigma\} = \overline{\mathbf{A}}_2\Sigma$, *where*

$$\Sigma = \begin{cases} \{\langle \ell], [r\rangle\} & \textit{if } \mathsf{h}(\mathbf{w}) \neq \mathsf{h}(\mathbf{w}') \textit{ and } \mathsf{t}(\mathbf{w}) \neq \mathsf{t}(\mathbf{w}'), \\ \{\langle \ell]\} & \textit{if } \mathsf{h}(\mathbf{w}) \neq \mathsf{h}(\mathbf{w}') \textit{ and } \mathsf{t}(\mathbf{w}) = \mathsf{t}(\mathbf{w}'), \\ \{[r\rangle\} & \textit{if } \mathsf{h}(\mathbf{w}) = \mathsf{h}(\mathbf{w}') \textit{ and } \mathsf{t}(\mathbf{w}) \neq \mathsf{t}(\mathbf{w}'). \end{cases}$$

Proof The case when $\mathsf{h}(\mathbf{w}) \neq \mathsf{h}(\mathbf{w}')$ and $\mathsf{t}(\mathbf{w}) \neq \mathsf{t}(\mathbf{w}')$ is covered by Lemma 3.17. The other cases can be similarly established. □

Proposition 3.19 *Every variety in the interval* $\mathfrak{I}_1 = [\mathbf{A_0} \vee \mathbf{B_2}, \overline{\mathbf{A}}_2]$ *is finitely based.*

Proof As shown in Sect. 3.1.2, each variety $\mathbf{V}$ in $\mathfrak{I}_1 = [\mathbf{A_0} \vee \mathbf{B_2}, \overline{\mathbf{A}}_2]$ is defined within $\overline{\mathbf{A}}_2$ by identities of the form

$$\sigma : \mathbf{a}_{(\ell)}\mathbf{w}\mathbf{b}_{(r)} \approx \mathbf{a}_{(\ell)}\mathbf{w}'\mathbf{b}_{(r)}$$

that satisfy all of the following properties:

(A) $\mathbf{w}, \mathbf{w}' \in \mathscr{A}^+$ are connected words with $\mathsf{con}(\mathbf{w}) = \mathsf{con}(\mathbf{w}')$ and $\mathcal{A}_0, \mathcal{B}_2 \models \mathbf{w} \approx \mathbf{w}'$;
(B) $\mathbf{a}_{(\ell)}, \mathbf{w}, \mathbf{b}_{(r)}$ are pairwise disjoint and $\mathbf{a}_{(\ell)}, \mathbf{w}', \mathbf{b}_{(r)}$ are pairwise disjoint;
(C) if $\mathsf{h}(\mathbf{w}) = \mathsf{h}(\mathbf{w}')$, then $\ell = 0$;
(D) if $\mathsf{t}(\mathbf{w}) = \mathsf{t}(\mathbf{w}')$, then $r = 0$.

Suppose that $\mathsf{h}(\mathbf{w}) = \mathsf{h}(\mathbf{w}')$ and $\mathsf{t}(\mathbf{w}) = \mathsf{t}(\mathbf{w}')$, so that by (C) and (D), the identity σ is $\mathbf{w} \approx \mathbf{w}'$. Then $\overline{\mathbf{B}}_2 \models \mathbf{w} \approx \mathbf{w}'$ by Lemma 3.8(i). Now since $\mathbf{B_2} \models \mathbf{w} \approx \mathbf{w}'$ by (A) and $\overline{\mathbf{A}}_2 = \overline{\mathbf{B}}_2 \vee \mathbf{B_2}$ by Lemma 3.14, it follows that $\overline{\mathbf{A}}_2 \models \mathbf{w} \approx \mathbf{w}'$, whence the identity σ is redundant in the definition of $\mathbf{V}$ and so can be omitted from its identity basis. Therefore, assume further that each identity σ satisfies either $\mathsf{h}(\mathbf{w}) \neq \mathsf{h}(\mathbf{w}')$ or $\mathsf{t}(\mathbf{w}) \neq \mathsf{t}(\mathbf{w}')$. It then follows from Lemma 3.18 that $\mathbf{V} = \overline{\mathbf{A}}_2(\Sigma_1 \cup \Sigma_2)$ for some sets $\Sigma_1 \subseteq \{\langle \ell] \mid \ell \in \widehat{\mathbb{N}}\}$ and $\Sigma_2 \subseteq \{[r\rangle \mid r \in \widehat{\mathbb{N}}\}$. By Lemma 3.10,

$$\overline{\mathbf{A}}_2\Sigma_1 = \begin{cases} \overline{\mathbf{A}}_2\{\langle \ell_0]\} & \text{if } \Sigma_1 \neq \varnothing \text{ and } \ell_0 = \min\{\ell \in \widehat{\mathbb{N}} \mid \langle \ell] \in \Sigma_1\}, \\ \overline{\mathbf{A}}_2 & \text{if } \Sigma_1 = \varnothing; \end{cases}$$

$$\overline{\mathbf{A}}_2\Sigma_2 = \begin{cases} \overline{\mathbf{A}}_2\{[r_0\rangle\} & \text{if } \Sigma_2 \neq \varnothing \text{ and } r_0 = \min\{r \in \widehat{\mathbb{N}} \mid [r\rangle \in \Sigma_2\}, \\ \overline{\mathbf{A}}_2 & \text{if } \Sigma_2 = \varnothing. \end{cases}$$

Since $\overline{\mathbf{A}}_2$ is finitely based by Proposition 3.5, the variety $\mathbf{V} = \overline{\mathbf{A}}_2\Sigma_1 \cap \overline{\mathbf{A}}_2\Sigma_2$ is also finitely based. □

3.2.3 Varieties in $\mathfrak{I}_3 = [\mathbf{B}_2, \overline{\mathbf{A}}_0]$ and $\mathfrak{I}_4 = [\mathbf{B}_0, \overline{\mathbf{A}}_0 \cap \overline{\mathbf{B}}_2]$

In the present subsection, all varieties in $\mathfrak{I}_3 = [\mathbf{B}_2, \overline{\mathbf{A}}_0]$ and $\mathfrak{I}_4 = [\mathbf{B}_0, \overline{\mathbf{A}}_0 \cap \overline{\mathbf{B}}_2]$ are shown to be finitely based.

Lemma 3.20

(i) *The identities* (3.1) *and*

$$x^2y^2x^2y^2 \approx x^2y^2 \tag{3.8}$$

constitute an identity basis for the variety $\overline{\mathbf{A}}_0$.

(ii) *The identities* $\{(3.1), (3.8), xy^2x \approx xyx\}$ *constitute an identity basis for the variety* $\overline{\mathbf{A}}_0 \cap \overline{\mathbf{B}}_2$.

Proof These identity bases are obtained from Lemma 3.1 and Proposition 3.5. □

A word is *semi-connected* if it is a product of pairwise disjoint connected words. Note that a connected word is vacuously semi-connected.

Lemma 3.21 *The identities* $\{(3.1), (3.8)\}$ *can be used to convert any semi-connected word into a connected word.*

Proof It suffices to verify the lemma for a semi-connected word $\mathbf{w}_1\mathbf{w}_2$, where $\mathbf{w}_1$ and $\mathbf{w}_2$ are disjoint connected words, since the general case can be established by induction. Since $\mathbf{w}_1$ is a connected word, the variable $t = \mathsf{t}(\mathbf{w}_1)$ occurs at least twice in $\mathbf{w}_1$. Therefore, $\mathbf{w}_1 = \mathbf{a}t\mathbf{b}t$ for some $\mathbf{a}, \mathbf{b} \in \mathscr{A}_\varnothing^+$, whence $\mathbf{w}_1 \overset{(3.1)}{\approx} \mathbf{w}_1(\mathbf{b}t)^2$. Similarly, the variable $h = \mathsf{h}(\mathbf{w}_2)$ occurs at least twice in $\mathbf{w}_2$, so that $\mathbf{w}_2 \overset{(3.1)}{\approx} (h\mathbf{c})^2\mathbf{w}_2$ for some $\mathbf{c} \in \mathscr{A}_\varnothing^+$. Consequently,

$$\begin{aligned}\mathbf{w}_1\mathbf{w}_2 &\overset{(3.1)}{\approx} \mathbf{w}_1(\mathbf{b}t)^2(h\mathbf{c})^2\mathbf{w}_2\\ &\overset{(3.8)}{\approx} \mathbf{w}_1(\mathbf{b}t)^2(h\mathbf{c})^2(\mathbf{b}t)^2(h\mathbf{c})^2\mathbf{w}_2\\ &\overset{(3.1)}{\approx} \mathbf{w}_1(h\mathbf{c})^2(\mathbf{b}t)^2\mathbf{w}_2,\end{aligned}$$

where the word $\mathbf{w}_1(h\mathbf{c})^2(\mathbf{b}t)^2\mathbf{w}_2$ is connected. □

Lemma 3.22 (Lee and Volkov 2007, Proposition 3.2(ii)) *Let S be any fixed semigroup from $\{\mathcal{B}_0, \mathcal{B}_2\}$ and $\mathbf{w} \approx \mathbf{w}'$ be any identity satisfied by S. Suppose that $\mathbf{w} = \mathbf{w}_1\mathbf{s}\mathbf{w}_2$ for some pairwise disjoint words $\mathbf{w}_1, \mathbf{s}, \mathbf{w}_2 \in \mathscr{A}^+$ such that $\mathbf{s}$ is simple. Then $\mathbf{w}' = \mathbf{w}_1'\mathbf{s}\mathbf{w}_2'$ for some $\mathbf{w}_1', \mathbf{w}_2' \in \mathscr{A}^+$ such that $\mathbf{w}_1', \mathbf{s}, \mathbf{w}_2'$ are pairwise disjoint. Further, the identities $\mathbf{w}_1 \approx \mathbf{w}_1'$ and $\mathbf{w}_2 \approx \mathbf{w}_2'$ are satisfied by S.*

Proposition 3.23 *Every variety in the interval $\mathfrak{I}_3 = [\mathbf{B_2}, \overline{\mathbf{A_0}}]$ is finitely based.*

Proof Let $\mathbf{V} \in \mathfrak{I}_3$. Since the variety $\overline{\mathbf{A_0}}$ is finitely based by Lemma 3.20(i), assume that $\mathbf{V} \neq \overline{\mathbf{A_0}}$. Then $\mathbf{V} = \overline{\mathbf{A_0}}\Sigma$ for some nonempty set Σ of nontrivial identities satisfied by $\mathcal{B}_2$. Consider any identity $\sigma : \mathbf{w} \approx \mathbf{w}'$ in Σ. If either $\mathbf{w}$ or $\mathbf{w}'$ is simple, then the contradiction $\mathbf{w} = \mathbf{w}'$ follows from Lemma 3.22. Therefore, both $\mathbf{w}$ and $\mathbf{w}'$ are non-simple words. The word $\mathbf{w}$ can be written in the form $\mathbf{w} = \mathbf{s}_0\mathbf{w}_1\mathbf{s}_1\mathbf{w}_2\mathbf{s}_2 \cdots \mathbf{w}_k\mathbf{s}_k$, where

- $\mathbf{s}_0, \mathbf{s}_k \in \mathscr{A}^+_\varnothing$ and $\mathbf{s}_1, \mathbf{s}_2, \ldots, \mathbf{s}_{k-1} \in \mathscr{A}^+$ are simple words;
- $\mathbf{w}_1, \mathbf{w}_2, \ldots, \mathbf{w}_k$ are semi-connected words;
- $\mathbf{s}_0, \mathbf{w}_1, \mathbf{s}_1, \mathbf{w}_2, \mathbf{s}_2, \ldots, \mathbf{w}_k, \mathbf{s}_k$ are pairwise disjoint.

Then by Lemma 3.22, the word $\mathbf{w}'$ is of the form $\mathbf{w}' = \mathbf{s}_0\mathbf{w}_1'\mathbf{s}_1\mathbf{w}_2'\mathbf{s}_2 \cdots \mathbf{w}_k'\mathbf{s}_k$, where

- $\mathbf{w}_1', \mathbf{w}_2', \ldots, \mathbf{w}_k'$ are semi-connected words;
- $\mathbf{s}_0, \mathbf{w}_1', \mathbf{s}_1, \mathbf{w}_2', \mathbf{s}_2, \ldots, \mathbf{w}_k', \mathbf{s}_k$ are pairwise disjoint;
- $\mathcal{B}_2 \models \mathbf{w}_i \approx \mathbf{w}_i'$ and $\mathsf{con}(\mathbf{w}_i) = \mathsf{con}(\mathbf{w}_i')$ for all i.

It follows from Lemma 3.21 that for each i, there exist connected words $\widetilde{\mathbf{w}}_i, \widetilde{\mathbf{w}}_i' \in \mathscr{A}^+$ such that the deduction $\{(3.1), (3.8)\} \vdash \{\mathbf{w}_i \approx \widetilde{\mathbf{w}}_i, \mathbf{w}_i' \approx \widetilde{\mathbf{w}}_i'\}$ holds. Since the semigroup $\mathcal{B}_2$ satisfies the identities $\{(3.1), (3.8)\}$, it also satisfies the identities $\widetilde{\mathbf{w}}_i \approx \widetilde{\mathbf{w}}_i'$. Further, $\mathcal{A}_0 \models \widetilde{\mathbf{w}}_i \approx \widetilde{\mathbf{w}}_i'$ by Lemma 2.1. Hence, the identity $\widetilde{\sigma} : \widetilde{\mathbf{w}} \approx \widetilde{\mathbf{w}}'$, where

$$\widetilde{\mathbf{w}} = \mathbf{s}_0\widetilde{\mathbf{w}}_1\mathbf{s}_1\widetilde{\mathbf{w}}_2\mathbf{s}_2 \cdots \widetilde{\mathbf{w}}_k\mathbf{s}_k \quad \text{and} \quad \widetilde{\mathbf{w}}' = \mathbf{s}_0\widetilde{\mathbf{w}}_1'\mathbf{s}_1\widetilde{\mathbf{w}}_2'\mathbf{s}_2 \cdots \widetilde{\mathbf{w}}_k'\mathbf{s}_k,$$

is satisfied by $\mathbf{A_0} \vee \mathbf{B_2}$. Since $\overline{\mathbf{A_0}} \models \{(3.1), (3.8)\}$ and the identities $\{(3.1), (3.8)\}$ can be used to construct $\widetilde{\sigma}$ from σ, it follows that $\overline{\mathbf{A_0}}\{\sigma\} = \overline{\mathbf{A_0}}\{\widetilde{\sigma}\}$.

Since the identity σ is arbitrarily chosen from Σ, the construction of $\widetilde{\sigma}$ from σ in the preceding paragraph can be repeated on every other identity in Σ to obtain the set $\widetilde{\Sigma} = \{\widetilde{\sigma} \mid \sigma \in \Sigma\}$ of identities satisfied by $\mathbf{A_0} \vee \mathbf{B_2}$ with the property that $\overline{\mathbf{A_0}}\widetilde{\Sigma} = \overline{\mathbf{A_0}}\Sigma = \mathbf{V}$. Now the variety $\overline{\mathbf{A_2}}\widetilde{\Sigma}$ belongs to the interval $\mathfrak{I}_1 = [\mathbf{A_0} \vee \mathbf{B_2}, \overline{\mathbf{A_2}}]$ and so is finitely based by Proposition 3.19. Consequently, $\mathbf{V} = \overline{\mathbf{A_0}}\widetilde{\Sigma} = \overline{\mathbf{A_2}}\widetilde{\Sigma} \cap \overline{\mathbf{A_0}}$ is also finitely based. □

Proposition 3.24 *Every variety in the interval* $\mathfrak{I}_4 = [\mathbf{B_0}, \overline{\mathbf{A_0}} \cap \overline{\mathbf{B_2}}]$ *is finitely based.*

Proof This can be established in the same manner as Proposition 3.23. □

3.2.4 Varieties in $\mathfrak{I}_5 = \mathfrak{L}(\overline{\mathbf{B_0}})$

In the present subsection, all varieties in $\mathfrak{I}_5 = \mathfrak{L}(\overline{\mathbf{B_0}})$ are shown to be finitely based.

Lemma 3.25

(i) *The identities* (3.1) *and*

$$x^2y^2z^2 \approx x^2yz^2 \tag{3.9}$$

constitute an identity basis for the variety $\overline{\mathbf{B_0}}$.

(ii) *Any connected identity* $\mathbf{w} \approx \mathbf{w}'$ *with* $\mathsf{con}(\mathbf{w}) = \mathsf{con}(\mathbf{w}')$, $\mathsf{h}(\mathbf{w}') = \mathsf{h}(\mathbf{w}')$, *and* $\mathsf{t}(\mathbf{w}) = \mathsf{t}(\mathbf{w}')$ *is deducible from* $\mathsf{id}\,\overline{\mathbf{B_0}}$. *Specifically, the identities*

$$x^2yx \approx xyx, \quad xyx^2 \approx xyx \tag{3.10}$$

are deducible from $\mathsf{id}\,\overline{\mathbf{B_0}}$.

Proof

(i) This follows from Lemma 3.1 and Proposition 3.5(iv).
(ii) Since $\overline{\mathbf{B_0}} \subseteq \overline{\mathbf{B_2}} \models \mathbf{w} \approx \mathbf{w}'$ by Lemmas 3.6 and 3.8(i), the identity $\mathbf{w} \approx \mathbf{w}'$ is satisfied by $\overline{\mathbf{B_0}}$ and so is deducible from $\mathsf{id}\,\overline{\mathbf{B_0}}$. □

The number $|\operatorname{con}(\mathbf{w})|$ of distinct variables in a word $\mathbf{w}$ is obviously bounded by its length $|\mathbf{w}|$. Define the *depth* of $\mathbf{w}$ to be the difference $|\mathbf{w}| - |\operatorname{con}(\mathbf{w})|$.

Lemma 3.26 (Volkov 1984a) *Let m be any fixed positive integer and Σ be any set of identities. Suppose that each identity in Σ is formed by a pair of words with depth at most m. Then Σ defines a finitely based variety.*

Lemma 3.27 *The identities from $\operatorname{id}\overline{\mathbf{B}}_0$ can be used to convert any word into a word with depth at most two.*

Proof It suffices to consider non-simple words since all simple words have depth zero. Any non-simple word $\mathbf{w}$ can be uniquely written in the form $\mathbf{w} = \mathbf{a}x\mathbf{u}y\mathbf{b}$, where

(A) the variables of $\mathbf{a}, \mathbf{b} \in \mathscr{A}_\varnothing^+$ are all simple in $\mathbf{w}$

and the variables x and y are non-simple in $\mathbf{w}$. (Note that $x \in \operatorname{con}(\mathbf{u}y)$, $y \in \operatorname{con}(x\mathbf{u})$, and it is possible that $x = y$.) Now since the word $x^2\mathbf{u}^2y^2$ is connected, it follows from Lemma 3.9 that the identities (3.4) can be used to convert it into some unique word $\mathbf{v}$ in canonical form such that $\mathsf{h}(\mathbf{v}) = x$, $\mathsf{t}(\mathbf{v}) = y$, and

(B) $\operatorname{con}(\mathbf{v}) = \operatorname{con}(x\mathbf{u}y)$.

Therefore, the deduction $\operatorname{id}\overline{\mathbf{B}}_0 \vdash x^2\mathbf{u}^2y^2 \approx \mathbf{v}$ holds by Lemma 3.25(ii), whence

$$\begin{aligned} \mathbf{w} &\overset{(3.10)}{\approx} \mathbf{a}x^2\mathbf{u}y^2\mathbf{b} \\ &\overset{(3.9)}{\approx} \mathbf{a}x^2\mathbf{u}^2y^2\mathbf{b} \\ &\overset{\operatorname{id}\overline{\mathbf{B}}_0}{\approx} \mathbf{a}\mathbf{v}\mathbf{b}. \end{aligned}$$

By (A) and (B), the words $\mathbf{a}, \mathbf{v}, \mathbf{b}$ are pairwise disjoint. It follows that the depth of $\mathbf{avb}$ is identical to the depth of $\mathbf{v}$. But since $\mathbf{v}$ is a word in canonical form, its depth is one or two depending on whether it is of the form (CF1) or (CF2). □

Proposition 3.28 *Every variety in the interval $\mathfrak{I}_5 = \mathfrak{L}(\overline{\mathbf{B}}_0)$ is finitely based.*

Proof Let $\mathbf{V} \in \mathfrak{I}_5$. Then $\mathbf{V} = \overline{\mathbf{B}}_0\Sigma$ for some set Σ of identities. By Lemma 3.27, the words forming the identities in Σ can be chosen to have depth at most two. Therefore, by Lemma 3.26, the variety defined by Σ is finitely based. Since the variety $\overline{\mathbf{B}}_0$ is finitely based by Lemma 3.25(i), the variety $\mathbf{V}$ is also finitely based. □

3.3 The Lattice $\mathfrak{L}(\mathbf{A_2})$

By Proposition 3.7, the lattice $\mathfrak{L}(\overline{\mathbf{A_2}})$ is the disjoint union of the following intervals:

$$\mathfrak{I}_1 = [\mathbf{A_0} \vee \mathbf{B_2}, \overline{\mathbf{A_2}}], \qquad \mathfrak{I}_2 = [\mathbf{A_0}, \overline{\mathbf{B_2}}], \qquad \mathfrak{I}_3 = [\mathbf{B_2}, \overline{\mathbf{A_0}}],$$

$$\mathfrak{I}_4 = [\mathbf{B_0}, \overline{\mathbf{A_0}} \cap \overline{\mathbf{B_2}}], \qquad \mathfrak{I}_5 = \mathfrak{L}(\overline{\mathbf{B_0}}).$$

It follows that the lattice $\mathfrak{L}(\mathbf{A_2})$ is the disjoint union of the intervals $\mathfrak{I}_5$ and

$$[\mathbf{B_0}, \mathbf{A_2}] = \mathfrak{I}_1 \cup \mathfrak{I}_2 \cup \mathfrak{I}_3 \cup \mathfrak{I}_4 \cup \{\mathbf{A_2}\}.$$

Recall that a variety is *finitely universal* if its lattice of subvarieties contains an isomorphic copy of every finite lattice. In Sect. 3.3.1, the varieties $\mathbf{B_0}$ and $\overline{\mathbf{B_0}}$ are shown to be finitely universal. Therefore, a complete description of the interval $\mathfrak{I}_5 = \mathfrak{L}(\overline{\mathbf{B_0}})$ is infeasible and every variety in the interval $[\mathbf{B_0}, \mathbf{A_2}]$ is finitely universal. But a complete description of the interval $[\mathbf{B_0}, \mathbf{A_2}]$ is possible.

In Sect. 3.3.2, several varieties are introduced that would assist in showing which varieties in the interval $[\mathbf{B_0}, \mathbf{A_2}]$ are finitely generated. In Sects. 3.3.3 and 3.3.4, the intervals $\mathfrak{I}_1$, $\mathfrak{I}_2$, $\mathfrak{I}_3$, and $\mathfrak{I}_4$ are each shown to be isomorphic to the direct product of two countably infinite chains:

$$\mathfrak{I}_1 \cong \mathfrak{I}_2 \cong \widehat{\mathbb{N}} \times \widehat{\mathbb{N}} \text{ and } \mathfrak{I}_3 \cong \mathfrak{I}_4 \cong (\widehat{\mathbb{N}} \backslash \{\widehat{\omega}\}) \times (\widehat{\mathbb{N}} \backslash \{\widehat{\omega}\}).$$

Finitely generated varieties in these four intervals are also identified. These results are then combined in Sect. 3.3.5 to give a complete description of the interval $[\mathbf{B_0}, \mathbf{A_2}]$. Given that the intervals $\mathfrak{I}_1$, $\mathfrak{I}_2$, $\mathfrak{I}_3$, and $\mathfrak{I}_4$ are all distributive, it is not surprising that the interval $[\mathbf{B_0}, \mathbf{A_2}]$ is also distributive.

3.3.1 The Interval $\mathfrak{I}_5 = \mathfrak{L}(\overline{\mathbf{B_0}})$

The variety $\mathbf{H}$ of semigroups defined by the identity

$$xyx \approx y^2 \tag{3.11}$$

is finitely universal (Volkov 1989a). Hence, any variety that contains $\mathbf{H}$ is also finitely universal.

Lemma 3.29 *For any non-simple word* $\mathbf{u} \in \mathscr{A}^+$ *and any* $z \in \mathscr{A}$, *the deduction* (3.11) $\vdash \mathbf{u} \approx z^2$ *holds.*

Proof The identity $\sigma : x^2 \approx y^2$ is deducible from (3.11) because

$$\begin{aligned} y^2 &\overset{(3.11)}{\approx} y^2 y \\ &\overset{(3.11)}{\approx} x(yxy) \\ &\overset{(3.11)}{\approx} xx^2 \\ &\overset{(3.11)}{\approx} x^2. \end{aligned}$$

Therefore, it suffices to establish the deduction $\{(3.11), \sigma\} \vdash \mathbf{u} \approx z^2$. By assumption, $\mathbf{u} = \mathbf{u}_1 x \mathbf{u}_2 x \mathbf{u}_3$ for some $x \in \mathscr{A}$ and $\mathbf{u}_1, \mathbf{u}_2, \mathbf{u}_3 \in \mathscr{A}_\varnothing^+$. If $\mathbf{u}_2 = \varnothing$, then

$$\begin{aligned} \mathbf{u} &= \mathbf{u}_1 x^2 \mathbf{u}_3 \\ &\overset{\sigma}{\approx} \mathbf{u}_1 (\mathbf{u}_1^2 x \mathbf{u}_3^2)^2 \mathbf{u}_3 \\ &\overset{(3.11)}{\approx} (\mathbf{u}_1^2 x \mathbf{u}_3^2)^2 \\ &\overset{\sigma}{\approx} z^2; \end{aligned}$$

if $\mathbf{u}_2 \neq \varnothing$, then

$$\begin{aligned} \mathbf{u} &= \mathbf{u}_1 x \mathbf{u}_2 x \mathbf{u}_3 \\ &\overset{(3.11)}{\approx} \mathbf{u}_1 \mathbf{u}_2^2 \mathbf{u}_3 \\ &\overset{\sigma}{\approx} \mathbf{u}_1 (\mathbf{u}_1^2 x \mathbf{u}_3^2)^2 \mathbf{u}_3 \\ &\overset{(3.11)}{\approx} (\mathbf{u}_1^2 x \mathbf{u}_3^2)^2 \\ &\overset{\sigma}{\approx} z^2 \end{aligned}$$

as required. □

Proposition 3.30 *Suppose that* $\mathbf{V}$ *is any variety defined by some set* Σ *of identities, each of which is formed by some pair of non-simple words. Then the inclusion* $\mathbf{H} \subseteq \mathbf{V}$ *holds.*

Proof For any identity $\mathbf{u} \approx \mathbf{v}$ in Σ, since the words $\mathbf{u}$ and $\mathbf{v}$ are non-simple, it follows from Lemma 3.29 that $(3.11) \vdash \{\mathbf{u} \approx z^2, \mathbf{v} \approx z^2\} \vdash \mathbf{u} \approx \mathbf{v}$. Therefore, the deduction $(3.11) \vdash \Sigma$ holds, whence the inclusion $\mathbf{H} \subseteq \mathbf{V}$ also holds. □

Corollary 3.31 *The varieties* $\mathbf{B}_0$ *and* $\overline{\mathbf{B}}_0$ *are finitely universal.*

Proof By Lemmas 3.1, 3.2(iii), and 3.25(i), the varieties $\mathbf{B_0}$ and $\overline{\mathbf{B_0}}$ are defined by identities formed by non-simple words. Hence, the result holds by Proposition 3.30. □

Since the inclusions $\mathbf{B_0} \subseteq \mathbf{A_0}$ and $\mathbf{B_0} \subseteq \mathbf{B_2}$ hold by Lemma 3.2, the semigroups $\mathcal{A}_0$, $\mathcal{A}_2$, $\mathcal{B}_0$, and $\mathcal{B}_2$ generate finitely universal varieties.

Remark 3.32 Up to isomorphism and anti-isomorphism, $\mathcal{A}_0$, $\mathcal{B}_0$, and two other semigroups of order four are the smallest semigroups that generate a finitely universal variety (Lee 2007a).

3.3.2 The Varieties $\mathbf{D}_\ell$, $\mathbf{E}$, $\mathbf{F}$, $\langle \ell]_i$, and $[r\rangle_i$

3.3.2.1 The Varieties $\mathbf{D}_\ell$

Recall that a *diverse identity* is an identity of the form $x_1x_2\cdots x_k \approx \mathbf{w}$ that is not a permutation identity.

Lemma 3.33 (Malyshev 1981) *Every variety that satisfies some nontrivial permutation identity and some diverse identity is small.*

Lemma 3.34 (Jackson 2005b, Lemma 6.1) *Every small locally finite variety is finitely generated.*

For each $\ell \in \{0, 1, 2, \ldots\}$, let $\mathbf{D}_\ell$ denote the subvariety of $\mathbf{A_2}$ defined by the diverse identity

$$x_1x_2\cdots x_\ell y \approx x_1x_2\cdots x_\ell z. \qquad (\delta_\ell)$$

Note that $\mathbf{D}_0 = \mathbf{0}$. Since the semigroup $\mathcal{B}_0$ violates the identity δ_ℓ, the inclusion $\mathbf{D}_\ell \subseteq \overline{\mathbf{B_0}}$ holds.

Lemma 3.35 *For each $\ell \in \{0, 1, 2, \ldots\}$, the variety $\mathbf{D}_\ell$ is Cross.*

Proof The variety $\mathbf{D}_\ell = \mathbf{A_2}\{\delta_\ell\}$ is finitely based because $\mathbf{A_2}$ is finitely based. It is clear that $\mathbf{D}_\ell$ satisfies the permutation identity $x_1x_2\cdots x_\ell yz \approx x_1x_2\cdots x_\ell zy$ and so is small by Lemma 3.33. Since $\mathbf{A_2}$ is locally finite, its subvariety $\mathbf{D}_\ell$ is also locally finite and so is finitely generated by Lemma 3.34. □

Lemma 3.36 *Suppose that $k, \ell \in \{0, 1, 2, \ldots\}$ with $k < \ell$. Then $\mathbf{D}_\ell \models \{\langle \ell], [0\rangle\}$ and $\mathbf{D}_\ell \not\models \langle k]$. Consequently, the inclusions $\mathbf{D}_0 \subset \mathbf{D}_1 \subset \mathbf{D}_2 \subset \cdots$ hold and are proper.*

Table 3.1 Multiplication tables of $\mathcal{E}$ and $\mathcal{F}$

$\mathcal{E}$	0	a	b	c	d	e
0	0	0	0	0	0	0
a	0	0	0	0	0	b
b	0	0	0	0	b	b
c	0	a	b	c	0	0
d	0	0	0	0	d	d
e	0	0	0	0	e	e

$\mathcal{F}$	0	a	b	c	d
0	0	0	0	0	0
a	0	0	0	a	a
b	0	a	b	0	a
c	0	0	0	c	c
d	0	0	0	d	d

Proof By Lemma 3.1, the variety $\mathbf{D}_\ell = \mathbf{A_2}\{\delta_\ell\}$ is defined by $\{(3.1), \delta_\ell\}$. The deduction $\delta_\ell \vdash \langle \ell]$ clearly holds, so that $\mathbf{D}_\ell \models \langle \ell]$. Since

$$\begin{aligned} x^2y^2x^2y^2 &\overset{(3.1)}{\approx} x^{2\ell}y^2x^2y^2 \\ &\overset{\delta_\ell}{\approx} x^{2\ell}y^2x^2 \\ &\overset{(3.1)}{\approx} x^2y^2x^2, \end{aligned}$$

the deduction $\{(3.1), \delta_\ell\} \vdash [0\rangle$ holds, whence $\mathbf{D}_\ell \models [0\rangle$.

Now suppose that the deduction $\{(3.1), \delta_\ell\} \vdash a_1a_2\cdots a_kx^2y^2x^2y^2 \approx \mathbf{w}$ holds for some $\mathbf{w} \in \mathscr{A}^+$. Then there exists a sequence

$$a_1a_2\cdots a_kx^2y^2x^2y^2 = \mathbf{w}_1, \mathbf{w}_2, \ldots, \mathbf{w}_r = \mathbf{w}$$

of words, where each identity $\mathbf{w}_i \approx \mathbf{w}_{i+1}$ is directly deducible from some identity in $\{(3.1), \delta_\ell\}$. It is easily seen that $a_1a_2\cdots a_kx$ remains a prefix of every $\mathbf{w}_i$, so that $\mathbf{w} \neq a_1a_2\cdots a_ky^2x^2y^2$. It follows that $\{(3.1), \delta_\ell\} \nvdash \langle k]$, whence $\mathbf{D}_\ell \not\models \langle k]$. □

3.3.2.2 The Varieties E and F

Let **E** and **F** denote the varieties generated by the semigroups $\mathcal{E}$ and $\mathcal{F}$ in Table 3.1, respectively. The semigroup $\mathcal{E}$ was first brought to my attention by Volkov (email communication, 19 Jan 2005), and the semigroup $\mathcal{F}$ appeared in Lee and Volkov (2007, Sect. 4) as LC_0.

Lemma 3.37

(i) **E** *is a variety in* $\mathfrak{I}_4 = [\mathbf{B_0}, \overline{\mathbf{A}}_\mathbf{0} \cap \overline{\mathbf{B}}_\mathbf{2}]$ *such that* $\mathbf{E} \models \{\langle\widehat{\omega}], [0\rangle\}$ *and* $\mathbf{E} \not\models \langle\omega]$.
(ii) **F** *is a variety in* $\mathfrak{I}_2 = [\mathbf{A_0}, \overline{\mathbf{B}}_\mathbf{2}]$ *such that* $\mathbf{F} \models [0\rangle$ *and* $\mathbf{F} \not\models \langle\widehat{\omega}]$.

Proof

(i) It is routinely checked that $\mathcal{E} \models \{(3.1), (3.8), xy^2x \approx xyx, \langle\widehat{\omega}], [0\rangle\}$; in particular, $\mathbf{E} \subseteq \overline{\mathbf{A_0}} \cap \overline{\mathbf{B_2}}$ by Lemma 3.20(ii). Since the subsemigroup $\{0, \mathsf{b}, \mathsf{c}, \mathsf{d}\}$ of $\mathcal{E}$ is isomorphic to $\mathcal{B}_0$, the inclusion $\mathbf{B_0} \subseteq \mathbf{E}$ holds, so that $\mathbf{E} \in \mathfrak{I}_4$. The semigroup $\mathcal{E}$ violates the identity $\langle\omega]$ under the substitution $(a_1, a_2, x, y) \mapsto (\mathsf{c}, \mathsf{a}, \mathsf{d}, \mathsf{e})$, so that $\mathbf{E} \not\models \langle\omega]$.

(ii) It is routinely checked that $\mathcal{F} \models \{(3.1), xy^2x \approx xyx, [0\rangle\}$; in particular, $\mathbf{F} \subseteq \overline{\mathbf{B_2}}$ by Proposition 3.5(ii). Since the subsemigroup $\{0, \mathsf{a}, \mathsf{b}, \mathsf{d}\}$ of $\mathcal{F}$ is isomorphic to $\mathcal{A}_0$, the inclusion $\mathbf{A_0} \subseteq \mathbf{F}$ holds, so that $\mathbf{F} \in \mathfrak{I}_2$. The semigroup $\mathcal{F}$ violates the identity $\langle\widehat{\omega}]$ under the substitution $(a_1, x, y) \mapsto (\mathsf{b}, \mathsf{c}, \mathsf{d})$, so that $\mathbf{F} \not\models \langle\widehat{\omega}]$. □

3.3.2.3 The Varieties $\langle\ell]_i$ and $[r\rangle_i$

For any $\ell, r \in \widehat{\mathbb{N}}$, define the varieties

$$\begin{aligned}
&\langle\ell]_1 = \overline{\mathbf{A_2}}\{\langle\ell]\}, && [r\rangle_1 = \overline{\mathbf{A_2}}\{[r\rangle\}, && \langle\ell, r\rangle_1 = \overline{\mathbf{A_2}}\{\langle\ell], [r\rangle\},\\
&\langle\ell]_2 = \overline{\mathbf{B_2}}\{\langle\ell]\}, && [r\rangle_2 = \overline{\mathbf{B_2}}\{[r\rangle\}, && \langle\ell, r\rangle_2 = \overline{\mathbf{B_2}}\{\langle\ell], [r\rangle\},\\
&\langle\ell]_3 = \overline{\mathbf{A_0}}\{\langle\ell]\}, && [r\rangle_3 = \overline{\mathbf{A_0}}\{[r\rangle\}, && \langle\ell, r\rangle_3 = \overline{\mathbf{A_0}}\{\langle\ell], [r\rangle\},\\
&\langle\ell]_4 = (\overline{\mathbf{A_0}} \cap \overline{\mathbf{B_2}})\{\langle\ell]\}, && [r\rangle_4 = (\overline{\mathbf{A_0}} \cap \overline{\mathbf{B_2}})\{[r\rangle\}, && \langle\ell, r\rangle_4 = (\overline{\mathbf{A_0}} \cap \overline{\mathbf{B_2}})\{\langle\ell], [r\rangle\}.
\end{aligned}$$

In other words, if $\mathbf{V}_i$ denotes the largest variety in the interval $\mathfrak{I}_i$, then

$$\langle\ell]_i = \mathbf{V}_i\{\langle\ell]\}, \quad [r\rangle_i = \mathbf{V}_i\{[r\rangle\}, \quad \text{and} \quad \langle\ell, r\rangle_i = \mathbf{V}_i\{\langle\ell], [r\rangle\} = \langle\ell]_i \cap [r\rangle_i.$$

Lemma 3.38 *The variety $\langle 0, 0\rangle_i$ coincides with the least variety in the interval $\mathfrak{I}_i$:*

(i) $\langle 0, 0\rangle_1 = \mathbf{A_0} \vee \mathbf{B_2}$;
(ii) $\langle 0, 0\rangle_2 = \mathbf{A_0}$;
(iii) $\langle 0, 0\rangle_3 = \mathbf{B_2}$;
(iv) $\langle 0, 0\rangle_4 = \mathbf{B_0}$.

Proof

(i) The inclusion $\mathbf{A_0} \vee \mathbf{B_2} \subseteq \langle 0, 0\rangle_1$ holds because $\mathcal{A}_0, \mathcal{B}_2 \models \{\langle 0], [0\rangle\}$. Conversely, the inclusion $\langle 0, 0\rangle_1 \subseteq \overline{\mathbf{A_2}}\{x^2y^2x^2 \approx y^2x^2y^2\}$ holds because

$$\begin{aligned}
x^2y^2x^2 &\overset{\langle 0]}{\approx} y^2x^2y^2x^2\\
&\overset{[0\rangle}{\approx} y^2x^2y^2;
\end{aligned}$$

since $\mathbf{A_0} \vee \mathbf{B_2} = \mathbf{A_2}\{x^2y^2x^2 \approx y^2x^2y^2\}$ by Lemma 3.2(iv), the result follows.

(ii) The inclusion $\mathbf{A_0} \subseteq \langle 0, 0\rangle_2$ holds because $\mathcal{A}_0 \models \{\langle 0], [0\rangle\}$. By Lemma 3.8(i), the variety $\overline{\mathbf{B}}_\mathbf{2}$ satisfies the identity $\sigma_1 : xyxy \approx x^2y^2x^2y^2$. Since

$$\begin{aligned} xyxy &\overset{\sigma_1}{\approx} x^2y^2x^2y^2 \\ &\overset{\langle 0]}{\approx} y^2x^2y^2 \\ &\overset{[0\rangle}{\approx} y^2x^2y^2x^2 \\ &\overset{\sigma_1}{\approx} yxyx, \end{aligned}$$

the inclusion $\langle 0, 0\rangle_2 \subseteq \mathbf{A_2}\{xyxy \approx yxyx\}$ holds. Since $\mathbf{A_0} = \mathbf{A_2}\{xyxy \approx yxyx\}$ by Lemma 3.2(i), the result follows.

(iii) The inclusion $\mathbf{B_2} \subseteq \langle 0, 0\rangle_3$ holds because $\mathcal{B}_2 \models \{\langle 0], [0\rangle\}$. By Lemma 3.20(i), the variety $\overline{\mathbf{A}}_\mathbf{0}$ satisfies the identity $\sigma_2 : x^2y^2x^2y^2 \approx x^2y^2$. Since

$$\begin{aligned} x^2y^2 &\overset{\sigma_2}{\approx} x^2y^2x^2y^2 \\ &\overset{\langle 0]}{\approx} y^2x^2y^2 \\ &\overset{[0\rangle}{\approx} y^2x^2y^2x^2 \\ &\overset{\sigma_2}{\approx} y^2x^2, \end{aligned}$$

the inclusion $\langle 0, 0\rangle_3 \subseteq \mathbf{A_2}\{x^2y^2 \approx y^2x^2\}$ holds. Since $\mathbf{B_2} = \mathbf{A_2}\{x^2y^2 \approx y^2x^2\}$ by Lemma 3.2(ii), the result follows.

(iv) It follows from parts (ii) and (iii) that $\mathbf{A_0} \cap \mathbf{B_2} = \langle 0, 0\rangle_2 \cap \langle 0, 0\rangle_3 = \langle 0, 0\rangle_4$. Then the result holds because $\mathbf{A_0} \cap \mathbf{B_2} = \mathbf{B_0}$ by Lemma 3.2(iii). □

Lemma 3.39 (Almeida 1994, Proposition 3.7.1) *Suppose that S is any finite semigroup of order n. Then for any $a_1, a_2, \ldots, a_n \in S$, there exist some $s_1, s_2, s_3 \in S$ such that $a_1a_2 \cdots a_n = s_1s_2^2s_3$.*

Lemma 3.40 *Suppose that S is any finite semigroup of order n such that $S \models \langle\omega]$. Then $S \models \langle n]$.*

Proof Let $a_1, a_2, \ldots, a_n, b, c \in S$. Then by Lemma 3.39, there exist $s_1, s_2, s_3 \in S$ such that $a_1a_2 \cdots a_n = s_1s_2^2s_3$. Since

$$\begin{aligned} a_1a_2 \cdots a_nb^2c^2b^2c^2 &= s_1s_2^2s_3b^2c^2b^2c^2 \\ &\overset{\langle\omega]}{=} s_1s_2^2s_3c^2b^2c^2 \end{aligned}$$

$$= a_1 a_2 \cdots a_n c^2 b^2 c^2,$$

it follows that $S \models \langle n]$. □

3.3.3 The Intervals $\mathfrak{J}_1 = [\mathbf{A}_0 \vee \mathbf{B}_2, \overline{\mathbf{A}}_2]$ and $\mathfrak{J}_2 = [\mathbf{A}_0, \overline{\mathbf{B}}_2]$

Proposition 3.41 *The interval* $\mathfrak{J}_1 = [\mathbf{A}_0 \vee \mathbf{B}_2, \overline{\mathbf{A}}_2]$ *is given in Fig. 3.2.*

Proof As shown in the proof of Proposition 3.19, each variety in the interval $\mathfrak{J}_1$ is of the form $\overline{\mathbf{A}}_2\Sigma$, where $\Sigma \subseteq \{\langle \ell], [r\rangle\}$ for some $\ell, r \in \widehat{\mathbb{N}}$. By Lemma 3.10, these varieties are as positioned in Fig. 3.2. It remains to verify that the varieties in Fig. 3.2 are distinct and identify those that are finitely generated.

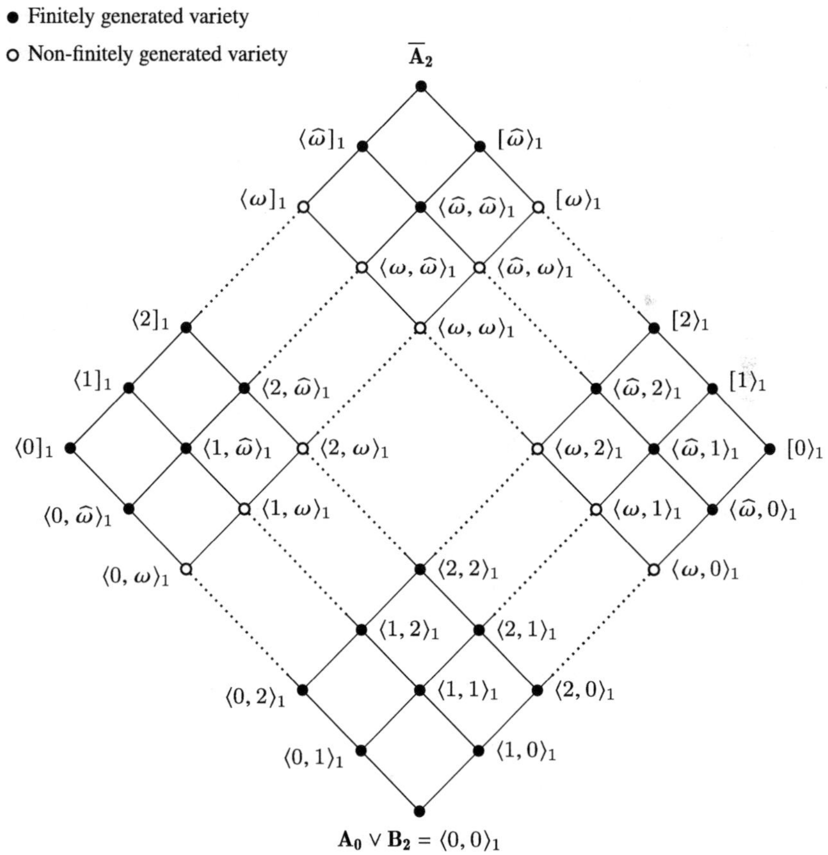

Fig. 3.2 The interval $\mathfrak{J}_1 = [\mathbf{A}_0 \vee \mathbf{B}_2, \overline{\mathbf{A}}_2]$

The equality $\mathbf{A_0} \vee \mathbf{B_2} = \langle 0, 0\rangle_1$ holds by Lemma 3.38(i). Consider the chain

$$\langle 0,0\rangle_1 \subseteq \langle 1,0\rangle_1 \subseteq \langle 2,0\rangle_1 \subseteq \cdots \subseteq \langle \omega,0\rangle_1 \subseteq \langle \widehat{\omega},0\rangle_1 \subseteq [0\rangle_1$$

in Fig. 3.2. Suppose that $1 \le \ell < \omega$. Then $\mathbf{D}_\ell \models \{\langle \ell], [0\rangle\}$ and $\mathbf{D}_\ell \not\models \langle \ell - 1]$ by Lemma 3.36, whence $\mathbf{A_0} \vee \mathbf{B_2} \vee \mathbf{D}_\ell \subseteq \langle \ell, 0\rangle_1$ and $\mathbf{A_0} \vee \mathbf{B_2} \vee \mathbf{D}_\ell \nsubseteq \langle \ell - 1, 0\rangle_1$. Hence, $\mathbf{A_0} \vee \mathbf{B_2} \vee \mathbf{D}_\ell = \langle \ell, 0\rangle_1$ is finitely generated by Lemma 3.35. It follows that the varieties $\langle 0,0\rangle_1, \langle 1,0\rangle_1, \langle 2,0\rangle_1, \ldots$ are distinct and finitely generated. Further, $\mathbf{D}_{\ell+1}$ is a subvariety of $\langle \omega, 0\rangle_1$ that is not contained in $\langle \ell, 0\rangle_1$. Therefore, $\langle \omega, 0\rangle_1 \nsubseteq \langle \ell, 0\rangle_1$ for all $\ell < \omega$. If the variety $\langle \omega, 0\rangle_1$ is finitely generated by some semigroup S of order n, then $S \models \langle n]$ by Lemma 3.40, so that the contradiction $\langle \omega, 0\rangle_1 \subseteq \langle n, 0\rangle_1$ is deduced. Hence, the variety $\langle \omega, 0\rangle_1$ is non-finitely generated. Now $\mathbf{E} \vee \mathbf{A_0} \vee \mathbf{B_2} \subseteq \langle \widehat{\omega}, 0\rangle_1$ and $\mathbf{E} \vee \mathbf{A_0} \vee \mathbf{B_2} \nsubseteq \langle \omega, 0\rangle_1$ by Lemma 3.37(i), so that $\langle \widehat{\omega}, 0\rangle_1 = \mathbf{E} \vee \mathbf{A_0} \vee \mathbf{B_2}$ is finitely generated. Similarly, $\mathbf{F} \vee \mathbf{A_0} \vee \mathbf{B_2} \subseteq [0\rangle_1$ and $\mathbf{F} \vee \mathbf{A_0} \vee \mathbf{B_2} \nsubseteq \langle \widehat{\omega}, 0\rangle_1$ by Lemma 3.37(ii), so that $[0\rangle_1 = \mathbf{F} \vee \mathbf{A_0} \vee \mathbf{B_2}$ is finitely generated. Consequently, the varieties

$$\langle 0,0\rangle_1 \subset \langle 1,0\rangle_1 \subset \langle 2,0\rangle_1 \subset \cdots \subset \langle \omega,0\rangle_1 \subset \langle \widehat{\omega},0\rangle_1 \subset [0\rangle_1,$$

with the exception of $\langle \omega, 0\rangle_1$, are all finitely generated. By symmetry, the varieties

$$\langle 0,0\rangle_1 \subset \langle 0,1\rangle_1 \subset \langle 0,2\rangle_1 \subset \cdots \subset \langle 0,\omega\rangle_1 \subset \langle 0,\widehat{\omega}\rangle_1 \subset \langle 0]_1,$$

with the exception of $\langle 0, \omega\rangle_1$, are all finitely generated.

Distinctness and the finite generation property of the varieties in Fig. 3.2 can now be easily verified. For instance, for any $r \in \{1, 2, 3, \ldots\} \cup \{\widehat{\omega}\}$, consider the varieties

$$\langle 0,r\rangle_1 \subseteq \langle 1,r\rangle_1 \subseteq \langle 2,r\rangle_1 \subseteq \cdots \subseteq \langle \omega,r\rangle_1 \subseteq \langle \widehat{\omega},r\rangle_1 \subseteq [r\rangle_1.$$

The procedure in the previous paragraph can be repeated to show that the varieties

$$\langle \ell,r\rangle_1 = \mathbf{D}_\ell \vee \langle 0,r\rangle_1, \quad \langle \widehat{\omega},r\rangle_1 = \mathbf{E} \vee \langle 0,r\rangle_1, \quad \text{and} \quad [r\rangle_1 = \mathbf{F} \vee \langle 0,r\rangle_1$$

are all distinct and finitely generated, while the variety $\langle \omega, r\rangle_1$ is non-finitely generated. □

Proposition 3.42 *The interval* $\mathfrak{I}_2 = [\mathbf{A_0}, \overline{\mathbf{B_2}}]$ *is given in Fig. 3.3.*

Proof This is established in the same manner as Proposition 3.41, with the minor difference in the least variety being $\mathbf{A_0} = \langle 0, 0\rangle_2$ instead of $\mathbf{A_0} \vee \mathbf{B_2} = \langle 0, 0\rangle_1$. □

Corollary 3.43 *The varieties* $\mathbf{F}$ *and* $[0\rangle_2$ *coincide.*

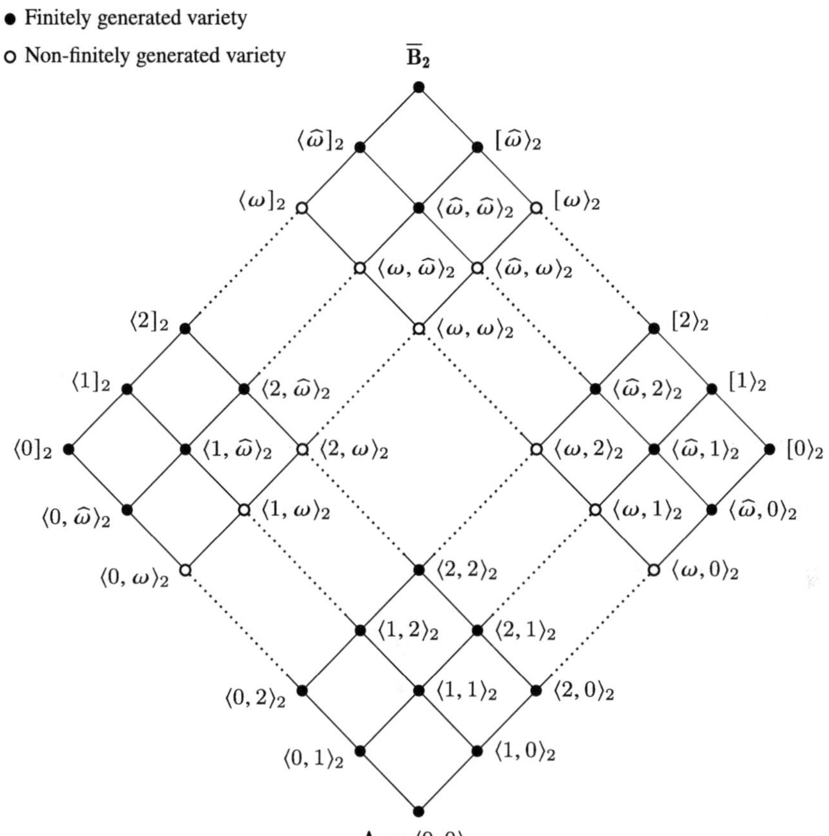

Fig. 3.3 The interval $\mathfrak{I}_2 = [\mathbf{A}_0, \overline{\mathbf{B}}_2]$

Proof This follows from Lemma 3.37(ii) and Proposition 3.42. □

3.3.4 The Intervals $\mathfrak{I}_3 = [\mathbf{B}_2, \overline{\mathbf{A}}_0]$ and $\mathfrak{I}_4 = [\mathbf{B}_0, \overline{\mathbf{A}}_0 \cap \overline{\mathbf{B}}_2]$

Lemma 3.44 *The variety* $\overline{\mathbf{A}}_0$ *satisfies the identities* $\{\langle\widehat{\omega}], [\widehat{\omega}\rangle\}$.

Proof By Lemmas 3.6, 3.16, and 3.20, the variety $\overline{\mathbf{A}}_0$ satisfies the identities (3.1), (3.5), and (3.8). Since

$$\begin{aligned} a_1^2x^2y^2x^2y^2 &\overset{(3.1)}{\approx} a_1^2(x^2y^2)^2y^2 \\ &\overset{(3.8)}{\approx} a_1^2(x^2y^2)^2a_1^2(x^2y^2)^2y^2 \end{aligned}$$

$$\stackrel{(3.5)}{\approx} a_1^2(y^2x^2)^2a_1^2(x^2y^2)^2y^2$$
$$\stackrel{(3.5)}{\approx} a_1^2(y^2x^2)^2a_1^2(y^2x^2)^2y^2$$
$$\stackrel{(3.8)}{\approx} a_1^2(y^2x^2)^2y^2$$
$$\stackrel{(3.1)}{\approx} a_1^2y^2x^2y^2,$$

it follows that $\overline{\mathbf{A_0}} \models \langle\widehat{\omega}]$. By symmetry, $\overline{\mathbf{A_0}} \models [\widehat{\omega}\rangle$. □

By Lemma 3.44, the identities $\{\langle\widehat{\omega}], [\widehat{\omega}\rangle\}$ do not define any proper subvariety of $\overline{\mathbf{A_0}}$. Therefore, the following descriptions of the intervals $\mathfrak{I}_3$ and $\mathfrak{I}_4$ are obtained when the appropriate arguments from the proof of Proposition 3.41 are repeated.

Proposition 3.45 *The interval* $\mathfrak{I}_3 = [\mathbf{B_2}, \overline{\mathbf{A_0}}]$ *is given in Fig. 3.4.*

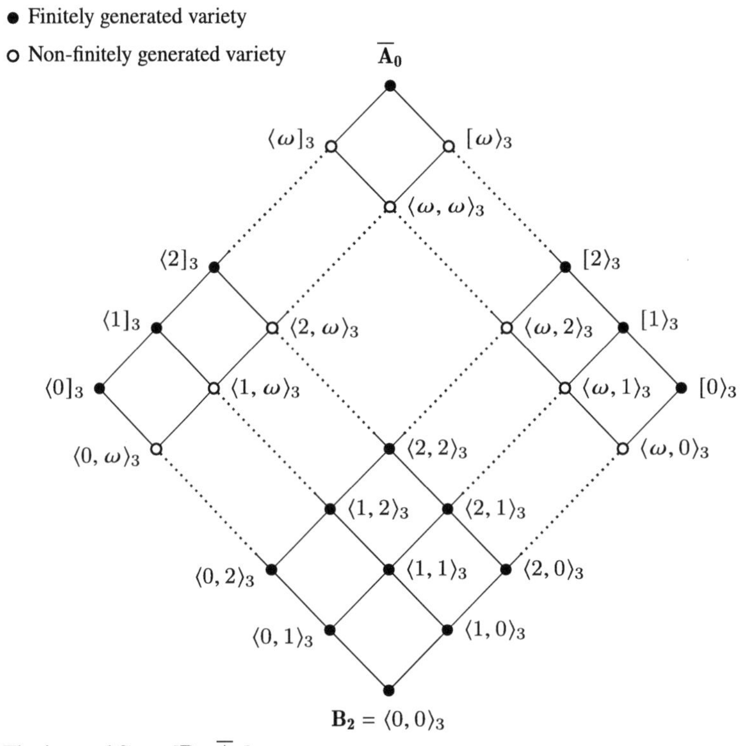

Fig. 3.4 The interval $\mathfrak{I}_3 = [\mathbf{B_2}, \overline{\mathbf{A_0}}]$

Proposition 3.46 *The interval* $\mathfrak{I}_4 = [\mathbf{B_0}, \overline{\mathbf{A_0}} \cap \overline{\mathbf{B_2}}]$ *is given in Fig. 3.5.*

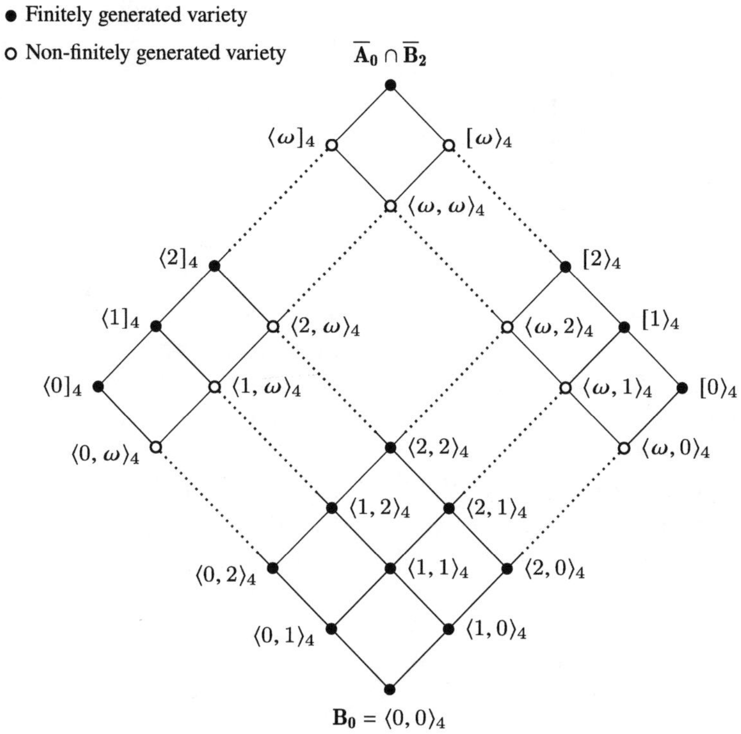

Fig. 3.5 The interval $\mathfrak{I}_4 = [\mathbf{B_0}, \overline{\mathbf{A_0}} \cap \overline{\mathbf{B_2}}]$

Corollary 3.47 *The varieties* $\mathbf{E}$ *and* $[0\rangle_4$ *coincide.*

Proof This follows from Lemma 3.37(i) and Proposition 3.46. □

3.3.5 The Interval $[\mathbf{B_0}, \mathbf{A_2}]$

Lemma 3.48

(i) *The variety* $\overline{\mathbf{A_2}}$ *is the only subvariety of* $\mathbf{A_2}$ *that covers* $\overline{\mathbf{B_2}}$.
(ii) *The variety* $\langle \widehat{\omega}, \widehat{\omega} \rangle_1$ *is the only subvariety of* $\mathbf{A_2}$ *that covers* $\overline{\mathbf{A_0}}$.
(iii) *The varieties* $\langle \widehat{\omega}, \widehat{\omega} \rangle_2$ *and* $\overline{\mathbf{A_0}}$ *are the only subvarieties of* $\mathbf{A_2}$ *that cover* $\overline{\mathbf{A_0}} \cap \overline{\mathbf{B_2}}$.

Proof

(i) The exclusion $\overline{\mathbf{B_2}} \not\subseteq \overline{\mathbf{A_0}}$ holds because $\mathcal{A}_0 \in \overline{\mathbf{B_2}}$. It follows from Lemma 3.6 and Proposition 3.7 that within the lattice $\mathfrak{L}(\mathbf{A_2})$, the variety $\overline{\mathbf{B_2}}$ can only be covered by some variety in the interval $\mathfrak{I}_1 = [\mathbf{A_0} \vee \mathbf{B_2}, \overline{\mathbf{A_2}}]$. The inclusion $\overline{\mathbf{B_2}} \subseteq \overline{\mathbf{A_2}}$ follows from Lemma 3.6. Since $\overline{\mathbf{B_2}} \nvDash \langle\widehat{\omega}]$ and $\overline{\mathbf{B_2}} \nvDash [\widehat{\omega}\rangle$ by Lemma 3.37(ii), it follows from Proposition 3.41 that $\overline{\mathbf{B_2}}$ is not contained in any proper subvariety of $\overline{\mathbf{A_2}}$ in $\mathfrak{I}_1$, whence $\overline{\mathbf{A_2}}$ is the only subvariety of $\mathbf{A_2}$ that covers $\overline{\mathbf{B_2}}$.

(ii) The exclusion $\overline{\mathbf{A_0}} \not\subseteq \overline{\mathbf{B_2}}$ holds because $\mathcal{B}_2 \in \overline{\mathbf{A_0}}$. It follows from Lemma 3.6 and Proposition 3.7 that within the lattice $\mathfrak{L}(\mathbf{A_2})$, the variety $\overline{\mathbf{A_0}}$ can only be covered by some variety in the interval $\mathfrak{I}_1 = [\mathbf{A_0} \vee \mathbf{B_2}, \overline{\mathbf{A_2}}]$. The inclusion $\overline{\mathbf{A_0}} \subseteq \langle\widehat{\omega}, \widehat{\omega}\rangle_1$ holds by Lemmas 3.6 and 3.44. Since $\overline{\mathbf{A_0}} \nvDash \langle\omega]$ and $\overline{\mathbf{A_0}} \nvDash [\omega\rangle$ by Lemma 3.37(i), it follows from Proposition 3.41 that $\overline{\mathbf{A_0}}$ is not contained in any proper subvariety of $\langle\widehat{\omega}, \widehat{\omega}\rangle_1$ in $\mathfrak{I}_1$, whence $\langle\widehat{\omega}, \widehat{\omega}\rangle_1$ is the only subvariety of $\mathbf{A_2}$ that covers $\overline{\mathbf{A_0}}$.

(iii) The exclusion $\overline{\mathbf{A_0}} \cap \overline{\mathbf{B_2}} \not\subseteq \mathbf{B_0}$ holds because $\mathcal{B}_0 \in \overline{\mathbf{A_0}} \cap \overline{\mathbf{B_2}}$. It follows from Lemma 3.6 and Proposition 3.7 that within the lattice $\mathfrak{L}(\mathbf{A_2})$, the variety $\overline{\mathbf{A_0}} \cap \overline{\mathbf{B_2}}$ can only be covered by some variety in $\mathfrak{I}_1 \cup \mathfrak{I}_2 \cup \mathfrak{I}_3$. If $\overline{\mathbf{A_0}} \cap \overline{\mathbf{B_2}}$ is a subvariety of some variety $\mathbf{V}$ in $\mathfrak{I}_1$, then $\overline{\mathbf{A_0}} \cap \overline{\mathbf{B_2}}$ is a subvariety of $\mathbf{V} \cap \overline{\mathbf{B_2}}$ in $\mathfrak{I}_2$. Hence, $\overline{\mathbf{A_0}} \cap \overline{\mathbf{B_2}}$ can only be covered by some variety in $\mathfrak{I}_2 \cup \mathfrak{I}_3$. The inclusion $\overline{\mathbf{A_0}} \cap \overline{\mathbf{B_2}} \subseteq \langle\widehat{\omega}, \widehat{\omega}\rangle_2$ holds by Lemmas 3.6 and 3.44, and the inclusion $\overline{\mathbf{A_0}} \cap \overline{\mathbf{B_2}} \subseteq \overline{\mathbf{A_0}}$ holds vacuously. Since $\overline{\mathbf{A_0}} \cap \overline{\mathbf{B_2}} \nvDash \langle\omega]$ and $\overline{\mathbf{A_0}} \cap \overline{\mathbf{B_2}} \nvDash [\omega\rangle$ by Lemma 3.37(i), it follows from Propositions 3.42 and 3.45 that the variety $\overline{\mathbf{A_0}} \cap \overline{\mathbf{B_2}}$ is neither contained in any proper subvariety of $\langle\widehat{\omega}, \widehat{\omega}\rangle_2$ in $\mathfrak{I}_2$ nor contained in any proper subvariety of $\overline{\mathbf{A_0}}$ in $\mathfrak{I}_3$. Consequently, $\langle\widehat{\omega}, \widehat{\omega}\rangle_2$ and $\overline{\mathbf{A_0}}$ are the only subvarieties of $\mathbf{A_2}$ that cover $\overline{\mathbf{A_0}} \cap \overline{\mathbf{B_2}}$. □

By Lemma 3.48 and methods similar to its proof, it is routinely shown how the four intervals $\mathfrak{I}_1$, $\mathfrak{I}_2$, $\mathfrak{I}_3$, and $\mathfrak{I}_4$, together with $\mathbf{A_2}$, combine to form the single interval $[\mathbf{B_0}, \mathbf{A_2}]$ in Fig. 3.6. To avoid overcrowding this figure, most of the varieties are unlabeled and most of the following coverings are not shown:

- $\overline{\mathbf{B_2}} \prec \overline{\mathbf{A_2}}$ and $\overline{\mathbf{A_0}} \cap \overline{\mathbf{B_2}} \prec \overline{\mathbf{A_0}}$;
- $\langle\ell]_2 \prec \langle\ell]_1$, $[r\rangle_2 \prec [r\rangle_1$, and $\langle\ell, r\rangle_2 \prec \langle\ell, r\rangle_1$ for all $\ell, r \in \widehat{\mathbb{N}}$;
- $\langle\ell]_4 \prec \langle\ell]_3$, $[r\rangle_4 \prec [r\rangle_3$, and $\langle\ell, r\rangle_4 \prec \langle\ell, r\rangle_3$ for all $\ell, r \in \widehat{\mathbb{N}}$.

Any unlabeled variety in Fig. 3.6 can be identified by referring to Figs. 3.2–3.5. It is easily seen that the interval $[\mathbf{B_0}, \mathbf{A_2}]$ is a distributive lattice.

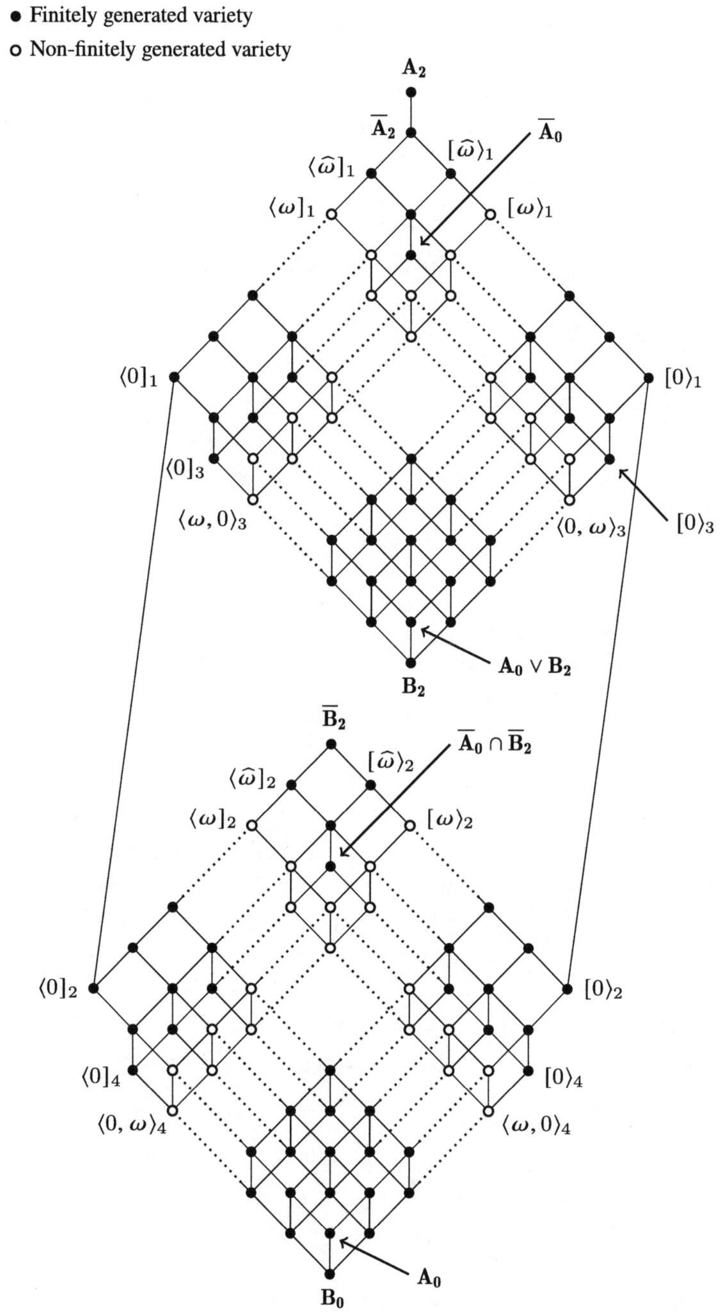

Fig. 3.6 The interval $[\mathbf{B}_0, \mathbf{A}_2]$

3.4 Subvarieties of A_2 That Are Cross, Finitely Generated, or Small

The proof of Theorem 1.7—a characterization of Cross subvarieties of $\mathbf{A_2}$—is given in Sect. 3.4.1. Some results on finitely generated subvarieties of $\mathbf{A_2}$ are established in Sect. 3.4.2.

3.4.1 Cross Subvarieties and Small Subvarieties of A_2

Lemma 3.49 (See Lee 2010b, Lemma 7)

(i) *Suppose that some diverse identity is deducible from some set Σ of identities. Then Σ contains some diverse identity.*
(ii) *An identity is non-diverse if and only if it is deducible from the identities*

$$xyx \approx y^2, \quad xy \approx yx. \tag{3.12}$$

Proof

(i) Suppose that every identity in Σ is non-diverse. Then Σ can only contain permutation identities or identities formed by non-simple words. It is then easily shown that no diverse identity can be deduced from Σ.
(ii) The identities (3.12) are non-diverse. Hence, by part (i), every identity deducible from (3.12) is non-diverse. Conversely, let $\mathbf{u} \approx \mathbf{v}$ be any non-diverse identity. It is clear that any permutation identity is deducible from the second identity in (3.12). Therefore, it suffices to assume that both $\mathbf{u}$ and $\mathbf{v}$ are non-simple words; in this case, the deduction $xyx \approx y^2 \vdash \mathbf{u} \approx \mathbf{v}$ holds by Lemma 3.29. □

Recall that $\mathbf{H_{com}}$ denotes the variety defined by the identities (3.12). By comparing the identities (3.11) and (3.12), it is easily seen that $\mathbf{H_{com}}$ is the largest commutative subvariety of $\mathbf{H}$.

Proof of Theorem 1.7 It suffices to establish the equivalence of the following statements for any subvariety $\mathbf{V}$ of $\mathbf{A_2}$: (a) $\mathbf{V}$ is Cross; (b) $\mathbf{V}$ is small; (c) $\mathbf{V}$ is finitely generated and $\mathbf{B_0} \nsubseteq \mathbf{V}$; (d) $\mathbf{V}$ satisfies some diverse identity; (e) $\mathbf{H_{com}} \nsubseteq \mathbf{V}$.

(a) $\Rightarrow$ (b). This holds by the definition of a Cross variety.
(b) $\Rightarrow$ (c). Suppose that $\mathbf{V}$ is small. Then since $\mathbf{V}$ is locally finite, it is also finitely generated by Lemma 3.34. Further, $\mathbf{B_0} \nsubseteq \mathbf{V}$ because the variety $\mathbf{B_0}$ is not small by Corollary 3.31.

(c) ⇒ (d). Let S be any finite semigroup that generates $\mathbf{V}$ with $n = |S|$. The assumption $\mathbf{B_0} \nsubseteq \mathbf{V}$ implies that $S \in \overline{\mathbf{B_0}}$, so that by Lemma 3.25, the semigroup S satisfies the identities $\{(3.1), (3.9), (3.10)\}$. In the following, it is shown that S, and so also $\mathbf{V}$, satisfy the diverse identity

$$x_1x_2\cdots x_n y z_1z_2\cdots z_n \approx x_1x_2\cdots x_n y^2 z_1z_2\cdots z_n.$$

Let $a_1, a_2, \ldots, a_n, b, c_1, c_2, \ldots, c_n \in S$. Then by Lemma 3.39, there exist $s_1, s_2, s_3, t_1, t_2, t_3 \in S$ such that $a_1a_2\cdots a_n = s_1s_2^2s_3$ and $c_1c_2\cdots c_n = t_1t_2^2t_3$. Since

$$\begin{aligned} a_1a_2\cdots a_nbc_1c_2\cdots c_n &= s_1s_2^2s_3bt_1t_2^2t_3 \\ &\overset{(3.9)}{=} s_1s_2^2(s_3bt_1)^2t_2^2t_3 \\ &\overset{(3.10)}{=} s_1s_2^2(s_3b^2t_1)^2t_2^2t_3 \\ &\overset{(3.9)}{=} s_1s_2^2s_3b^2t_1t_2^2t_3 \\ &= a_1a_2\cdots a_nb^2c_1c_2\cdots c_n, \end{aligned}$$

the semigroup S indeed satisfies the diverse identity.

(d) ⇒ (a). Suppose that $\mathbf{V}$ satisfies a diverse identity $\sigma : \mathbf{u} \approx \mathbf{v}$. Then $\mathbf{B_0} \nsubseteq \mathbf{V}$ because the semigroup $\mathcal{B}_0$ violates every diverse identity. Hence, $\mathbf{V} \subseteq \overline{\mathbf{B_0}} \subseteq \overline{\mathbf{A_2}}$ by Lemma 3.6, so that $\mathbf{V}$ satisfies the identities $\{(3.5), (3.10)\}$ by Lemmas 3.16 and 3.25(ii). By symmetry, there are two cases.

CASE 1: $\mathbf{u}$ is simple and $\mathbf{v}$ is non-simple. Then

$$\mathbf{v} = \mathbf{v}_0 \prod_{i=1}^{n} (y\mathbf{v}_i)$$

for some $\mathbf{v}_0, \mathbf{v}_1, \ldots, \mathbf{v}_n \in \mathscr{A}_\varnothing^+$ with $n \geq 2$ and $y \notin \mathsf{con}(\mathbf{v}_0\mathbf{v}_1\cdots\mathbf{v}_n)$.

1.1: $y \in \mathsf{con}(\mathbf{u})$. Choose any variables $x_1, x_2, x_3, x_4 \notin \mathsf{con}(\mathbf{uv})$. Let φ and χ denote the substitutions $y \mapsto x_1x_2x_3x_4$ and $y \mapsto x_1x_3x_2x_4$, respectively. Since

$$\begin{aligned} \mathbf{u}\varphi &\overset{\sigma}{\approx} \mathbf{v}\varphi \\ &= \mathbf{v}_0 \prod_{i=1}^{n} (x_1x_2x_3x_4\mathbf{v}_i) \\ &\overset{(3.10)}{\approx} \mathbf{v}_0 \prod_{i=1}^{n} (x_1x_2^2x_3^2x_4\mathbf{v}_i) \end{aligned}$$

$$\begin{aligned}
&\overset{(3.5)}{\approx} \mathbf{v}_0 \prod_{i=1}^{n} (x_1 x_3^2 x_2^2 x_4 \mathbf{v}_i) \\
&\overset{(3.10)}{\approx} \mathbf{v}_0 \prod_{i=1}^{n} (x_1 x_3 x_2 x_4 \mathbf{v}_i) \\
&= \mathbf{v}\chi \\
&\overset{\sigma}{\approx} \mathbf{u}\chi,
\end{aligned}$$

the variety $\mathbf{V}$ satisfies the nontrivial permutation identity $\mathbf{u}\varphi \approx \mathbf{u}\chi$.

1.2: $y \notin \mathsf{con}(\mathbf{u})$. Then $\mathbf{u}y \approx \mathbf{v}y$ is a diverse identity satisfied by the variety $\mathbf{V}$, where $\mathbf{u}y$ is simple and $\mathbf{v}y$ is non-simple. Repeat the procedure in Case 1.1 on $\mathbf{u}y \approx \mathbf{v}y$ to obtain a nontrivial permutation identity satisfied by $\mathbf{V}$.

CASE 2: $\mathbf{u}$ and $\mathbf{v}$ are simple. Then $\mathsf{con}(\mathbf{u}) \neq \mathsf{con}(\mathbf{v})$ because σ is not a permutation identity. Generality is not lost by assuming that $x \in \mathsf{con}(\mathbf{v}) \backslash \mathsf{con}(\mathbf{u})$. Then $\mathbf{u}x \approx \mathbf{v}x$ is a diverse identity satisfied by $\mathbf{V}$, where $\mathbf{u}x$ is simple and $\mathbf{v}x$ is non-simple. Repeat the procedure in Case 1 on $\mathbf{u}x \approx \mathbf{v}x$ to obtain a nontrivial permutation identity satisfied by $\mathbf{V}$.

In any case, the variety $\mathbf{V}$ satisfies some nontrivial permutation identity. Hence, $\mathbf{V}$ is small by Lemma 3.33. Since $\mathbf{V}$ is locally finite, it is also finitely generated by Lemma 3.34. By Theorem 1.6, all subvarieties of $\mathbf{A_2}$, which include $\mathbf{V}$, are finitely based. Consequently, $\mathbf{V}$ is Cross.

(d) $\Leftrightarrow$ (e). It follows from Lemma 3.49(ii) that the variety $\mathbf{V}$ satisfies some diverse identity if and only if (3.12) $\nvdash \mathsf{id}\,\mathbf{V}$, that is, $\mathbf{H}_{\mathsf{com}} \nsubseteq \mathbf{V}$. □

Corollary 3.50 *The non-small subvarieties of* $\mathbf{A_2}$ *constitute the interval* $[\mathbf{H}_{\mathsf{com}}, \mathbf{A_2}]$.

Proposition 3.51 *The Cross subvarieties of* $\mathbf{A_2}$ *constitute an incomplete sublattice of* $\mathfrak{L}(\mathbf{A_2})$.

Proof Let $\mathbf{U}$ and $\mathbf{V}$ be any Cross subvarieties of $\mathbf{A_2}$. Then $\mathbf{B_0} \nsubseteq \mathbf{U}, \mathbf{V}$ by Theorem 1.7, so that $\mathbf{U}, \mathbf{V} \subseteq \overline{\mathbf{B_0}}$. Therefore, $\mathbf{U} \vee \mathbf{V} \subseteq \overline{\mathbf{B_0}}$, whence $\mathbf{B_0} \nsubseteq \mathbf{U} \vee \mathbf{V}$. Since the variety $\mathbf{U} \vee \mathbf{V}$ is finitely generated, it is also Cross by Theorem 1.7. Now the variety $\mathbf{U} \cap \mathbf{V}$ is vacuously Cross, so the Cross subvarieties of $\mathbf{A_2}$ form a sublattice of $\mathfrak{L}(\mathbf{A_2})$; this sublattice is incomplete because by Lemmas 3.35 and 3.36, the varieties $\mathbf{D}_0 \subset \mathbf{D}_1 \subset \mathbf{D}_2 \subset \cdots$ are Cross, while their complete varietal join $\bigvee_{\ell \geq 0} \mathbf{D}_\ell$ is not small and so also not Cross. □

3.4.2 Finitely Generated Subvarieties of $\mathbf{A}_2$

By Theorem 1.7, every small subvariety of $\mathbf{A}_2$ is Cross and so also finitely generated. But the converse obviously does not hold since the finitely generated variety $\mathbf{A}_2$ is not small.

Proposition 3.52 *The finitely generated subvarieties of* $\mathbf{A}_2$ *constitute an incomplete sublattice of* $\mathfrak{L}(\mathbf{A}_2)$.

Proof Let $\mathbf{U}$ and $\mathbf{V}$ be any finitely generated subvarieties of $\mathbf{A}_2$. Then the join $\mathbf{U} \vee \mathbf{V}$ is clearly finitely generated. To show that the intersection $\mathbf{U} \cap \mathbf{V}$ is also finitely generated, consider the lattice $\mathfrak{L}(\mathbf{A}_2)$ as the disjoint union of $[\mathbf{B}_0, \mathbf{A}_2]$ and $\mathfrak{L}(\overline{\mathbf{B}}_0)$. There are two cases.

CASE 1: $\mathbf{U}, \mathbf{V} \in [\mathbf{B}_0, \mathbf{A}_2]$. Then $\mathbf{U} \cap \mathbf{V}$ is finitely generated by Fig. 3.6.
CASE 2: $\mathbf{U} \in \mathfrak{L}(\overline{\mathbf{B}}_0)$ or $\mathbf{V} \in \mathfrak{L}(\overline{\mathbf{B}}_0)$. Then either $\mathbf{B}_0 \not\subseteq \mathbf{U}$ or $\mathbf{B}_0 \not\subseteq \mathbf{V}$, so that by Theorem 1.7, either $\mathbf{U}$ or $\mathbf{V}$ is Cross. Therefore, $\mathbf{U} \cap \mathbf{V}$ is Cross and so also finitely generated.

Hence, the finitely generated subvarieties of $\mathbf{A}_2$ form a sublattice of $\mathfrak{L}(\mathbf{A}_2)$; this lattice is incomplete since it is easily seen from Fig. 3.2 that the complete varietal join of the finitely generated varieties $\langle 0, 0\rangle_1, \langle 1, 0\rangle_1, \langle 2, 0\rangle_1, \ldots$ is the non-finitely generated variety $\langle \omega, 0\rangle_1$. □

It follows from Figs. 3.2 and 3.5 that

$$\langle \omega]_1 = \overline{\mathbf{A}}_2\{\langle \omega]\} \text{ and } [\omega\rangle_1 = \overline{\mathbf{A}}_2\{[\omega\rangle\}$$

are maximal non-finitely generated varieties in $\mathfrak{I}_1 = [\mathbf{A}_0 \vee \mathbf{B}_2, \overline{\mathbf{A}}_2]$, and

$$\langle 0, \omega\rangle_4 = (\overline{\mathbf{A}}_0 \cap \overline{\mathbf{B}}_2)\{\langle 0], [\omega\rangle\} \text{ and } \langle \omega, 0\rangle_4 = (\overline{\mathbf{A}}_0 \cap \overline{\mathbf{B}}_2)\{\langle \omega], [0\rangle\}$$

are minimal non-finitely generated varieties in $\mathfrak{I}_4 = [\mathbf{B}_0, \overline{\mathbf{A}}_0 \cap \overline{\mathbf{B}}_2]$.

Proposition 3.53 *The non-finitely generated subvarieties of* $\mathbf{A}_2$ *form the intervals*

$$[\mathbf{H}_{\mathsf{com}}, \overline{\mathbf{B}}_0], \ [\langle 0, \omega\rangle_4, [\omega\rangle_1], \ \textit{and} \ [\langle \omega, 0\rangle_4, \langle \omega]_1]. \tag{3.13}$$

Proof The lattice $\mathfrak{L}(\mathbf{A}_2)$ is the disjoint union of $\mathfrak{L}(\overline{\mathbf{B}}_0)$ and $[\mathbf{B}_0, \mathbf{A}_2]$. By Theorem 1.7, the non-finitely generated varieties in $\mathfrak{L}(\overline{\mathbf{B}}_0)$ constitute the interval $[\mathbf{H}_{\mathsf{com}}, \overline{\mathbf{B}}_0]$. By Fig. 3.6, the non-finitely generated varieties in $[\mathbf{B}_0, \mathbf{A}_2]$ form the second and third intervals in (3.13). □

Since $\mathfrak{L}(\mathbf{A}_2) = \mathfrak{L}(\overline{\mathbf{B}}_0) \cup [\mathbf{B}_0, \mathbf{A}_2]$ is a disjoint union, it is of interest to know which varieties in $[\mathbf{B}_0, \mathbf{A}_2]$ contain $\overline{\mathbf{B}}_0$. It turns out that $\overline{\mathbf{B}}_0$ is covered by some variety in the interval $\mathfrak{I}_4 = [\mathbf{B}_0, \overline{\mathbf{A}}_0 \cap \overline{\mathbf{B}}_2]$.

Proposition 3.54 *The variety $\overline{\mathbf{B}}_0$ is the unique maximal subvariety of $\langle \omega, \omega\rangle_4$.*

Proof Recall from Lemma 3.25(i) that the identities {(3.1), (3.9)} constitute an identity basis for $\overline{\mathbf{B}}_0$. Alternatively, the identities (3.1) and $\sigma : x^2yzt^2 \approx x^2zyt^2$ also constitute an identity basis for $\overline{\mathbf{B}}_0$ (Lee 2007d, Lemma 6). Then $\overline{\mathbf{B}}_0 \models \langle \omega]$ because

$$\begin{aligned} a_1^2a_2y^2x^2y^2 &\overset{\sigma}{\approx} a_1^2y^2a_2x^2y^2 \\ &\overset{(3.1)}{\approx} (a_1^2y^2a_2x^2x^2)y^2 \\ &\overset{\sigma}{\approx} a_1^2a_2x^2y^2x^2y^2; \end{aligned}$$

by symmetry, $\overline{\mathbf{B}}_0 \models [\omega\rangle$. Hence, by Lemma 3.6, the inclusion

$$\overline{\mathbf{B}}_0 \subseteq (\overline{\mathbf{A}}_0 \cap \overline{\mathbf{B}}_2)\{\langle\omega], [\omega\rangle\} = \langle\omega, \omega\rangle_4$$

holds; this inclusion is proper because $\mathcal{B}_0 \in \langle\omega, \omega\rangle_4 \backslash \overline{\mathbf{B}}_0$. Let $\mathbf{V}$ be any variety such that $\overline{\mathbf{B}}_0 \subset \mathbf{V} \subset \langle\omega, \omega\rangle_4$. Since $\mathbf{V} \nsubseteq \overline{\mathbf{B}}_0$, the variety $\mathbf{V}$ contains $\mathbf{B}_0$ and so belongs to the interval $\mathfrak{I}_4 = [\mathbf{B}_0, \overline{\mathbf{A}}_0 \cap \overline{\mathbf{B}}_2]$. Now since $\mathbf{V}$ is a proper subvariety of $\langle\omega, \omega\rangle_4$, it is easily seen from Fig. 3.5 that $\mathbf{V}$ satisfies $\langle k]$ or $[k\rangle$ for some finite k. It follows that $\overline{\mathbf{B}}_0$ satisfies either $\langle k]$ or $[k\rangle$, but this is impossible because it is easily shown that the identities $\langle k]$ and $[k\rangle$ are not deducible from the identity basis {(3.1), (3.9)} for $\overline{\mathbf{B}}_0$. Therefore, the variety $\mathbf{V}$ does not exist, whence $\overline{\mathbf{B}}_0$ is a maximal subvariety of $\langle\omega, \omega\rangle_4$. The uniqueness of $\overline{\mathbf{B}}_0$ follows because the variety $\langle\omega, \omega\rangle_4$ has no maximal subvarieties containing $\mathbf{B}_0$; see Fig. 3.5. □

Based on Fig. 3.6 and Proposition 3.54, the interval $[\overline{\mathbf{B}}_0, \mathbf{A}_2]$ is given in Fig. 3.7. To avoid overcrowding this figure, the following coverings are not shown:

- $\langle \ell]_2 \prec \langle \ell]_1$, $[r\rangle_2 \prec [r\rangle_1$, and $\langle \ell, r\rangle_2 \prec \langle \ell, r\rangle_1$ for all $\ell, r \in \{\omega, \widehat{\omega}\}$;
- $\overline{\mathbf{A}}_0 \cap \overline{\mathbf{B}}_2 \prec \overline{\mathbf{A}}_0$, $\langle\omega]_4 \prec \langle\omega]_3$, and $[\omega\rangle_4 \prec [\omega\rangle_3$.

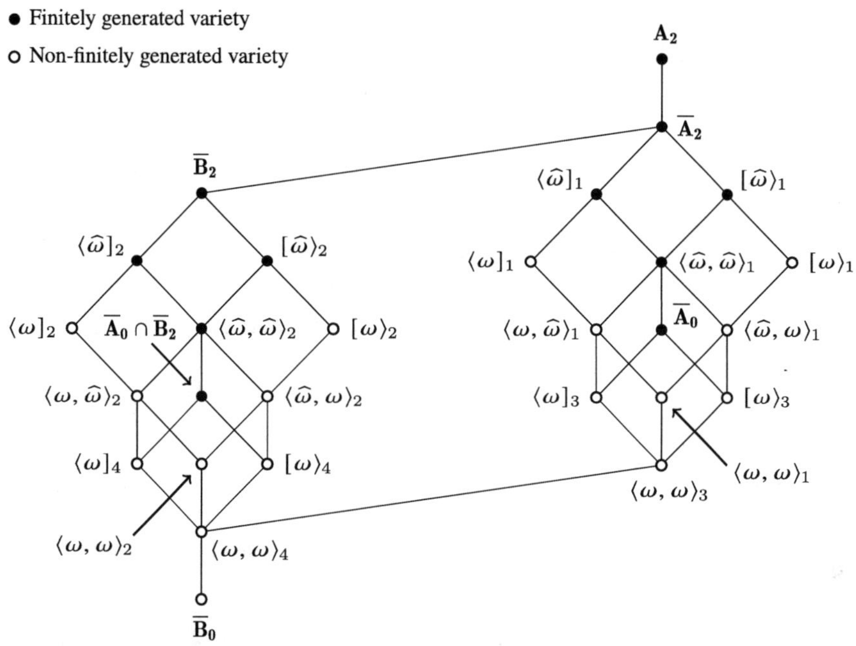

Fig. 3.7 The interval $[\overline{\mathbf{B}}_0, \mathbf{A}_2]$

3.5 Summary

The identity (3.3) that precisely excludes $\mathcal{A}_2$ from any subvariety of $\mathbf{P}_n$ was first found in Lee (2007b) within the context of finite semigroups. The finite basis property of all subvarieties of $\mathbf{A}_2$—the main result of Sect. 3.2—was published in Lee (2008a). The description in Sect. 3.3 of the lattice $\mathfrak{L}(\mathbf{A}_2)$ and the results in Sect. 3.4 on subvarieties of $\mathbf{A}_2$ that are Cross, finitely generated, or small were published in Lee (2010a).

4 Pseudo-Simple Hereditarily Finitely Based Identities

Recall that an identity is *hereditarily finitely based* if it defines a hereditarily finitely based variety. A word is *almost simple* if it contains one variable twice and any other variable at most once, and an identity is *pseudo-simple* if it is formed by a simple word and an almost simple word. As described in Sect. 1.3.2, the present chapter is concerned with pseudo-simple hereditarily finitely based identities.

Theorem 4.1 *Identities of the form*

$$x_1x_2\cdots x_\ell y^2 z_1z_2\cdots z_r \approx x_1x_2\cdots x_\ell y z_1z_2\cdots z_r, \tag{4.1}$$

where $\ell, r \geq 0$ *and* $(\ell, r) \neq (0, 0)$, *are precisely all pseudo-simple identities that are not hereditarily finitely based.*

Each identity from (4.1) is satisfied by continuum many varieties (Kad'ourek 2000) and so cannot be hereditarily finitely based.

Recall that an identity $\mathbf{w} \approx \mathbf{w}'$ is *homotypical* if $\mathsf{con}(\mathbf{w}) = \mathsf{con}(\mathbf{w}')$. In Sect. 4.1, it is shown that any non-homotypical pseudo-simple identity is hereditarily finitely based. In Sect. 4.2, it is shown that any homotypical pseudo-simple identity that is not of the form (4.1) is hereditarily finitely based. Consequently, the proof of Theorem 4.1 is complete.

4.1 Non-Homotypical Identities

Lemma 4.2 *Any pseudo-simple identity can be used to deduce a periodicity identity.*

E. W. H. Lee, *Advances in the Theory of Varieties of Semigroups*, Frontiers in Mathematics, https://doi.org/10.1007/978-3-031-16497-2_4

Proof Let $\mathbf{w} \approx \mathbf{w}'$ be any pseudo-simple identity, where $\mathbf{w}$ is a simple word and $\mathbf{w}'$ is an almost simple word. If $n = |\mathbf{w}|$ and $n' = |\mathbf{w}'|$ are different, then $x^n \approx x^{n'}$ is a periodicity identity deducible from $\mathbf{w} \approx \mathbf{w}'$. Hence, assume that $n = n'$. Let y be the variable that occurs twice in $\mathbf{w}'$. Then

(A) $\mathbf{w}' \stackrel{\circ}{=} \mathbf{a}y^2$ for some $\mathbf{a} \in \mathscr{A}_\varnothing^+$ such that $y \notin \mathsf{con}(\mathbf{a})$ and $|\mathbf{a}| = n - 2$;
(B) $y \notin \mathsf{con}(\mathbf{w})$ or $\mathbf{w} \stackrel{\circ}{=} \mathbf{b}y$ for some $\mathbf{b} \in \mathscr{A}_\varnothing^+$ such that $y \notin \mathsf{con}(\mathbf{b})$ and $|\mathbf{b}| = n - 1$.

Let φ denote the substitution that maps y to x^2 and every other variable to x. Then $\mathbf{w}\varphi \in \{x^n, x^{n+1}\}$ and $\mathbf{w}'\varphi = x^{n+2}$. In any case, the periodicity identity $x^{n+3} \approx x^{n+1}$ is deducible from $\mathbf{w} \approx \mathbf{w}'$. □

Lemma 4.3 *Any simple semigroup that satisfies a pseudo-simple identity is completely simple.*

Proof This follows from Lemma 4.2 since any simple semigroup that satisfies a periodicity identity is completely simple; see Howie (1995, Theorem 3.2.11). □

Lemma 4.4 (Pollák 1982; See Also Volkov 2002, Proposition 1.6B) *Let $\mathbf{w} \approx \mathbf{w}'$ be any non-homotypical identity, where $\mathbf{w}$ is a simple word. Suppose that any simple semigroup that satisfies the identity $\mathbf{w} \approx \mathbf{w}'$ is finitely based. Then the identity $\mathbf{w} \approx \mathbf{w}'$ is hereditarily finitely based.*

Proposition 4.5 *Any non-homotypical pseudo-simple identity is hereditarily finitely based.*

Proof Let $\mathbf{w} \approx \mathbf{w}'$ be any non-homotypical pseudo-simple identity, where $\mathbf{w}$ is a simple word and $\mathbf{w}'$ is an almost simple word. It is shown in the following that any simple semigroup S that satisfies $\mathbf{w} \approx \mathbf{w}'$ is finitely based. The identity $\mathbf{w} \approx \mathbf{w}'$ is thus hereditarily finitely based by Lemma 4.4.

By Lemma 4.3, the semigroup S is completely simple. Let y be the variable that occurs exactly twice in $\mathbf{w}'$. Since the word $\mathbf{w}$ is simple, y occurs at most once in $\mathbf{w}$. Therefore, any subgroup G of S satisfies either $y \approx y^2$ or $1 \approx y^2$, whence G is either trivial or of exponent two. It is well known and easily shown that any group of exponent two is commutative. Hence, S is a completely simple semigroup with commutative subgroups, and such a semigroup is finitely based (Mashevitzky 1978; Rasin 1979). □

4.2 Homotypical Identities

Proposition 4.6 *Any homotypical pseudo-simple identity that is not of the form* (4.1) *is hereditarily finitely based.*

The remainder of the section is devoted to the proof of Proposition 4.6. Reference will often be made to a homotypical pseudo-simple identity

$$\mathbf{z} \approx \mathbf{z}',$$

where $\mathbf{z}$ is a simple word and $\mathbf{z}'$ is an almost simple word. Therefore, assume that

$$\mathbf{z} \stackrel{\circ}{=} x_1 x_2 \cdots x_n y \quad \text{and} \quad \mathbf{z}' \stackrel{\circ}{=} x_1 x_2 \cdots x_n y^2$$

for some distinct $x_1, x_2, \ldots, x_n, y \in \mathscr{A}$ with $n \geq 0$. Note that if $n = 0$, then $\mathbf{z} = y$ and $\mathbf{z}' = y^2$.

Lemma 4.7 *Let* $\mathbf{w} \approx \mathbf{w}'$ *be any pseudo-simple identity. Suppose that some nontrivial permutation identity is deducible from* $\mathbf{w} \approx \mathbf{w}'$. *Then* $\mathbf{w} \approx \mathbf{w}'$ *is hereditarily finitely based.*

Proof Let S be any semigroup that satisfies the identity $\mathbf{w} \approx \mathbf{w}'$, so that S also satisfies some nontrivial permutation identity. Then S is uniformly periodic by Lemma 4.2. Every uniformly periodic semigroup that satisfies some nontrivial permutation identity is finitely based (Perkins 1969, Theorem 22), so that S is also finitely based. □

Lemma 4.8 *Suppose that* $x_i x_j$ *is a factor of* $\mathbf{z}'$ *that is not a factor of* $\mathbf{z}$. *Then the identity* $\mathbf{z} \approx \mathbf{z}'$ *is hereditarily finitely based.*

Proof No generality is lost—and more convenience is gained—by letting $x_i x_j = x_1 x_2$. By assumption, the identity $\mathbf{z} \approx \mathbf{z}'$ is one of the following:

(A) $\mathbf{a} x_1 \mathbf{b} x_2 \mathbf{c} \approx \mathbf{a}' x_1 x_2 \mathbf{b}'$ with $\mathbf{abc} \stackrel{\circ}{=} x_3 x_4 \cdots x_n y$, $\mathbf{b} \neq \varnothing$, and $\mathbf{a}'\mathbf{b}' \stackrel{\circ}{=} x_3 x_4 \cdots x_n y^2$;
(B) $\mathbf{a} x_2 \mathbf{b} x_1 \mathbf{c} \approx \mathbf{a}' x_1 x_2 \mathbf{b}'$ with $\mathbf{abc} \stackrel{\circ}{=} x_3 x_4 \cdots x_n y$ and $\mathbf{a}'\mathbf{b}' \stackrel{\circ}{=} x_3 x_4 \cdots x_n y^2$.

First assume that $\mathbf{z} \approx \mathbf{z}'$ is the identity in (A). Choose any variable z that does not occur in $\mathbf{z} \approx \mathbf{z}'$. Then performing the substitution $x_1 \mapsto x_1 z$ in $\mathbf{z} \approx \mathbf{z}'$ results in the identity

$$\mathbf{a} x_1 z \mathbf{b} x_2 \mathbf{c} \approx \mathbf{a}' x_1 z x_2 \mathbf{b}',$$

and performing the substitution $x_2 \mapsto z x_2$ in $\mathbf{z} \approx \mathbf{z}'$ results in the identity

$$\mathbf{a} x_1 \mathbf{b} z x_2 \mathbf{c} \approx \mathbf{a}' x_1 z x_2 \mathbf{b}'.$$

Hence, the deduction $\mathbf{z} \approx \mathbf{z}' \vdash \mathbf{a}x_1z\mathbf{b}x_2\mathbf{c} \approx \mathbf{a}x_1\mathbf{b}zx_2\mathbf{c}$ holds. Now $\mathbf{a}x_1z\mathbf{b}x_2\mathbf{c} \approx \mathbf{a}x_1\mathbf{b}zx_2\mathbf{c}$ is a nontrivial permutation identity because $\mathbf{b} \neq \varnothing$. Therefore, the identity $\mathbf{z} \approx \mathbf{z}'$ is hereditarily finitely based by Lemma 4.7.

Now assume that $\mathbf{z} \approx \mathbf{z}'$ is the identity in (B). By performing the same substitutions from the previous paragraph, the nontrivial permutation identity $\mathbf{a}x_2\mathbf{b}x_1z\mathbf{c} \approx \mathbf{a}zx_2\mathbf{b}x_1\mathbf{c}$ is deduced from $\mathbf{z} \approx \mathbf{z}'$. Therefore, the identity $\mathbf{z} \approx \mathbf{z}'$ is hereditarily finitely based by Lemma 4.7. □

Lemma 4.9 *Suppose that any of the following conditions holds*:

(a) *there exists some i such that* $\mathsf{h}(\mathbf{z}') = x_i \neq \mathsf{h}(\mathbf{z})$;
(b) *there exists some i such that* $\mathsf{t}(\mathbf{z}') = x_i \neq \mathsf{t}(\mathbf{z})$.

Then the identity $\mathbf{z} \approx \mathbf{z}'$ is hereditarily finitely based.

Proof By symmetry, it suffices to assume that (a) holds. Choose any variable z that does not occur in the identity $\mathbf{z} \approx \mathbf{z}'$. Then zx_i is a factor of $z\mathbf{z}'$ that is not a factor of $z\mathbf{z}$. Since the variables z and x_i are simple in both $z\mathbf{z}$ and $z\mathbf{z}'$, the proof of Lemma 4.8 can be repeated to show that the identity $z\mathbf{z} \approx z\mathbf{z}'$ is hereditarily finitely based. Therefore, the identity $\mathbf{z} \approx \mathbf{z}'$ is also hereditarily finitely based. □

Lemma 4.10 (Pollák 1986, Theorem A) *Suppose that for some i, the word yx_iy is a factor of $\mathbf{z}'$. Then the identity $\mathbf{z} \approx \mathbf{z}'$ is hereditarily finitely based.*

Lemma 4.11 *Suppose that the following conditions hold*:

(a) $\mathbf{z} \approx \mathbf{z}'$ *is not of the form* (4.1);
(b) $n \leq 1$ *in* $\mathbf{z} \approx \mathbf{z}'$.

Then the identity $\mathbf{z} \approx \mathbf{z}'$ is hereditarily finitely based.

Proof If $n = 0$, then the identity $\mathbf{z} \approx \mathbf{z}'$ is the idempotency identity $y \approx y^2$ and so is hereditarily finitely based (Birjukov 1970; Fennemore 1970; Gerhard 1970). Therefore, assume that $n = 1$, whence $\mathbf{z} \in \{x_1y, yx_1\}$ and $\mathbf{z}' \in \{x_1y^2, yx_1y, y^2x_1\}$. Since the identities $x_1y \approx x_1y^2$ and $yx_1 \approx y^2x_1$ are of the form (4.1), it follows from (a) that the identity $\mathbf{z} \approx \mathbf{z}'$ is one of the following: $x_1y \approx yx_1y$, $x_1y \approx y^2x_1$, $yx_1 \approx x_1y^2$, and $yx_1 \approx yx_1y$. The identity $\mathbf{z} \approx \mathbf{z}'$ is then hereditarily finitely based by Lemmas 4.9 and 4.10. □

Lemma 4.12 *For any $\ell, r \geq 0$ and $m \geq 1$, the following identities are hereditarily finitely based*:

$$a_1a_2\cdots a_\ell yb_1b_2\cdots b_m yc_1c_2\cdots c_r \approx a_1a_2\cdots a_\ell yb_1b_2\cdots b_m c_1c_2\cdots c_r, \qquad \langle \ell; m; r]$$

$$a_1a_2\cdots a_\ell yb_1b_2\cdots b_m yc_1c_2\cdots c_r \approx a_1a_2\cdots a_\ell b_1b_2\cdots b_m yc_1c_2\cdots c_r. \qquad [\ell; m; r\rangle$$

Proof By symmetry, it suffices to consider only the identity $\langle \ell; m; r]$. If $m = 1$, then $\langle \ell; m; r]$ is hereditarily finitely based by Lemma 4.10. Therefore, assume that $m \geq 2$. Since

$$\begin{aligned} & a_1a_2\cdots a_{\ell+1} yb_1b_2\cdots b_{m-1} c_1c_2\cdots c_r \\ \stackrel{\langle \ell; m; r]}{\approx}\; & a_1a_2\cdots a_{\ell+1} yb_1b_2\cdots b_{m-1} a_{\ell+1} c_1c_2\cdots c_r \\ \stackrel{\langle \ell; m; r]}{\approx}\; & a_1a_2\cdots a_{\ell+1} yb_1b_2\cdots b_{m-1} a_{\ell+1} yc_1c_2\cdots c_r \\ \stackrel{\langle \ell; m; r]}{\approx}\; & a_1a_2\cdots a_{\ell+1} yb_1b_2\cdots b_{m-1} yc_1c_2\cdots c_r, \end{aligned}$$

the deduction $\langle \ell; m; r] \vdash \langle \ell + 1; m - 1; r]$ holds. If $m - 1 \geq 2$, then the same argument establishes the deduction $\langle \ell + 1; m - 1; r] \vdash \langle \ell + 2; m - 2; r]$. This can be repeated so that

$$\begin{aligned} \langle \ell; m; r] & \vdash \langle \ell + 1; m - 1; r] \\ & \vdash \langle \ell + 2; m - 2; r] \\ & \;\;\vdots \\ & \vdash \langle \ell + m - 1; 1; r]. \end{aligned}$$

Since the identity $\langle \ell+m-1; 1; r]$ is hereditarily finitely based by Lemma 4.10, the identity $\langle \ell; m; r]$ is also hereditarily finitely based. □

Proof of Proposition 4.6 Suppose that

(A) the identity $\mathbf{z} \approx \mathbf{z}'$ is not hereditarily finitely based.

Then by Lemma 4.11, either $\mathbf{z} \approx \mathbf{z}'$ is of the form (4.1) or $n \geq 2$ in $\mathbf{z} \approx \mathbf{z}'$. If the identity $\mathbf{z} \approx \mathbf{z}'$ is of the form (4.1), then the proof is complete. Hence, it remains to assume that $n \geq 2$ in $\mathbf{z} \approx \mathbf{z}'$. By assumption,

$$\mathbf{z}' = a_1a_2\cdots a_\ell yb_1b_2\cdots b_m yc_1c_2\cdots c_r,$$

where $a_1a_2\cdots a_\ell b_1b_2\cdots b_m c_1c_2\cdots c_r \stackrel{\circ}{=} x_1x_2\cdots x_n$ with $\ell + m + r = n \geq 2$ and $\ell, m, r \geq 0$. By Lemmas 4.8 and 4.9, any of the following implies that $\mathbf{z} \approx \mathbf{z}'$ is hereditarily finitely based:

(B) $\ell \geq 1$ and $a_1a_2\cdots a_\ell$ is not a prefix of $\mathbf{z}$;
(C) $r \geq 1$ and $c_1c_2\cdots c_r$ is not a suffix of $\mathbf{z}$.

Therefore, by (A), neither (B) nor (C) holds, that is,

(D) either $\ell = 0$ or $a_1a_2\cdots a_\ell$ is a prefix of $\mathbf{z}$;
(E) either $r = 0$ or $c_1c_2\cdots c_r$ is a suffix of $\mathbf{z}$.

It follows that $\mathbf{z} = a_1a_2\cdots a_\ell \mathbf{b} c_1c_2\cdots c_r$ with $\mathbf{b} \overset{\circ}{=} b_1b_2\cdots b_m y$.

CASE 1: $m = 1$. Then the identity $\mathbf{z} \approx \mathbf{z}'$ is either $\langle \ell; 1; r]$ or $[\ell; 1; r\rangle$.
CASE 2: $m \geq 2$. Then by (A) and Lemma 4.8, the factor $b_1b_2\cdots b_m$ of $\mathbf{z}'$ is also a factor of $\mathbf{z}$. Therefore, either

$$\mathbf{z} = a_1a_2\cdots a_\ell y b_1b_2\cdots b_m c_1c_2\cdots c_r$$
$$\text{or } \mathbf{z} = a_1a_2\cdots a_\ell b_1b_2\cdots b_m y c_1c_2\cdots c_r,$$

whence the identity $\mathbf{z} \approx \mathbf{z}'$ is either $\langle \ell; m; r]$ or $[\ell; m; r\rangle$.

By Lemma 4.12, any of the two cases above contradicts (A). Therefore, $m = 0$ and the identity $\mathbf{z} \approx \mathbf{z}'$ is

$$a_1a_2\cdots a_\ell y c_1c_2\cdots c_r \approx a_1a_2\cdots a_\ell y^2 c_1c_2\cdots c_r.$$

Since $\ell + r = n \geq 2$ implies that $(\ell, r) \neq (0, 0)$, the identity $\mathbf{z} \approx \mathbf{z}'$ is of the form (4.1). □

4.3 Summary

Theorem 4.1, the main result of the present chapter, is a complete description of pseudo-simple identities that are not hereditarily finitely based. This result was published in Lee (2014b).

Sufficient Conditions for the Non-Finite Basis Property

5

The main result of the present chapter is the following restatement of Theorem 1.25.

Theorem 5.1 *Suppose that S is any semigroup that satisfies the identities*

$$x^n y_1^n y_2^n \cdots y_m^n x^n \approx x^n y_m^n y_{m-1}^n \cdots y_1^n x^n, \quad m \in \{2, 3, 4, \ldots\} \tag{5.1}$$

for some fixed $n \geq 2$ and that $\mathcal{L}_3 \in \mathscr{V}_{\mathsf{sem}}\{S\}$. Then S is non-finitely based.

Restrictions on identities satisfied by the semigroup $\mathcal{L}_3$ are established in Sect. 5.1. Based on these results, the proof of Theorem 5.1 is given in Sect. 5.2. In Sect. 5.3, an identity that precisely excludes $\mathcal{L}_3$ from any subvariety of $\mathbf{P}_n$ is given. It follows that Theorem 5.1 can be stated more concisely when applied to only finite semigroups.

Corollary 5.2 *For each $\ell \geq 3$, the semigroup $\mathcal{L}_\ell$ is non-finitely based.*

Proof Identifying the product efe in the semigroup $\mathcal{L}_\ell$ with $\mathtt{0}$ results in the semigroup $\mathcal{L}_3$, so that $\mathcal{L}_3 \in \mathscr{V}_{\mathsf{sem}}\{\mathcal{L}_\ell\}$. Let n be any integer greater than $\ell/2$. Then for any element $a \in \mathcal{L}_\ell$ that is not an idempotent, $a^n = \mathtt{0}$. Based on this observation, it is routinely verified that the semigroup $\mathcal{L}_\ell$ satisfies the identities (5.1) and so is non-finitely based by Theorem 5.1. □

More generally, for any $\ell \geq 3$ and any semigroup T that satisfies the identities (5.1), the direct product $\mathcal{L}_\ell \times T$ is non-finitely based. Examples of such semigroups T include periodic groups and commutative semigroups.

E. W. H. Lee, *Advances in the Theory of Varieties of Semigroups*, Frontiers in Mathematics, https://doi.org/10.1007/978-3-031-16497-2_5

Corollary 5.3 *For each $n \geq 1$, the direct product $\mathcal{L}_{\ell,n} = \mathcal{L}_\ell \times \mathcal{Z}_n$ is finitely based if and only if $\ell = 2$.*

Proof This follows from the preceding observation and the finite basis property of the semigroup $\mathcal{L}_2 \times \mathcal{Z}_n = \mathcal{A}_0 \times \mathcal{Z}_n$ (Lee and Volkov 2011). □

5.1 Identities Satisfied by $\mathcal{L}_3$

Lemma 5.4 *For any $\mathbf{w} \approx \mathbf{w}' \in \mathsf{id}\{\mathcal{L}_3\}$, the following conditions hold:*

(i) $\mathsf{sim}(\mathbf{w}) = \mathsf{sim}(\mathbf{w}')$ *and* $\mathsf{non}(\mathbf{w}) = \mathsf{non}(\mathbf{w}')$;
(ii) *if either* $\mathbf{w}$ *or* $\mathbf{w}'$ *is simple, then* $\mathbf{w} = \mathbf{w}'$;
(iii) *if* $\mathbf{w}$ *and* $\mathbf{w}'$ *are non-simple, then*

$$\mathbf{w} = \mathbf{s}_0\mathbf{w}_1\mathbf{s}_1\mathbf{w}_2\mathbf{s}_2\cdots\mathbf{w}_m\mathbf{s}_m \quad \textit{and} \quad \mathbf{w}' = \mathbf{s}_0\mathbf{w}'_1\mathbf{s}_1\mathbf{w}'_2\mathbf{s}_2\cdots\mathbf{w}'_m\mathbf{s}_m,$$

where

- $\mathbf{s}_0, \mathbf{s}_1, \ldots, \mathbf{s}_m \in \mathscr{A}^+_\varnothing$ *are simple words;*
- $\mathbf{w}_1, \mathbf{w}_2, \ldots, \mathbf{w}_m, \mathbf{w}'_1, \mathbf{w}'_2, \ldots, \mathbf{w}'_m \in \mathscr{A}^+$ *are connected words;*
- $\mathbf{s}_0, \mathbf{w}_1, \mathbf{s}_1, \mathbf{w}_2, \mathbf{s}_2, \ldots, \mathbf{w}_m, \mathbf{s}_m$ *are pairwise disjoint;*
- $\mathbf{s}_0, \mathbf{w}'_1, \mathbf{s}_1, \mathbf{w}'_2, \mathbf{s}_2, \ldots, \mathbf{w}'_m, \mathbf{s}_m$ *are pairwise disjoint;*
- $\mathsf{con}(\mathbf{w}_i) = \mathsf{con}(\mathbf{w}'_i)$ *and* $\mathcal{L}_3 \models \mathbf{w}_i \approx \mathbf{w}'_i$ *for all* i.

Proof Part (i) is routinely verified. By identifying the elements 0 and ef in $\mathcal{L}_3$, the semigroup $\mathcal{A}_0$ is obtained. Hence, $\mathcal{A}_0 \in \mathscr{V}_{\mathsf{sem}}\{\mathcal{L}_3\}$. Since the semigroup $\mathcal{L}_3$ is idempotent-separable by Lemma 2.2(i), parts (ii) and (iii) hold by Lemma 2.3. □

For each $m \geq 2$, define the sets

$$\mathscr{P}^{\uparrow}_m = \left\{ x^{s_0}\left(\prod_{i=1}^{m} y_i^{t_i}\right)x^{s_1} \;\middle|\; s_0, s_1 \geq 1 \text{ and } t_1, t_2, \ldots, t_m \geq 2 \right\},$$

$$\mathscr{P}^{\downarrow}_m = \left\{ x^{s_0}\left(\prod_{i=m}^{1} y_i^{t_i}\right)x^{s_1} \;\middle|\; s_0, s_1 \geq 1 \text{ and } t_1, t_2, \ldots, t_m \geq 2 \right\},$$

$$\mathscr{Q}_m^{\uparrow} = \left\{ x^{s_0}\left(\prod_{i=1}^{m}(y_i^{s_i} h_i y_i^{t_i})\right)x^{t_0} \,\middle|\, s_0, t_0, s_1, t_1, \ldots, s_m, t_m \geq 1 \right\},$$

$$\text{and } \mathscr{Q}_m^{\downarrow} = \left\{ x^{s_0}\left(\prod_{i=m}^{1}(y_i^{s_i} h_i y_i^{t_i})\right)x^{t_0} \,\middle|\, s_0, t_0, s_1, t_1, \ldots, s_m, t_m \geq 1 \right\}.$$

Note that the identities in (5.1) are formed by words from $\mathscr{P}_m^{\uparrow}$ and $\mathscr{P}_m^{\downarrow}$. Identities formed by words from $\mathscr{Q}_m^{\uparrow}$ and $\mathscr{Q}_m^{\downarrow}$ are also required in Chaps. 6 and 9.

Lemma 5.5 *Let* $\mathbf{w} \approx \mathbf{w}' \in \mathsf{id}\{\mathcal{L}_3\}$.

(i) *If* $\mathbf{w} \in \mathscr{P}_m^{\uparrow}$, *then* $\mathbf{w}' \in \mathscr{P}_m^{\uparrow} \cup \mathscr{P}_m^{\downarrow}$.
(ii) *If* $\mathbf{w} \in \mathscr{Q}_m^{\uparrow}$, *then* $\mathbf{w}' \in \mathscr{Q}_m^{\uparrow} \cup \mathscr{Q}_m^{\downarrow}$.

Proof It suffices to establish part (ii) since part (i) is similar. Suppose that $\mathbf{w} \in \mathscr{Q}_m^{\uparrow}$. Then

$$\mathbf{w} = x^{s_0} \cdot y_1^{s_1} h_1 y_1^{t_1} \cdot y_2^{s_2} h_2 y_2^{t_2} \cdots y_m^{s_m} h_m y_m^{t_m} \cdot x^{t_0}$$

for some $s_0, t_0, s_1, t_2, \ldots, s_m, t_m \geq 1$. Further, it follows from Lemma 5.4(i) that

(A) $\mathsf{sim}(\mathbf{w}') = \mathsf{sim}(\mathbf{w}) = \{h_1, \ldots, h_m\}$ and $\mathsf{non}(\mathbf{w}') = \mathsf{non}(\mathbf{w}) = \{x, y_1, \ldots, y_m\}$.

Let $\varphi_1 : \mathscr{A} \to \mathcal{L}_3$ denote the substitution

$$z \mapsto \begin{cases} \mathsf{f} & \text{if } z = x, \\ \mathsf{e} & \text{otherwise.} \end{cases}$$

Then $\mathbf{w}'\varphi_1 = \mathbf{w}\varphi_1 = \mathsf{fef}$, so that $\mathbf{w}'\varphi_1$ is a product of the form

$$\mathsf{f} \cdot \mathsf{f} \cdots \mathsf{f} \cdot \mathsf{e} \cdot \mathsf{e} \cdots \mathsf{e} \cdot \mathsf{f} \cdot \mathsf{f} \cdots \mathsf{f}. \tag{5.2}$$

Therefore, by (A),

(B) $\mathbf{w}' = x^{s_0'} \mathbf{u} x^{t_0'}$ for some $\mathbf{u} \in \{h_1, y_1, h_2, y_2, \ldots, h_m, y_m\}^+$ such that $\mathsf{sim}(\mathbf{u}) = \{h_1, h_2, \ldots, h_m\}$ and some $s_0', t_0' \geq 1$.

Consider any fixed $i \in \{1, 2, \ldots, m\}$. Let $\varphi_2 : \mathscr{A} \to \mathcal{L}_3$ denote the substitution

$$z \mapsto \begin{cases} \mathsf{e} & \text{if } z \in \{h_i, y_i\}, \\ \mathsf{f} & \text{otherwise.} \end{cases}$$

Then $\mathbf{w}'\varphi_2 = \mathbf{w}\varphi_2 = \mathsf{fef}$, so that $\mathbf{w}'\varphi_2$ is a product of the form (5.2), whence all occurrences of h_i and y_i in $\mathbf{w}'$ form a single factor. Therefore, by (A) and (B),

(C) $\mathbf{w}' = x^{s'_0}\mathbf{u}_1 y_i^{s'_i} h_i y_i^{t'_i} \mathbf{u}_2 x^{t'_0}$ for some $\mathbf{u}_1, \mathbf{u}_2 \in \mathscr{A}_\varnothing^+$ and $s'_i, t'_i \geq 0$ with $h_i, y_i \notin \mathsf{con}(\mathbf{u}_1\mathbf{u}_2)$ and $s'_i + t'_i \geq 2$.

Suppose that $s'_i = 0$. Let $\varphi_3 : \mathscr{A} \to \mathcal{L}_3$ denote the substitution

$$z \mapsto \begin{cases} \mathsf{fe} & \text{if } z = h_i, \\ \mathsf{e} & \text{if } z = y_i, \\ \mathsf{f} & \text{otherwise.} \end{cases}$$

Then $\mathbf{w}\varphi_3 = 0 \neq \mathsf{fef} = \mathbf{w}'\varphi_3$, which is impossible. Therefore, $s'_i \geq 1$. A symmetrical argument yields $t'_i \geq 1$. Now since i is arbitrary, it follows from (C) that

$$\mathbf{w}' = x^{s'_0}\mathbf{u}x^{t'_0},$$

where $\mathbf{u}$ is a product of $y_1^{s'_1} h_1 y_1^{t'_1}, y_2^{s'_2} h_2 y_2^{t'_2}, \ldots, y_m^{s'_m} h_m y_m^{t'_m}$ in some order. Hence,

(D) there exists some permutation π of $\{1, 2, \ldots, m\}$ such that

$$\mathbf{w}' = x^{s'_0} \cdot y_{1\pi}^{s'_{1\pi}} h_{1\pi} y_{1\pi}^{t'_{1\pi}} \cdot y_{2\pi}^{s'_{2\pi}} h_{2\pi} y_{2\pi}^{t'_{2\pi}} \cdots y_{m\pi}^{s'_{m\pi}} h_{m\pi} y_{m\pi}^{t'_{m\pi}} \cdot x^{t'_0}.$$

Suppose there exists some $i \in \{1, 2, \ldots, m-1\}$ such that neither $y_i^{s'_i} h_i y_i^{t'_i} \cdot y_{i+1}^{s'_{i+1}} h_{i+1} y_{i+1}^{t'_{i+1}}$ nor $y_{i+1}^{s'_{i+1}} h_{i+1} y_{i+1}^{t'_{i+1}} \cdot y_i^{s'_i} h_i y_i^{t'_i}$ is a factor of $\mathbf{w}'$. Let $\varphi_4 : \mathscr{A} \to \mathcal{L}_3$ denote the substitution

$$z \mapsto \begin{cases} \mathsf{e} & \text{if } z \in \{y_i, h_i, y_{i+1}, h_{i+1}\}, \\ \mathsf{f} & \text{otherwise.} \end{cases}$$

Then $\mathbf{w}\varphi_4 = \mathsf{fef} \neq 0 = \mathbf{w}'\varphi_4$, which is impossible. Hence,

(E) for each $i \in \{1, 2, \ldots, m-1\}$, either $y_i^{s'_i} h_i y_i^{t'_i} \cdot y_{i+1}^{s'_{i+1}} h_{i+1} y_{i+1}^{t'_{i+1}}$ or $y_{i+1}^{s'_{i+1}} h_{i+1} y_{i+1}^{t'_{i+1}} \cdot y_i^{s'_i} h_i y_i^{t'_i}$ is a factor of $\mathbf{w}'$.

It follows from (D) and (E) that the sequence $(1\pi, 2\pi, \ldots, m\pi)$ coincides with either $(1, 2, \ldots, m)$ or $(m, m-1, \ldots, 1)$, whence either $\mathbf{w}' \in \mathscr{Q}_m^{\uparrow}$ or $\mathbf{w}' \in \mathscr{Q}_m^{\downarrow}$. □

Lemma 5.6 *Let* $\mathbf{w} \approx \mathbf{w}'$ *be any identity directly deducible from some identity in* $\mathsf{id}_m\{\mathcal{L}_3\}$. *Suppose that* $\mathbf{w} \in \mathscr{P}_m^{\uparrow}$. *Then* $\mathbf{w}' \in \mathscr{P}_m^{\uparrow}$.

Proof Suppose that $\mathbf{w} \approx \mathbf{w}'$ is directly deducible from $\mathbf{u} \approx \mathbf{u}' \in \mathsf{id}_m\{\mathcal{L}_3\}$. Then there exist words $\mathbf{e}, \mathbf{f} \in \mathscr{A}_\varnothing^+$ and a substitution $\varphi : \mathscr{A} \to \mathscr{A}^+$ such that $\mathbf{w} = \mathbf{e}(\mathbf{u}\varphi)\mathbf{f}$ and $\mathbf{w}' = \mathbf{e}(\mathbf{u}'\varphi)\mathbf{f}$. Since $\mathbf{w} \in \mathscr{P}_m^{\uparrow}$,

$$\mathbf{e}(\mathbf{u}\varphi)\mathbf{f} = \mathbf{w} = x^{s_0} y_1^{t_1} y_2^{t_2} \cdots y_m^{t_m} x^{s_1}$$

for some $s_0, s_1 \geq 1$ and $t_1, t_2, \ldots, t_m \geq 2$. There are three cases.

CASE 1: $\mathbf{u}\varphi$ is simple. Since $\mathbf{u}\varphi \approx \mathbf{u}'\varphi \in \mathsf{id}_m\{\mathcal{L}_3\}$, it follows from Lemma 5.4(ii) that $\mathbf{u}\varphi = \mathbf{u}'\varphi$. Hence, $\mathbf{w}' = \mathbf{w} \in \mathscr{P}_m^{\uparrow}$.

CASE 2: $\mathbf{u}\varphi$ is connected. The connected factors of $\mathbf{w}$ are exhausted by the following:

(A) the non-singleton factors of $x^{s_0}, y_1^{t_1}, y_2^{t_2}, \ldots, y_m^{t_m}, x^{s_1}$;

(B) $x^{r_0} y_1^{t_1} y_2^{t_2} \cdots y_m^{t_m} x^{r_1}$ where $1 \leq r_0 \leq s_0$ and $1 \leq r_1 \leq s_1$.

2.1: $\mathbf{u}\varphi$ belongs to (A). It suffices to assume that $\mathbf{u}\varphi$ is a factor of some $y_i^{t_i}$ since the argument is similar when $\mathbf{u}\varphi$ is a factor of either x^{s_0} or x^{s_1}. Then $\mathbf{u}\varphi = y_i^p$ for some $p \in \{2, 3, \ldots, t_i\}$, so that

$$\mathbf{w} = \underbrace{x^{s_0} y_1^{t_1} y_2^{t_2} \cdots y_{i-1}^{t_{i-1}} y_i^{p'}}_{\mathbf{e}} \cdot y_i^p \cdot \underbrace{y_i^{p''} y_{i+1}^{t_{i+1}} \cdots y_m^{t_m} x^{s_1}}_{\mathbf{f}}$$

for some $p', p'' \geq 0$ with $p' + p + p'' = t_i$. Since $\mathbf{u}\varphi \approx \mathbf{u}'\varphi \in \mathsf{id}\{\mathcal{L}_3\}$, it follows from Lemma 5.4(i) that $\mathsf{sim}(\mathbf{u}\varphi) = \mathsf{sim}(\mathbf{u}'\varphi)$ and $\mathsf{non}(\mathbf{u}\varphi) = \mathsf{non}(\mathbf{u}'\varphi)$. Thus, $\mathbf{u}'\varphi = y_i^q$ for some $q \geq 2$, whence $\mathbf{w}' = \mathbf{e}(\mathbf{u}'\varphi)\mathbf{f} \in \mathscr{P}_m^{\uparrow}$.

2.2: $\mathbf{u}\varphi$ belongs to (B). Then

$$\mathbf{w} = \underbrace{x^{s_0 - r_0}}_{\mathbf{e}} \cdot \underbrace{x^{r_0} y_1^{t_1} y_2^{t_2} \cdots y_m^{t_m} x^{r_1}}_{\mathbf{u}\varphi} \cdot \underbrace{x^{s_1 - r_1}}_{\mathbf{f}}.$$

By assumption, $|\mathsf{con}(\mathbf{u})| < m + 1 = |\mathsf{con}(\mathbf{u}\varphi)|$. Hence, there exists some variable z of $\mathbf{u}$ such that the factor $z\varphi$ of $\mathbf{u}\varphi$ contains at least two distinct variables. It follows that one of the following words is a factor of $z\varphi$:

$$xy_1,\ y_1y_2,\ y_2y_3,\ \ldots,\ y_{m-1}y_m,\ y_mx. \tag{5.3}$$

Since $z \in \mathsf{con}(\mathbf{u}) = \mathsf{con}(\mathbf{u}')$ by Lemma 5.4(i), the word $z\varphi$ is a factor of $\mathbf{u}'\varphi$. Therefore, some word from (5.3) is also a factor of $\mathbf{u}'\varphi$, whence $\mathbf{w}' = \mathbf{e}(\mathbf{u}'\varphi)\mathbf{f} \notin \mathscr{P}_m^{\downarrow}$. Since $\mathbf{w} \approx \mathbf{w}' \in \mathsf{id}\{\mathcal{L}_3\}$ and $\mathbf{w} \in \mathscr{P}_m^{\uparrow}$, it follows from Lemma 5.5(i) that $\mathbf{w}' \in \mathscr{P}_m^{\uparrow} \cup \mathscr{P}_m^{\downarrow}$. Consequently, $\mathbf{w}' \in \mathscr{P}_m^{\uparrow}$.

CASE 3: $\mathbf{u}\varphi$ is neither simple nor connected. Then $\mathbf{u}\varphi$ is a product of two or more pairwise disjoint words, at least one of which is connected. Hence, $\mathbf{u}\varphi$ is a factor of some word from the set

$$\{x^{s_0} y_1^{t_1} y_2^{t_2} \cdots y_m^{t_m},\, y_1^{t_1} y_2^{t_2} \cdots y_m^{t_m} x^{s_1} \mid s_0, s_1 \geq 1 \text{ and } t_1, t_2, \ldots, t_m \geq 2\}$$

with $|\mathsf{con}(\mathbf{u}\varphi)| \geq 2$. It then follows from Lemma 5.4(iii) that $\mathbf{u}'\varphi$ is also a factor of some word from the above set with $|\mathsf{con}(\mathbf{u}'\varphi)| \geq 2$. Therefore, $\mathbf{u}'\varphi$ contains a factor from (5.3). Hence, the same argument from Case 2 yields $\mathbf{w}' \in \mathscr{P}_m^{\uparrow}$. □

5.2 Proof of Theorem 5.1

Let S be any semigroup that satisfies the identities (5.1) for some fixed $n \geq 2$ with $\mathcal{L}_3 \in \mathscr{V}_{\mathsf{sem}}\{S\}$. Seeking a contradiction, suppose that S is finitely based. Then there exists some $m \geq 2$ such that $\mathsf{id}_m\{S\}$ is an identity basis for S. By assumption, S satisfies the identity $\mathbf{p} \approx \mathbf{p}'$ from (5.1), where

$$\mathbf{p} = x^n y_1^n y_2^n \cdots y_m^n x^n \in \mathscr{P}_m^{\uparrow} \text{ and } \mathbf{p}' = x^n y_m^n y_{m-1}^n \cdots y_1^n x^n \in \mathscr{P}_m^{\downarrow}.$$

Therefore, $\mathbf{p} \approx \mathbf{p}'$ is deducible from $\mathsf{id}_m\{S\}$, whence there exists a sequence

$$\mathbf{p} = \mathbf{w}_1, \mathbf{w}_2, \ldots, \mathbf{w}_r = \mathbf{p}'$$

of words where each identity $\mathbf{w}_i \approx \mathbf{w}_{i+1}$ is directly deducible from some identity in $\mathsf{id}_m\{S\}$. It is clear that $\mathbf{w}_1 = \mathbf{p} \in \mathscr{P}_m^{\uparrow}$. Suppose that $\mathbf{w}_i \in \mathscr{P}_m^{\uparrow}$ for some $i \geq 1$. Then since $\mathbf{w}_i \approx \mathbf{w}_{i+1}$ is directly deducible from an identity in $\mathsf{id}_m\{S\} \subseteq \mathsf{id}_m\{\mathcal{L}_3\}$, it follows from Lemma 5.6 that $\mathbf{w}_{i+1} \in \mathscr{P}_m^{\uparrow}$. Therefore, $\mathbf{w}_1, \mathbf{w}_2, \ldots, \mathbf{w}_r \in \mathscr{P}_m^{\uparrow}$ by induction, whence the contradiction $\mathbf{p}' \in \mathscr{P}_m^{\uparrow}$ is obtained.

5.3 Specialized Versions of Theorem 5.1

Lemma 5.7 *Let S be any semigroup in $\mathbf{P}_n$ for some $n \geq 2$. Then $\mathcal{L}_3 \notin \mathscr{V}_{\mathsf{sem}}\{S\}$ if and only if S satisfies the identity*

$$(x^n y^n x^n)^{n+1} \approx x^n y^n x^n. \tag{5.4}$$

Proof The semigroup $\mathcal{L}_3$ violates the identity (5.4) under the substitution $(x, y) \mapsto (\mathsf{f}, \mathsf{e})$. Hence, if S satisfies the identity (5.4), then $\mathcal{L}_3 \notin \mathscr{V}_{\mathsf{sem}}\{S\}$. Conversely, suppose that S violates the identity (5.4). Then there exist idempotents $a, b \in S$ such that $(bab)^{n+1} \neq bab$. It follows that the elements $a, b, ab, ba, bab \in S$ are distinct. Let T denote the subsemigroup of S generated by a and b. Seeking a contradiction, suppose the existence of an element $z \in \{a, b, ab, ba, bab\} \cap TabaT$. If $z = b$, then

$$bab = baz \in baTabaT \subseteq TabaT.$$

If $z \in \{a, ab, ba, bab\}$, then

$$bab = bzb \in bTabaTb \subseteq TabaT.$$

Thus, $bab \in TabaT$ in any case. Since $bab = bbabb \in bTabaTb$, it follows that $bab = (bab)^{k+1}$ for some $k \geq 1$. But this implies that $bab = (bab)^{nk+1} \stackrel{\wp_n}{=} (bab)^{n+1}$, which contradicts the assumption on a and b. Therefore, the element z does not exist, whence $\{a, b, ab, ba, bab\} \cap TabaT = \varnothing$. It follows that $TabaT$ is an ideal of T. Now the quotient $T/(TabaT)$ is isomorphic to the semigroup $\mathcal{L}_3$, so that $\mathcal{L}_3 \in \mathscr{V}_{\mathsf{sem}}\{S\}$. □

Theorem 1.26 is easily seen to be a consequence of Theorem 5.1 and Lemma 5.7.

For each element a in a finite semigroup S, there exists precisely one idempotent of S in the set $\{a^r \mid r = 1, 2, 3, \ldots\}$, commonly denoted by a^ω. Identities that involve factors of the form $\mathbf{w}^\omega$ are call *pseudoidentities*. Refer to the monograph of Almeida (1994) for more information on pseudoidentities and finite semigroup theory. In general, a finite algebra has a finite identity basis if and only if it has a finite pseudoidentity basis (Almeida 1994, Corollary 4.3.8).

Given any finite semigroup S, let $\mathsf{PV}\{S\}$ denote the pseudovariety generated by S. A pseudoidentity $\mathbf{u} \approx \mathbf{v}$ is an *exclusion pseudoidentity* for a finite semigroup S if the following equivalence holds for every finite semigroup T:

$$T \models \mathbf{u} \approx \mathbf{v} \quad \Longleftrightarrow \quad S \notin \mathsf{PV}\{T\}.$$

For example, it follows from Lemma 5.7 that

$$(x^\omega y^\omega x^\omega)^{\omega+1} \approx x^\omega y^\omega x^\omega$$

is an exclusion pseudoidentity for $\mathcal{L}_3$. Similarly, it follows from Lemma 3.4 that

$$((x^\omega y)^\omega (yx^\omega)^\omega)^\omega \approx (x^\omega yx^\omega)^\omega$$

is an exclusion pseudoidentity for $\mathcal{A}_2$. Refer to Rhodes and Steinberg (2009) for more information on exclusion pseudoidentities and to Almeida (1994), Lee (2017b), and Lee et al. (2019, 2022) for more examples.

Theorem 1.26 can be stated more concisely if it is only applied to finite semigroups.

Corollary 5.8 *Any finite semigroup that satisfies the pseudoidentities*

$$x^\omega y_1^\omega y_2^\omega \cdots y_m^\omega x^\omega \approx x^\omega y_m^\omega y_{m-1}^\omega \cdots y_1^\omega x^\omega, \quad m \in \{2, 3, 4, \ldots\}$$

but violates the pseudoidentity $(x^\omega y^\omega x^\omega)^{\omega+1} \approx x^\omega y^\omega x^\omega$ *is non-finitely based.*

5.4 Summary

The identity (5.4) that precisely excludes $\mathcal{L}_3$ from any subvariety of $\mathbf{P}_n$ was first found in Lee (2012a). The sufficient condition in Theorem 5.1 for the non-finite basis property was published in Lee (2018a).

Semigroups Without Irredundant Identity Bases 6

As described in Sect. 1.4.3, the present chapter is concerned with semigroups without irredundant identity bases. In Sect. 6.1, a sufficient condition is established under which a non-finitely based finite algebra of any signature type has no irredundant identity bases. Recall from Corollary 5.3 that the semigroup

$$\mathcal{L}_{3,n} = \mathcal{L}_3 \times \mathcal{Z}_n$$

is non-finitely based. In Sects. 6.2–6.5, identities satisfied by the semigroup $\mathcal{L}_{3,n}$ are investigated and an explicit identity basis for it is exhibited. In Sect. 6.6, the sufficient condition from Sect. 6.1 is applied to show that $\mathcal{L}_{3,n}$ has no irredundant identity bases.

6.1 Sufficient Condition for the Nonexistence of Irredundant Identity Bases

For any identity $\sigma : \mathbf{t} \approx \mathbf{t}'$, let $\#\sigma$ denote the number of distinct variables occurring in σ, that is, $\#\sigma = |\,\mathsf{con}(\mathbf{tt}')|$. For any set Σ of identities, define $\#\Sigma = \max\{\#\sigma \mid \sigma \in \Sigma\}$. Write $\#\Sigma = \infty$ to indicate that $\#\Sigma$ does not exist.

Lemma 6.1 (Birkhoff 1935) *Suppose that Σ is any identity basis for a non-finitely based finite algebra. Then $\#\Sigma = \infty$.*

Remark 6.2 Lemma 6.1 need not hold for an infinite algebra. For instance, the set Σ of identities $(x^p y^p)^2 \approx (y^p x^p)^2$, where p ranges over the primes, is an identity basis for some non-finitely based infinite monoid (Isbell 1970), but $\#\Sigma < \infty$.

E. W. H. Lee, *Advances in the Theory of Varieties of Semigroups*, Frontiers in Mathematics, https://doi.org/10.1007/978-3-031-16497-2_6

Theorem 6.3 *Let A be any non-finitely based finite algebra. Suppose that*

(a) *some identity basis $\Lambda = \Lambda_0 \cup \{\lambda_1, \lambda_2, \lambda_3, \ldots\}$ for A exists, where Λ_0 is some finite subset of $\mathsf{id}\{A\}$ and $\lambda_1, \lambda_2, \lambda_3, \ldots$ are identities such that $\Lambda_0 \cup \{\lambda_{k+1}\} \vdash \lambda_k$ for all $k \geq 1$;*
(b) *any identity basis for A contains identities $\theta_1, \theta_2, \theta_3, \ldots$ such that $\Lambda_0 \cup \{\theta_k\} \vdash \lambda_k$ for all $k \geq 1$.*

Then A has no irredundant identity bases.

Proof Let Θ be any identity basis for A. Then it suffices to show that some proper subset of Θ is also an identity basis for A. Since $\Lambda_0 \subseteq \mathsf{id}\{A\}$ is finite, there exists some finite $\Theta_0 \subseteq \Theta$ such that

(A) $\Theta_0 \vdash \Lambda_0$.

By (b), there exist identities $\theta_1, \theta_2, \theta_3, \ldots$ in Θ such that $\Lambda_0 \cup \{\theta_k\} \vdash \lambda_k$ for all $k \geq 1$. Since

$$\begin{aligned}\Theta_0 \cup \{\theta_1, \theta_2, \theta_3, \ldots\} &\vdash \Lambda_0 \cup \{\theta_1, \theta_2, \theta_3, \ldots\} && \text{by (A)}\\ &\vdash \Lambda_0 \cup \{\lambda_1, \lambda_2, \lambda_3, \ldots\},\end{aligned}$$

the subset $\Theta_0 \cup \{\theta_1, \theta_2, \theta_3, \ldots\}$ of Θ is also an identity basis for A. The proof is complete if $\Theta \neq \Theta_0 \cup \{\theta_1, \theta_2, \theta_3, \ldots\}$. Therefore, assume that $\Theta = \Theta_0 \cup \{\theta_1, \theta_2, \theta_3, \ldots\}$. Now since the algebra A is finite and non-finitely based, it follows from Lemma 6.1 that the sequence $\#\theta_1, \#\theta_2, \#\theta_3, \ldots$ is unbounded from above and so contains a subsequence $\#\theta_{k_1}, \#\theta_{k_2}, \#\theta_{k_3}, \ldots$ such that

(B) $\#\Theta_0 < \#\theta_{k_1} < \#\theta_{k_2} < \#\theta_{k_3} < \cdots$.

Then the subset $\Theta_0 \cup \{\theta_{k_2}, \theta_{k_3}, \theta_{k_4}, \ldots\}$ of Θ is an identity basis for A because

$$\begin{aligned}\Theta_0 \cup \{\theta_{k_2}, \theta_{k_3}, \theta_{k_4}, \ldots\} &\vdash \Lambda_0 \cup \{\theta_{k_2}, \theta_{k_3}, \theta_{k_4}, \ldots\} && \text{by (A)}\\ &\vdash \Lambda_0 \cup \{\lambda_{k_2}, \lambda_{k_3}, \lambda_{k_4}, \ldots\} && \text{by (b)}\\ &\vdash \Lambda_0 \cup \{\lambda_1, \lambda_2, \lambda_3, \ldots\} && \text{by (a).}\end{aligned}$$

By (B), the subset $\Theta_0 \cup \{\theta_{k_2}, \theta_{k_3}, \theta_{k_4}, \ldots\}$ of Θ does not contain θ_{k_1} and so is proper. □

6.2 Identities Satisfied by $\mathcal{L}_{3,n}$

In this section, some important identities satisfied by the semigroup $\mathcal{L}_{3,n}$ are presented. The first lemma contains identities that are eventually shown in Sect. 6.5 to form an identity basis for $\mathcal{L}_{3,n}$.

Lemma 6.4 *The semigroup $\mathcal{L}_{3,n}$ satisfies the identities*

$$x^{n+2} \approx x^2, \quad x^{n+1}yx \approx xyx, \quad xyx^{n+1} \approx xyx, \tag{6.1a}$$

$$x^2yx \approx xyx^2, \tag{6.1b}$$

$$xyxzx \approx xzxyx, \tag{6.1c}$$

$$(\mathbf{h}_1 xy\mathbf{h}_2)^{n+1} \approx (\mathbf{h}_1 yx\mathbf{h}_2)^{n+1}, \tag{6.1d}$$

$$x^p\mathbf{h}_1 y^q\mathbf{h}_2 x\mathbf{h}_3 y \approx y^q\mathbf{h}_1 x^p\mathbf{h}_2 y\mathbf{h}_3 x, \tag{6.1e}$$

$$x\mathbf{h}_1 y\mathbf{h}_2 x\mathbf{h}_3 y \approx xy^n\mathbf{h}_1 y\mathbf{h}_2 x\mathbf{h}_3 y, \tag{6.1f}$$

$$x\mathbf{h}_1 y\mathbf{h}_2 x\mathbf{h}_3 y \approx y^n x\mathbf{h}_1 y\mathbf{h}_2 x\mathbf{h}_3 y, \tag{6.1g}$$

$$x\mathbf{h}_1 y\mathbf{h}_2 x\mathbf{h}_3 y \approx x\mathbf{h}_1 x^n y\mathbf{h}_2 x\mathbf{h}_3 y, \tag{6.1h}$$

$$x\mathbf{h}_1 y\mathbf{h}_2 x\mathbf{h}_3 y \approx x\mathbf{h}_1 yx^n\mathbf{h}_2 x\mathbf{h}_3 y, \tag{6.1i}$$

$$x\mathbf{h}_1 y\mathbf{h}_2 x\mathbf{h}_3 y \approx x\mathbf{h}_1 y\mathbf{h}_2 xy^n\mathbf{h}_3 y, \tag{6.1j}$$

$$x\mathbf{h}_1 y\mathbf{h}_2 x\mathbf{h}_3 y \approx x\mathbf{h}_1 y\mathbf{h}_2 y^n x\mathbf{h}_3 y, \tag{6.1k}$$

$$x\mathbf{h}_1 y\mathbf{h}_2 x\mathbf{h}_3 y \approx x\mathbf{h}_1 y\mathbf{h}_2 x\mathbf{h}_3 x^n y, \tag{6.1l}$$

$$x\mathbf{h}_1 y\mathbf{h}_2 x\mathbf{h}_3 y \approx x\mathbf{h}_1 y\mathbf{h}_2 x\mathbf{h}_3 yx^n, \tag{6.1m}$$

$$x\left(\prod_{i=1}^{m}(y_i h_i y_i)\right)x \approx x\left(\prod_{i=m}^{1}(y_i h_i y_i)\right)x, \tag{6.1n}$$

where $\mathbf{h}_i \in \{\varnothing, h_i\}$, $p, q \in \{1, 2, 3, \ldots\}$, *and* $m \in \{2, 3, 4, \ldots\}$.

Note that the identities (6.1n) are formed by words from the sets $\mathscr{Q}_m^{\uparrow}$ and $\mathscr{Q}_m^{\downarrow}$ introduced in Sect. 5.1. It is convenient to write

$$\mathfrak{q}_m^{\uparrow} = x\left(\prod_{i=1}^{m}(y_i h_i y_i)\right)x \text{ and } \mathfrak{q}_m^{\downarrow} = x\left(\prod_{i=m}^{1}(y_i h_i y_i)\right)x.$$

Proof of Lemma 6.4 It is routinely checked that the group $\mathcal{Z}_n$ satisfies the identities (6.1) and that the semigroup $\mathcal{L}_3$ satisfies the identities (6.1a)–(6.1m). To verify that $\mathcal{L}_3$ satisfies the identity $\mathfrak{q}_m^{\uparrow} \approx \mathfrak{q}_m^{\downarrow}$ from (6.1n), consider any substitution $\varphi : \mathscr{A} \to \mathcal{L}_3$. If $x\varphi \in$

$\{\mathsf{0}, \mathsf{ef}, \mathsf{fe}, \mathsf{fef}\}$, then it is routinely checked that $\mathsf{q}_m^{\uparrow}\varphi = \mathsf{0} = \mathsf{q}_m^{\downarrow}\varphi$. Further, if $x\varphi = \mathsf{e}$, then

$$\mathsf{q}_m^{\uparrow}\varphi = \mathsf{q}_m^{\downarrow}\varphi = \begin{cases} \mathsf{e} & \text{if } y_i\varphi = h_i\varphi = \mathsf{e} \text{ for all } i, \\ \mathsf{0} & \text{otherwise.} \end{cases}$$

Therefore, it suffices to assume that $x\varphi = f$, so that

$$\mathsf{q}_m^{\uparrow}\varphi = f\bigg(\prod_{i=1}^{m}((y_i\varphi)(h_i\varphi)(y_i\varphi))\bigg)f \quad \text{and} \quad \mathsf{q}_m^{\downarrow}\varphi = f\bigg(\prod_{i=m}^{1}((y_i\varphi)(h_i\varphi)(y_i\varphi))\bigg)f.$$

It then follows that $\mathsf{q}_m^{\uparrow}\varphi, \mathsf{q}_m^{\downarrow}\varphi \in \mathsf{f}\mathcal{L}_3\mathsf{f} = \{\mathsf{0}, \mathsf{f}, \mathsf{fef}\}$, whence there are three cases.

CASE 1: $\mathsf{q}_m^{\uparrow}\varphi = \mathsf{f}$. Then $y_i\varphi = h_i\varphi = \mathsf{f}$ for all i, so that $\mathsf{q}_m^{\downarrow}\varphi = \mathsf{f}$.
CASE 2: $\mathsf{q}_m^{\uparrow}\varphi = \mathsf{fef}$. Then $\prod_{i=1}^{m}((y_i\varphi)(h_i\varphi)(y_i\varphi)) = \mathsf{f}^{s_1}\mathsf{e}^t\mathsf{f}^{s_2}$ for some $s_1, s_2 \geq 0$ and $t \geq 1$. Hence, $\mathsf{q}_m^{\downarrow}\varphi = \mathsf{f}(\mathsf{f}^{s_2}\mathsf{e}^t\mathsf{f}^{s_1})\mathsf{f} = \mathsf{fef}$.
CASE 3: $\mathsf{q}_m^{\uparrow}\varphi = 0$. If $\mathsf{q}_m^{\downarrow}\varphi \in \{\mathsf{f}, \mathsf{fef}\}$, then following the arguments in Cases 1 and 2 yields the contradiction $\mathsf{q}_m^{\uparrow}\varphi \in \{\mathsf{f}, \mathsf{fef}\}$. Therefore, $\mathsf{q}_m^{\downarrow}\varphi = \mathsf{0}$. □

Lemma 6.5 *The following identities are deducible from* $\{(6.1b), (6.1e)\}$:

$$\bigg(\prod_{i=1}^{m} x_i^{c_i}\bigg)\mathbf{h}\bigg(\prod_{i=1}^{m} x_i\bigg)^2 \approx \bigg(\prod_{i=1}^{m} x_i^{c_i+1}\bigg)\mathbf{h}\bigg(\prod_{i=1}^{m} x_i\bigg), \tag{6.2a}$$

$$\bigg(\prod_{i=1}^{m} x_i^{c_i}\bigg)\bigg(\prod_{i=1}^{m} x_i\bigg)\mathbf{h}\bigg(\prod_{i=1}^{m} x_i\bigg) \approx \bigg(\prod_{i=1}^{m} x_i^{c_i+1}\bigg)\mathbf{h}\bigg(\prod_{i=1}^{m} x_i\bigg), \tag{6.2b}$$

where $\mathbf{h} \in \{\varnothing, h\}$ *and* $c_i, m \in \{1, 2, 3, \ldots\}$. *Hence, the semigroup* $\mathcal{L}_{3,n}$ *satisfies the identities* (6.2).

Proof This is divided into two cases.

CASE 1: $\{(6.1b), (6.1e)\} \vdash (6.2a)$. It suffices to show that for each $m \geq 1$, the identities

$$(\mathtt{A}_m) \quad \bigg(\prod_{i=1}^{m} x_i^{c_i}\bigg)\mathbf{h}\bigg(\prod_{i=1}^{m} x_i\bigg)^2 \approx \bigg(\prod_{i=1}^{m} x_i^{c_i+1}\bigg)\mathbf{h}\bigg(\prod_{i=1}^{m} x_i\bigg),$$

where $\mathbf{h} \in \{\varnothing, h\}$ and $c_i \in \{1, 2, 3, \ldots\}$, are deducible from $\{(6.1b), (6.1e)\}$. The deduction $(6.1b) \vdash (\mathtt{A}_1)$ clearly holds. Suppose that the deduction $\{(6.1b), (6.1e)\} \vdash (\mathtt{A}_{m-1})$ holds. Then

$$
\begin{aligned}
\left(\prod_{i=1}^{m} x_i^{c_i}\right)\mathbf{h}\left(\prod_{i=1}^{m} x_i\right)^2 &= x_1^{c_1}\left(\prod_{i=2}^{m-1} x_i^{c_i}\right)x_m^{c_m}\mathbf{h}\left(\prod_{i=1}^{m-1} x_i\right)x_m x_1\left(\prod_{i=2}^{m-1} x_i\right)x_m \\
&\overset{(6.1e)}{\approx} x_m^{c_m}\left(\prod_{i=2}^{m-1} x_i^{c_i}\right)x_1^{c_1}\mathbf{h}\left(\prod_{i=1}^{m-1} x_i\right)x_m^2\left(\prod_{i=2}^{m-1} x_i\right)x_1 \\
&\overset{(6.1b)}{\approx} x_m^{c_m+1}\left(\prod_{i=2}^{m-1} x_i^{c_i}\right)x_1^{c_1}\mathbf{h}\left(\prod_{i=1}^{m-1} x_i\right)x_m\left(\prod_{i=2}^{m-1} x_i\right)x_1 \\
&\overset{(6.1e)}{\approx} x_1^{c_1}\left(\prod_{i=2}^{m-1} x_i^{c_i}\right)x_m^{c_m+1}\mathbf{h}\left(\prod_{i=1}^{m-1} x_i\right)x_1\left(\prod_{i=2}^{m-1} x_i\right)x_m \\
&= \left(\prod_{i=1}^{m-1} x_i^{c_i}\right)x_m^{c_m+1}\mathbf{h}\left(\prod_{i=1}^{m-1} x_i\right)^2 x_m \\
&\overset{(\mathtt{A}_{m-1})}{\approx} \left(\prod_{i=1}^{m-1} x_i^{c_i+1}\right)x_m^{c_m+1}\mathbf{h}\left(\prod_{i=1}^{m-1} x_i\right)x_m \\
&= \left(\prod_{i=1}^{m} x_i^{c_i+1}\right)\mathbf{h}\left(\prod_{i=1}^{m} x_i\right).
\end{aligned}
$$

Therefore, the deduction $\{(6.1b), (6.1e)\} \vdash (\mathtt{A}_m)$ also holds.

CASE 2: $\{(6.1b), (6.1e)\} \vdash (6.2b)$. By an argument symmetrical to Case 1, for each $m \geq 1$, the identities

$$
(\mathtt{B}_m) \quad \left(\prod_{i=1}^{m} x_i\right)^2\mathbf{h}\left(\prod_{i=1}^{m} x_i^{c_i}\right) \approx \left(\prod_{i=1}^{m} x_i\right)\mathbf{h}\left(\prod_{i=1}^{m} x_i^{c_i+1}\right),
$$

where $\mathbf{h} \in \{\varnothing, h\}$ and $c_i \in \{1, 2, 3, \ldots\}$, are deducible from $\{(6.1b), (6.1e)\}$. Since

$$
\begin{aligned}
\left(\prod_{i=1}^{m} x_i^{c_i}\right)\left(\prod_{i=1}^{m} x_i\right)\mathbf{h}\left(\prod_{i=1}^{m} x_i\right) &\overset{(6.1b)}{\approx} \left(\prod_{i=1}^{m} x_i\right)\left(\prod_{i=1}^{m} x_i\right)\mathbf{h}\left(\prod_{i=1}^{m} x_i^{c_i}\right) \\
&\overset{(\mathtt{B}_m)}{\approx} \left(\prod_{i=1}^{m} x_i\right)\mathbf{h}\left(\prod_{i=1}^{m} x_i^{c_i+1}\right) \\
&\overset{(6.1b)}{\approx} \left(\prod_{i=1}^{m} x_i^{c_i+1}\right)\mathbf{h}\left(\prod_{i=1}^{m} x_i\right),
\end{aligned}
$$

the deductions $\{(6.1b), (6.1e)\} \vdash \{(6.1b), (\mathtt{B}_m)\} \vdash (6.2b)$ hold. □

6.3 Sandwich Identities

In the present section, a specific type of identities, called *sandwich identities*, is introduced. It is shown in Lemma 6.10 that the semigroup $\mathcal{L}_{3,n}$ has an identity basis that consists of (6.1) and sandwich identities.

Let $\ll$ be a fixed total order on the alphabet $\mathscr{A}$. For any finite nonempty subset $\mathscr{X}$ of $\mathscr{A}$, write $\mathscr{X} = \{x_1 \ll x_2 \ll \cdots \ll x_r\}$ to indicate that

$$\mathscr{X} = \{x_1, x_2, \ldots, x_r\} \text{ and } x_1 \ll x_2 \ll \cdots \ll x_r.$$

For such a set $\mathscr{X}$, define

$$\mathscr{X}^{\boxplus} = \{x_1^{c_1} x_2^{c_2} \cdots x_r^{c_r} \mid 1 \leq c_1, c_2, \ldots, c_r \leq n\}.$$

Then the shortest word in $\mathscr{X}^{\boxplus}$ is

$$\vec{\mathscr{X}} = x_1 x_2 \cdots x_r.$$

For any word $\mathbf{w} \in \mathscr{A}^+$ with $\mathsf{con}(\mathbf{w}) = \{x_1 \ll x_2 \ll \cdots \ll x_r\}$, define $\min(\mathbf{w}) = x_1$.

Any word $\mathbf{w} \in \mathscr{A}^+$ can be uniquely decomposed into a product

$$\mathbf{w} = \mathbf{w}_1 \mathbf{w}_2 \cdots \mathbf{w}_k,$$

where $\mathbf{w}_1, \mathbf{w}_2, \ldots, \mathbf{w}_k \in \mathscr{A}^+$ are pairwise disjoint words, each of which is either a singleton or a connected word; this is called the *natural decomposition* of $\mathbf{w}$.

Lemma 6.6 *Let* $\mathbf{w} \approx \mathbf{w}' \in \mathsf{id}\{\mathcal{L}_{3,n}\}$.

(i) *If either* $\mathbf{w}$ *or* $\mathbf{w}'$ *is simple, then* $\mathbf{w} = \mathbf{w}'$.
(ii) *If* $\mathbf{w}$ *and* $\mathbf{w}'$ *are non-simple words, then the natural decompositions of these words are*

$$\mathbf{w} = \mathbf{w}_1 \mathbf{w}_2 \cdots \mathbf{w}_k \ \textit{and} \ \mathbf{w}' = \mathbf{w}_1' \mathbf{w}_2' \cdots \mathbf{w}_k',$$

where $\mathbf{w}_i \approx \mathbf{w}_i' \in \mathsf{id}\{\mathcal{L}_{3,n}\}$ *for all* i.

Consequently, the connected identities in $\mathsf{id}\{\mathcal{L}_{3,n}\}$ *constitute an identity basis for* $\mathcal{L}_{3,n}$.

Proof Since the semigroup $\mathcal{A}_0$ is obtained by identifying the elements 0 and ef from $\mathcal{L}_3$, the inclusion $\mathcal{A}_0 \in \mathscr{V}_{\mathsf{sem}}\{\mathcal{L}_3\}$ holds. By Lemma 2.2, the semigroup $\mathcal{L}_{3,n}$ is idempotent-separable. Hence, the result follows from Lemma 2.3. □

In this chapter, a connected word $\mathbf{s} \in \mathscr{A}^+$ is called a *sandwich* if

$$\mathbf{s} = \mathbf{x} \prod_{i=1}^{\ell} (\mathbf{u}_i \vec{\mathscr{X}}) \tag{6.3}$$

for some $\ell \geq 1$, finite nonempty $\mathscr{X} \subseteq \mathscr{A}$, $\mathbf{x} \in \mathscr{X}^{\boxplus}$, and $\mathbf{u}_i \in \mathscr{A}_{\varnothing}^+$ such that

(S1) $\vec{\mathscr{X}}, \mathbf{u}_1, \mathbf{u}_2, \ldots, \mathbf{u}_\ell$ are pairwise disjoint;
(S2) if $\ell \geq 2$, then $\mathbf{u}_i \neq \varnothing$ for all i and $\min(\mathbf{u}_1) \ll \min(\mathbf{u}_2) \ll \cdots \ll \min(\mathbf{u}_\ell)$.

The *level* of the sandwich $\mathbf{s}$ in (6.3) is the positive integer ℓ.

Remark 6.7 In (S2), the only scenario in which some of the factors $\mathbf{u}_1, \mathbf{u}_2, \ldots, \mathbf{u}_\ell$ of $\mathbf{s}$ can be empty is when $\ell = 1$ and $\mathbf{u}_1 = \varnothing$, that is, $\mathbf{s} = \mathbf{x}\vec{\mathscr{X}}$.

An identity $\mathbf{s} \approx \mathbf{s}'$ is a *sandwich identity* if $\mathbf{s}$ and $\mathbf{s}'$ are sandwiches. Denote by

$$\mathsf{id}_{\mathrm{SAN}}\{\mathcal{L}_{3,n}\}$$

the set of all sandwich identities satisfied by the semigroup $\mathcal{L}_{3,n}$.

Lemma 6.8 *Suppose that* $\mathbf{w}$ *is any connected word. Then the identities* (6.1) *can be used to convert* $\mathbf{w}$ *into some word* $\mathbf{w}'$ *such that* $\mathsf{h}(\mathbf{w}') = \mathsf{t}(\mathbf{w}')$.

Proof Suppose that $\mathsf{h}(\mathbf{w}) \neq \mathsf{t}(\mathbf{w})$. Then since the word $\mathbf{w}$ is connected, there exist distinct variables $\mathsf{h}(\mathbf{w}) = x_1, x_2, \ldots, x_r = \mathsf{t}(\mathbf{w})$ occurring in $\mathbf{w}$ in the following overlapping manner:

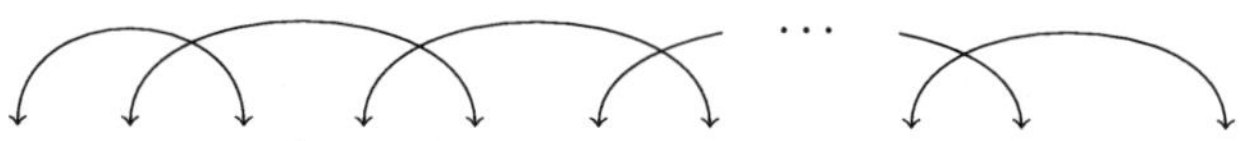

$$\mathbf{w} = x_1\, \mathbf{a}_1\, x_2\, \mathbf{a}_2\, x_1\, \mathbf{b}_1\, x_3\, \mathbf{a}_3\, x_2\, \mathbf{b}_2\, x_4\, \mathbf{a}_4\, x_3\, \mathbf{b}_3 \cdots x_r\, \mathbf{a}_r\, x_{r-1}\, \mathbf{b}_{r-1}\, x_r,$$

where $\mathbf{a}_i, \mathbf{b}_i \in \mathscr{A}_{\varnothing}^+$. (Note that $\mathbf{a}_i$ follows the first x_i and $\mathbf{b}_i$ follows the last x_i.) Hence,

$$\begin{aligned}
\mathbf{w} &\overset{(6.1\mathrm{m})}{\approx} x_1\mathbf{a}_1x_2\mathbf{a}_2x_1\mathbf{b}_1x_3\mathbf{a}_3x_2x_1^n\mathbf{b}_2x_4\mathbf{a}_4x_3\mathbf{b}_3\cdots x_r\mathbf{a}_rx_{r-1}\mathbf{b}_{r-1}x_r \\
&\overset{(6.1\mathrm{m})}{\approx} x_1\mathbf{a}_1x_2\mathbf{a}_2x_1\mathbf{b}_1x_3\mathbf{a}_3x_2x_1^n\mathbf{b}_2x_4\mathbf{a}_4x_3x_1^n\mathbf{b}_3\cdots x_r\mathbf{a}_rx_{r-1}\mathbf{b}_{r-1}x_r \\
&\ \ \vdots \\
&\overset{(6.1\mathrm{m})}{\approx} x_1\mathbf{a}_1x_2\mathbf{a}_2x_1\mathbf{b}_1x_3\mathbf{a}_3x_2x_1^n\mathbf{b}_2x_4\mathbf{a}_4x_3x_1^n\mathbf{b}_3\cdots x_r\mathbf{a}_rx_{r-1}x_1^n\mathbf{b}_{r-1}x_r
\end{aligned}$$

$$\begin{aligned} &\overset{(6.1\text{m})}{\approx} x_1\mathbf{a}_1x_2\mathbf{a}_2x_1\mathbf{b}_1x_3\mathbf{a}_3x_2x_1^n\mathbf{b}_2x_4\mathbf{a}_4x_3x_1^n\mathbf{b}_3\cdots x_r\mathbf{a}_rx_{r-1}x_1^n\mathbf{b}_{r-1}x_rx_1^n \\ &= \mathbf{w}', \end{aligned}$$

where $\mathsf{h}(\mathbf{w}') = \mathsf{t}(\mathbf{w}')$. □

Lemma 6.9 *Suppose that* $\mathbf{w}$ *is any connected word. Then the identities* (6.1) *can be used to convert* $\mathbf{w}$ *into some sandwich* $\mathbf{s}$ *with* $\mathsf{sim}(\mathbf{w}) = \mathsf{sim}(\mathbf{s})$ *and* $\mathsf{non}(\mathbf{w}) = \mathsf{non}(\mathbf{s})$.

Proof By Lemma 6.5, it suffices to convert the word $\mathbf{w}$, using the identities (6.1) and (6.2), into some sandwich $\mathbf{s}$ with $\mathsf{sim}(\mathbf{w}) = \mathsf{sim}(\mathbf{s})$ and $\mathsf{non}(\mathbf{w}) = \mathsf{non}(\mathbf{s})$. By Lemma 6.8, it can be assumed that $x_1 = \mathsf{h}(\mathbf{w}) = \mathsf{t}(\mathbf{w})$. Then $\mathbf{w}$ can be written as

$$\mathbf{w} = x_1\prod_{i=1}^{m_1}(\mathbf{w}_{1,i}x_1),$$

where $m_1 \geq 1$ and $\mathbf{w}_{1,i} \in \mathscr{A}_\varnothing^+$ with $x_1 \notin \mathsf{con}(\mathbf{w}_{1,i})$ for all i. If $m_1 = 1$, then $\mathbf{w}$ is already a sandwich. Therefore, assume that $m_1 \geq 2$.

Suppose that $\mathbf{w}_{1,i}$ and $\mathbf{w}_{1,j}$ are not disjoint with $i \neq j$, say $x_2 \in \mathsf{con}(\mathbf{w}_{1,i}) \cap \mathsf{con}(\mathbf{w}_{1,j})$. Then $\mathbf{w} = x_1\mathbf{h}_1x_2\mathbf{h}_2x_1\mathbf{h}_3x_2\mathbf{h}_4x_1$ for some $\mathbf{h}_i \in \mathscr{A}_\varnothing^+$, whence the identities (6.1f)–(6.1m) can be applied to perform the replacement $(x_1, x_2) \mapsto (x_1x_2^n, x_1^nx_2)$. The resulting word has the form

$$x_1^{c_{0,1}}x_2^{c_{0,2}}\prod_{i=1}^{m_2}(\mathbf{w}_{2,i}x_1^{c_{i,1}}x_2^{c_{i,2}}),$$

where $m_2 \geq 2$, $c_{i,j} \geq 1$, and $\mathbf{w}_{2,i} \in \mathscr{A}_\varnothing^+$ with $x_1, x_2 \notin \mathsf{con}(\mathbf{w}_{2,i})$ for all i. Similarly, if $\mathbf{w}_{2,i}$ and $\mathbf{w}_{2,j}$ are not disjoint with $i \neq j$, say $x_3 \in \mathsf{con}(\mathbf{w}_{2,i}) \cap \mathsf{con}(\mathbf{w}_{2,j})$, then the identities (6.1f)–(6.1m) can be applied to perform the replacement $(x_2^{c_{i,2}}, x_3) \mapsto (x_2^{c_{i,2}}x_3^n, x_1^nx_2^nx_3)$. The resulting word has the form

$$x_1^{c_{0,1}}x_2^{c_{0,2}}x_3^{c_{0,3}}\prod_{i=1}^{m_3}(\mathbf{w}_{3,i}x_1^{c_{i,1}}x_2^{c_{i,2}}x_3^{c_{i,3}}),$$

where $m_3 \geq 2$, $c_{i,j} \geq 1$, and $\mathbf{w}_{3,i} \in \mathscr{A}_\varnothing^+$ with $x_1, x_2, x_3 \notin \mathsf{con}(\mathbf{w}_{3,i})$ for all i. This can be repeated until a word of the form

$$x_1^{c_{0,1}}x_2^{c_{0,2}}\cdots x_r^{c_{0,r}}\prod_{i=1}^{m_r}(\mathbf{w}_{r,i}x_1^{c_{i,1}}x_2^{c_{i,2}}\cdots x_r^{c_{i,r}})$$

is obtained, where $m_r \geq 2$, $c_{i,j} \geq 1$, and $\mathbf{w}_{r,1}, \mathbf{w}_{r,2}, \ldots, \mathbf{w}_{r,m_r} \in \mathscr{A}_\varnothing^+$ are pairwise disjoint words with $x_1, x_2, \ldots, x_r \notin \mathsf{con}(\mathbf{w}_{r,i})$ for all i. It is easily seen that the identities (6.1a) and (6.1b) can be used to convert this word into

$$\mathbf{w}' = x_1^{c_1} x_2^{c_2} \cdots x_r^{c_r} \prod_{i=1}^{m_r} (\mathbf{w}_{r,i}\, x_1 x_2 \cdots x_r),$$

where $1 \leq c_i \leq n$. Let π be the permutation of $\{1, 2, \ldots, r\}$ such that $x_{1\pi} \ll x_{2\pi} \ll \cdots \ll x_{r\pi}$. Then

$$\begin{aligned}
\mathbf{w}' &\overset{(6.1a)}{\approx} x_1^{c_1} x_2^{c_2} \cdots x_r^{c_r} \mathbf{w}_{r,1} x_1 x_2 \cdots x_r \prod_{i=2}^{m_r} (\mathbf{w}_{r,i} (x_1 x_2 \cdots x_r)^{n+1}) \\
&\overset{(6.1e)}{\approx} x_{1\pi}^{c_{1\pi}} x_{2\pi}^{c_{2\pi}} \cdots x_{r\pi}^{c_{r\pi}} \mathbf{w}_{r,1} x_{1\pi} x_{2\pi} \cdots x_{r\pi} \prod_{i=2}^{m_r} (\mathbf{w}_{r,i} (x_1 x_2 \cdots x_r)^{n+1}) \\
&\overset{(6.1d)}{\approx} x_{1\pi}^{c_{1\pi}} x_{2\pi}^{c_{2\pi}} \cdots x_{r\pi}^{c_{r\pi}} \mathbf{w}_{r,1} x_{1\pi} x_{2\pi} \cdots x_{r\pi} \prod_{i=2}^{m_r} (\mathbf{w}_{r,i} (x_{1\pi} x_{2\pi} \cdots x_{r\pi})^{n+1}) \\
&\overset{(6.1a)}{\approx} x_{1\pi}^{c_{1\pi}} x_{2\pi}^{c_{2\pi}} \cdots x_{r\pi}^{c_{r\pi}} \prod_{i=1}^{m_r} (\mathbf{w}_{r,i}\, x_{1\pi} x_{2\pi} \cdots x_{r\pi}).
\end{aligned}$$

In summary, the identities (6.1) can be used to convert $\mathbf{w}$ into a word of the form

$$\mathbf{s} = \mathbf{x} \prod_{i=1}^{\ell} (\mathbf{u}_i \vec{\mathscr{X}}),$$

where $\ell \geq 1$, $\mathscr{X} = \{x_1 \ll x_2 \ll \cdots \ll x_r\} \subseteq \mathscr{A}$, $\mathbf{x} = x_1^{c_1} x_2^{c_2} \cdots x_r^{c_r} \in \mathscr{X}^{\boxplus}$, and $\mathbf{u}_i \in \mathscr{A}_\varnothing^+$ are such that $\vec{\mathscr{X}}, \mathbf{u}_1, \mathbf{u}_2, \ldots, \mathbf{u}_\ell$ are pairwise disjoint. Let c_i' denote the number in $\{1, 2, \ldots, n\}$ such that $c_i + 1 \equiv c_i' \pmod{n}$. If $\mathbf{u}_1 = \varnothing$, then

$$\begin{aligned}
\mathbf{s} &= x_1^{c_1} x_2^{c_2} \cdots x_r^{c_r} \vec{\mathscr{X}} \mathbf{u}_2 \vec{\mathscr{X}} \prod_{i=3}^{\ell} (\mathbf{u}_i \vec{\mathscr{X}}) \\
&\overset{(6.2b)}{\approx} x_1^{c_1+1} x_2^{c_2+1} \cdots x_r^{c_r+1} \mathbf{u}_2 \vec{\mathscr{X}} \prod_{i=3}^{\ell} (\mathbf{u}_i \vec{\mathscr{X}}) \\
&\overset{(6.1a)}{\approx} x_1^{c_1'} x_2^{c_2'} \cdots x_r^{c_r'} \mathbf{u}_2 \vec{\mathscr{X}} \prod_{i=3}^{\ell} (\mathbf{u}_i \vec{\mathscr{X}});
\end{aligned}$$

if $\mathbf{u}_k = \varnothing$ for some $k \in \{2, 3, \ldots, \ell\}$, then

$$\begin{aligned}
\mathbf{s} &= x_1^{c_1} x_2^{c_2} \cdots x_r^{c_r} \left(\prod_{i=1}^{k-1} (\mathbf{u}_i \vec{\mathscr{X}}) \right) \vec{\mathscr{X}} \left(\prod_{i=k+1}^{\ell} (\mathbf{u}_i \vec{\mathscr{X}}) \right) \\
&\overset{(6.2a)}{\approx} x_1^{c_1+1} x_2^{c_2+1} \cdots x_r^{c_r+1} \left(\prod_{i=1}^{k-1} (\mathbf{u}_i \vec{\mathscr{X}}) \right) \left(\prod_{i=k+1}^{\ell} (\mathbf{u}_i \vec{\mathscr{X}}) \right) \\
&\overset{(6.1a)}{\approx} x_1^{c_1'} x_2^{c_2'} \cdots x_r^{c_r'} \left(\prod_{i=1}^{k-1} (\mathbf{u}_i \vec{\mathscr{X}}) \right) \left(\prod_{i=k+1}^{\ell} (\mathbf{u}_i \vec{\mathscr{X}}) \right).
\end{aligned}$$

Hence, for any i, if the factor $\mathbf{u}_i$ is empty, then the $\vec{\mathscr{X}}$ that follows it can be "combined" with the prefix $\mathbf{x}$. Therefore, it can further be assumed that either $\mathbf{s} = \mathbf{x}\vec{\mathscr{X}}$ or $\mathbf{u}_i \neq \varnothing$ for all i. If $\mathbf{s} = \mathbf{x}\vec{\mathscr{X}}$, then the word $\mathbf{s}$ is already a sandwich. Hence, assume that $\mathbf{u}_i \neq \varnothing$ for all i; in this case, it remains to show that if $\ell \geq 2$, then the identities (6.1) can be used to rearrange the factors $\mathbf{u}_1, \mathbf{u}_2, \ldots, \mathbf{u}_\ell$ until $\mathbf{s}$ satisfies (S2) and so is a sandwich. To interchange $\mathbf{u}_i$ and $\mathbf{u}_{i+1}$ when $i \geq 2$, the identity (6.1c) can clearly be used. To interchange $\mathbf{u}_1$ and $\mathbf{u}_2$,

$$\begin{aligned}
\mathbf{s} &\overset{(6.1a)}{\approx} \mathbf{x}\mathbf{u}_1 \vec{\mathscr{X}} \mathbf{u}_2 \vec{\mathscr{X}} \vec{\mathscr{X}} \vec{\mathscr{X}}^{n-1} \prod_{i=3}^{\ell} (\mathbf{u}_i \vec{\mathscr{X}}) \\
&\overset{(6.1b)}{\approx} \vec{\mathscr{X}} \mathbf{u}_1 \vec{\mathscr{X}} \mathbf{u}_2 \vec{\mathscr{X}} \mathbf{x} \vec{\mathscr{X}}^{n-1} \prod_{i=3}^{\ell} (\mathbf{u}_i \vec{\mathscr{X}}) \\
&\overset{(6.1c)}{\approx} \vec{\mathscr{X}} \mathbf{u}_2 \vec{\mathscr{X}} \mathbf{u}_1 \vec{\mathscr{X}} \mathbf{x} \vec{\mathscr{X}}^{n-1} \prod_{i=3}^{\ell} (\mathbf{u}_i \vec{\mathscr{X}}) \\
&\overset{(6.1b)}{\approx} \mathbf{x}\mathbf{u}_2 \vec{\mathscr{X}} \mathbf{u}_1 \vec{\mathscr{X}} \vec{\mathscr{X}} \vec{\mathscr{X}}^{n-1} \prod_{i=3}^{\ell} (\mathbf{u}_i \vec{\mathscr{X}}) \\
&\overset{(6.1a)}{\approx} \mathbf{x}\mathbf{u}_2 \vec{\mathscr{X}} \mathbf{u}_1 \vec{\mathscr{X}} \prod_{i=3}^{\ell} (\mathbf{u}_i \vec{\mathscr{X}}).
\end{aligned}$$

Throughout this proof, the identities (6.1) and (6.2) have been used to convert the word $\mathbf{w}$ into some sandwich $\mathbf{s}$. Therefore, $\mathbf{w} \approx \mathbf{s} \in \mathsf{id}\{\mathcal{L}_3\}$ by Lemmas 6.4 and 6.5, so that $\mathsf{sim}(\mathbf{w}) = \mathsf{sim}(\mathbf{s})$ and $\mathsf{non}(\mathbf{w}) = \mathsf{non}(\mathbf{s})$ by Lemma 5.4(i). □

Lemma 6.10 *The identities* $\{(6.1)\} \cup \mathsf{id}_{\mathrm{SAN}}\{\mathcal{L}_{3,n}\}$ *constitute an identity basis for* $\mathcal{L}_{3,n}$.

Proof It follows from Lemmas 6.4 and 6.6 that (6.1) and the connected identities in $\mathsf{id}\{\mathcal{L}_{3,n}\}$ constitute an identity basis for the semigroup $\mathcal{L}_{3,n}$. The result then follows from Lemma 6.9. □

6.4 Restrictions on Sandwich Identities

This section establishes some properties of sandwich identities satisfied by the semigroup $\mathcal{L}_{3,n}$. In Sect. 6.4.1, it is shown that any two sandwiches that form such an identity must share the same level. In Sect. 6.4.2, *refined sandwich identities* are introduced; these are identities formed by certain sandwiches of level one. It is shown in Lemma 6.15 that refined sandwich identities satisfied by the semigroup $\mathcal{L}_{3,n}$ are of a very specific form.

6.4.1 Level of Sandwiches Forming Sandwich Identities

Lemma 6.11 *Suppose that* $\mathbf{s} \approx \mathbf{s}' \in \mathsf{id}_{\mathrm{SAN}}\{\mathcal{L}_{3,n}\}$, *where*

$$\mathbf{s} = \mathbf{x} \prod_{i=1}^{\ell} (\mathbf{u}_i \vec{\mathscr{X}})$$

is the sandwich in (6.3) *of level* ℓ. *Then*

$$\mathbf{s}' = \mathbf{x} \prod_{i=1}^{\ell} (\mathbf{u}_i' \vec{\mathscr{X}})$$

for some $\mathbf{u}_1', \mathbf{u}_2', \ldots, \mathbf{u}_\ell' \in \mathscr{A}_\varnothing^+$ *such that* $\mathsf{con}(\mathbf{u}_i) = \mathsf{con}(\mathbf{u}_i')$ *for all* i.

Proof By Lemma 5.4(i),

(A) $\mathsf{sim}(\mathbf{s}) = \mathsf{sim}(\mathbf{s}')$ and $\mathsf{non}(\mathbf{s}) = \mathsf{non}(\mathbf{s}')$.

Since $\mathbf{s}'$ is a sandwich, it can be written as

(B) $\mathbf{s}' = \mathbf{y} \prod_{i=1}^{\ell'} (\mathbf{u}_i' \vec{\mathscr{Y}})$ for some $\ell' \geq 1$, finite nonempty $\mathscr{Y} \subseteq \mathscr{A}$, $\mathbf{y} \in \mathscr{Y}^{\boxplus}$, and $\mathbf{u}_i' \in \mathscr{A}_\varnothing^+$ such that

(a) $\vec{\mathscr{Y}}, \mathbf{u}_1', \mathbf{u}_2', \ldots, \mathbf{u}_{\ell'}'$ are pairwise disjoint;
(b) if $\ell' \geq 2$, then $\mathbf{u}_i' \neq \varnothing$ for all i and $\min(\mathbf{u}_1') \ll \min(\mathbf{u}_2') \ll \cdots \ll \min(\mathbf{u}_{\ell'}')$.

Suppose that $\mathscr{X} \neq \mathscr{Y}$, say $x \in \mathscr{X} \backslash \mathscr{Y}$. Then $x \in \mathscr{X} \subseteq \mathsf{non}(\mathbf{s}) = \mathsf{non}(\mathbf{s}')$ by (A), so that (Ba) implies that $x \in \mathsf{non}(\mathbf{u}_k')$ for some $k \in \{1, 2, \ldots, \ell'\}$. Let $\varphi : \mathscr{A} \to \mathcal{L}_3$ denote the substitution

$$z \mapsto \begin{cases} \mathsf{e} & \text{if } z \in \mathsf{con}(\mathbf{u}_k'), \\ \mathsf{f} & \text{otherwise.} \end{cases}$$

Then $\mathbf{s}'\varphi = \mathsf{fef}$ by (Ba). On the other hand, $\mathbf{x}\varphi = \vec{\mathscr{X}}\varphi \neq \mathsf{f}$ due to $x\varphi = \mathsf{e}$, so that the contradiction $\mathbf{s}\varphi \neq \mathsf{fef}$ is deduced. Hence, $\mathscr{X} = \mathscr{Y}$. It then follows from (A) and (B) that

(C) $\mathbf{s} = \mathbf{y} \prod_{i=1}^{\ell'} (\mathbf{u}_i' \vec{\mathscr{X}})$ with $\mathbf{y} \in \mathscr{X}^{\boxplus}$;
(D) $\bigcup_{i=1}^{\ell} \mathsf{con}(\mathbf{u}_i) = \bigcup_{i=1}^{\ell'} \mathsf{con}(\mathbf{u}_i')$.

Suppose that i is such that $\mathsf{con}(\mathbf{u}_i) \nsubseteq \mathsf{con}(\mathbf{u}_j')$ for any j. Then in view of (D), there exist distinct $x, y \in \mathsf{con}(\mathbf{u}_i)$ such that $x \in \mathsf{con}(\mathbf{u}_j')$ and $y \in \mathsf{con}(\mathbf{u}_{j'}')$ with $j < j'$. Let $\chi : \mathscr{A} \to \mathcal{L}_3$ denote the substitution

$$z \mapsto \begin{cases} \mathsf{e} & \text{if } z \in \mathsf{con}(\mathbf{u}_i), \\ \mathsf{f} & \text{otherwise.} \end{cases}$$

Then $\mathbf{s}\chi = \mathsf{fef}$ by (S1). Since $\mathscr{X} \cap \mathsf{con}(\mathbf{u}_i) = \varnothing$ by (S1), it follows that $z\chi = \mathsf{f}$ for all $z \in \mathscr{X}$. Hence,

$$\begin{aligned} \mathbf{s}'\chi &= \cdots \mathbf{u}_j'\chi \cdots \vec{\mathscr{X}}\chi \cdots \mathbf{u}_{j'}'\chi \cdots \\ &= \cdots \mathsf{e} \cdots \mathsf{f} \cdots \mathsf{e} \cdots \\ &= 0 \neq \mathbf{s}\chi, \end{aligned}$$

which is a contradiction. Therefore, the inclusion $\mathsf{con}(\mathbf{u}_i) \subseteq \mathsf{con}(\mathbf{u}_j')$ holds for some j. By a symmetrical argument, the inclusion $\mathsf{con}(\mathbf{u}_j') \subseteq \mathsf{con}(\mathbf{u}_k)$ holds for some k, so that $\mathsf{con}(\mathbf{u}_i) \subseteq \mathsf{con}(\mathbf{u}_j') \subseteq \mathsf{con}(\mathbf{u}_k)$. By (S1), the words $\mathbf{u}_i$ and $\mathbf{u}_k$ are either equal or disjoint. Hence, $\mathbf{u}_i = \mathbf{u}_k$ is the only possibility, so that $\mathsf{con}(\mathbf{u}_i) = \mathsf{con}(\mathbf{u}_j')$. It has just been shown that for each $i \in \{1, 2, \ldots, \ell\}$, there exists some $j \in \{1, 2, \ldots, \ell'\}$ such that $\mathsf{con}(\mathbf{u}_i) = \mathsf{con}(\mathbf{u}_j')$. Therefore, $\ell = \ell'$ by (Ba) and (D), and it follows from (C) that

$$\mathbf{s}' = \mathbf{y} \prod_{i=1}^{\ell} (\mathbf{u}_i' \vec{\mathscr{X}})$$

with $\{\mathsf{con}(\mathbf{u}_i) \mid 1 \leq i \leq \ell\} = \{\mathsf{con}(\mathbf{u}_i') \mid 1 \leq i \leq \ell\}$. If $\ell \geq 2$, then (S2) and (Bb) imply

(E) $(\mathsf{con}(\mathbf{u}_1), \mathsf{con}(\mathbf{u}_2), \ldots, \mathsf{con}(\mathbf{u}_\ell)) = (\mathsf{con}(\mathbf{u}'_1), \mathsf{con}(\mathbf{u}'_2), \ldots, \mathsf{con}(\mathbf{u}'_\ell))$;

if $\ell = 1$, then (E) holds vacuously.

It remains to show that $\mathbf{x} = \mathbf{y}$. Suppose that $\mathscr{X} = \{x_1 \ll x_2 \ll \cdots \ll x_r\}$. Then by definition, $\vec{\mathscr{X}} = x_1 x_2 \cdots x_r$. Since $\mathbf{x}, \mathbf{y} \in \mathscr{X}^{\boxplus}$, there exist

(F) $c_1, c_2, \ldots, c_r, d_1, d_2, \ldots, d_r \in \{1, 2, \ldots, n\}$

such that $\mathbf{x} = x_1^{c_1} x_2^{c_2} \cdots x_r^{c_r}$ and $\mathbf{y} = x_1^{d_1} x_2^{d_2} \cdots x_r^{d_r}$. For any $j \in \{1, 2, \ldots, r\}$,

$$\begin{aligned} \mathsf{occ}(x_j, \mathbf{s}) &= \mathsf{occ}(x_j, \mathbf{x}) + \sum_{i=1}^{\ell} (\mathsf{occ}(x_j, \mathbf{u}_i) + \mathsf{occ}(x_j, \vec{\mathscr{X}})) \\ &= c_j + \sum_{i=1}^{\ell} (0 + 1) \\ &= c_j + \ell; \end{aligned}$$

similarly, $\mathsf{occ}(x_j, \mathbf{s}') = d_j + \ell$. Since $\mathsf{occ}(x_j, \mathbf{s}) \equiv \mathsf{occ}(x_j, \mathbf{s}') \pmod{n}$ because of the assumption $\mathcal{Z}_n \models \mathbf{s} \approx \mathbf{s}'$, it follows that $c_j \equiv d_j \pmod{n}$, whence $c_j = d_j$ by (F). Consequently, $\mathbf{x} = \mathbf{y}$. □

Lemma 6.12 *Suppose that* $\mathbf{s} \approx \mathbf{s}' \in \mathsf{id}_{\mathrm{SAN}}\{\mathcal{L}_{3,n}\}$, *where*

$$\mathbf{s} = \mathbf{x} \prod_{i=1}^{\ell} (\mathbf{u}_i \vec{\mathscr{X}}) \quad \text{and} \quad \mathbf{s}' = \mathbf{x} \prod_{i=1}^{\ell} (\mathbf{u}'_i \vec{\mathscr{X}})$$

are the sandwiches in Lemma 6.11. Then

$$\{(6.1), \mathbf{s} \approx \mathbf{s}'\} \sim \{(6.1)\} \cup \{x\mathbf{u}_i x \approx x\mathbf{u}'_i x \mid 1 \le i \le \ell\}.$$

Proof For each $i \in \{1, 2, \ldots, \ell\}$, let $\varphi_i : \mathscr{A} \to \mathscr{A}$ denote the substitution

$$z \mapsto \begin{cases} z & \text{if } z \in \mathsf{con}(\mathbf{u}_i) = \mathsf{con}(\mathbf{u}'_i), \\ x^n & \text{otherwise.} \end{cases}$$

Then by (S1), the deduction (6.1a) $\vdash \{x(\mathbf{s}\varphi_i)x \approx x\mathbf{u}_i x, x(\mathbf{s}'\varphi_i)x \approx x\mathbf{u}_i' x\}$ holds, so that the deduction $\{(6.1), \mathbf{s} \approx \mathbf{s}'\} \vdash x\mathbf{u}_i x \approx x\mathbf{u}_i' x$ follows. Conversely,

$$\begin{aligned}
\mathbf{s} &= \overbrace{x_1^{c_1} x_2^{c_2} \cdots x_r^{c_r}}^{\mathbf{x}}\, \mathbf{u}_1 \overbrace{x_1 x_2 \cdots x_r}^{\vec{\mathscr{X}}} \prod_{i=2}^{\ell} (\mathbf{u}_i \vec{\mathscr{X}}) \\
&\overset{(6.1\mathrm{k})}{\approx} x_1^{c_1} x_2^{c_2} \cdots x_r^{c_r} \mathbf{u}_1 x_r^n x_1 x_2 \cdots x_r \prod_{i=2}^{\ell} (\mathbf{u}_i \vec{\mathscr{X}}) \\
&\approx x_1^{c_1} x_2^{c_2} \cdots x_r^{c_r} \mathbf{u}_1' x_r^n x_1 x_2 \cdots x_r \prod_{i=2}^{\ell} (\mathbf{u}_i' \vec{\mathscr{X}}) \qquad \text{by } x\mathbf{u}_i x \approx x\mathbf{u}_i' x \\
&\overset{(6.1\mathrm{k})}{\approx} x_1^{c_1} x_2^{c_2} \cdots x_r^{c_r} \mathbf{u}_1' x_1 x_2 \cdots x_r \prod_{i=2}^{\ell} (\mathbf{u}_i' \vec{\mathscr{X}}) \\
&= \mathbf{s}'.
\end{aligned}$$

Therefore, the deduction $\{(6.1)\} \cup \{x\mathbf{u}_i x \approx x\mathbf{u}_i' x \mid 1 \leq i \leq \ell\} \vdash \mathbf{s} \approx \mathbf{s}'$ also holds. □

6.4.2 Refined Sandwich Identities

By Lemma 6.11, any identity in $\mathsf{id}_{\mathrm{SAN}}\{\mathcal{L}_{3,n}\}$ is formed by a pair of sandwiches that share the same level. Hence, it is unambiguous to define the *level* of a sandwich identity $\mathbf{s} \approx \mathbf{s}'$ to be the level shared by the sandwiches $\mathbf{s}$ and $\mathbf{s}'$. The present subsection examines identities in $\mathsf{id}_{\mathrm{SAN}}\{\mathcal{L}_{3,n}\}$ of level one.

Consider a word

$$\mathbf{r} = x\left(\prod_{i=1}^{k} \mathbf{p}_i\right)x, \tag{6.4}$$

where $k \geq 1$, $x \in \mathscr{A}$, and $\mathbf{p}_1, \mathbf{p}_2, \ldots, \mathbf{p}_k \in \mathscr{A}^+$ are such that $x, \mathbf{p}_1, \mathbf{p}_2, \ldots, \mathbf{p}_k$ are pairwise disjoint. Then $\mathbf{r}$ is a sandwich of level one. This sandwich is *refined* if it satisfies all of the following conditions:

(R1) each $\mathbf{p}_i$ is either a singleton or a sandwich;
(R2) if $k \geq 2$ and $\mathbf{p}_1, \mathbf{p}_2, \ldots, \mathbf{p}_k$ are all sandwiches, then $\min(\mathbf{p}_1) \ll \min(\mathbf{p}_k)$.

A sandwich identity $\mathbf{r} \approx \mathbf{r}'$ is *refined* if the sandwiches $\mathbf{r}$ and $\mathbf{r}'$ are refined. Denote by

$$\mathsf{id}_{\mathrm{REF}}\{\mathcal{L}_{3,n}\}$$

the set of all refined sandwich identities satisfied by the semigroup $\mathcal{L}_{3,n}$.

Lemma 6.13 *Suppose that* $\mathbf{s} \approx \mathbf{s}' \in \mathsf{id}_{\mathrm{SAN}}\{\mathcal{L}_{3,n}\}$. *Then there exists some finite subset* Σ *of* $\mathsf{id}_{\mathrm{REF}}\{\mathcal{L}_{3,n}\}$ *such that*

$$\{(6.1), \mathbf{s} \approx \mathbf{s}'\} \sim \{(6.1)\} \cup \Sigma.$$

Further, each identity $\mathbf{r} \approx \mathbf{r}'$ *in* Σ *can be chosen so that* $|\,\mathsf{sim}(\mathbf{r})| \leq |\,\mathsf{sim}(\mathbf{s})|$ *and* $|\,\mathsf{non}(\mathbf{r})| \leq |\,\mathsf{non}(\mathbf{s})|$.

Proof By Lemma 6.11,

$$\mathbf{s} = \mathbf{x}\prod_{i=1}^{\ell}(\mathbf{u}_i\vec{\mathscr{X}}) \text{ and } \mathbf{s}' = \mathbf{x}\prod_{i=1}^{\ell}(\mathbf{u}_i'\vec{\mathscr{X}})$$

for some $\ell \geq 1$, finite nonempty $\mathscr{X} \subseteq \mathscr{A}$, $\mathbf{x} \in \mathscr{X}^{\boxplus}$, and $\mathbf{u}_i, \mathbf{u}_i' \in \mathscr{A}_{\varnothing}^{+}$ such that $\mathsf{con}(\mathbf{u}_i) = \mathsf{con}(\mathbf{u}_i')$ for each i and $\vec{\mathscr{X}}, \mathbf{u}_1, \mathbf{u}_2, \ldots, \mathbf{u}_\ell$ are pairwise disjoint. By Lemma 6.12, the equivalence

(A) $\{(6.1), \mathbf{s} \approx \mathbf{s}'\} \sim \{(6.1)\} \cup \{x\mathbf{u}_i x \approx x\mathbf{u}_i' x \mid 1 \leq i \leq \ell\}$

holds, and it is easily seen that

(B) $|\,\mathsf{sim}(x\mathbf{u}_i x)| \leq |\,\mathsf{sim}(\mathbf{s})|$ and $|\,\mathsf{non}(x\mathbf{u}_i x)| \leq |\,\mathsf{non}(\mathbf{s})|$

for each i. In what follows, it is shown that the identities (6.1) can be used to convert each $x\mathbf{u}_i x$ into some refined sandwich $\mathbf{r}_i$ with

(C) $|\,\mathsf{sim}(x\mathbf{u}_i x)| = |\,\mathsf{sim}(\mathbf{r}_i)|$ and $|\,\mathsf{non}(x\mathbf{u}_i x)| = |\,\mathsf{non}(\mathbf{r}_i)|$.

Similarly, the identities (6.1) can be used to convert each $x\mathbf{u}_i' x$ into some refined sandwich $\mathbf{r}_i'$. Therefore, by letting

$$\Sigma = \{\mathbf{r}_i \approx \mathbf{r}_i' \mid 1 \leq i \leq \ell\},$$

the equivalence $\{(6.1), \mathbf{s} \approx \mathbf{s}'\} \sim \{(6.1)\} \cup \Sigma$ holds by (A), where $|\,\mathsf{sim}(\mathbf{r}_i)| \leq |\,\mathsf{sim}(\mathbf{s})|$ and $|\,\mathsf{non}(\mathbf{r}_i)| \leq |\,\mathsf{non}(\mathbf{s})|$ by (B) and (C), whence the lemma is proved.

Write $x\mathbf{u}_i x = x\mathbf{p}_1\mathbf{p}_2\cdots\mathbf{p}_k x$, where $\mathbf{u}_i = \mathbf{p}_1\mathbf{p}_2\cdots\mathbf{p}_k$ is the natural decomposition of $\mathbf{u}_i$ (so that $x, \mathbf{p}_1, \mathbf{p}_2, \ldots, \mathbf{p}_k$ are pairwise disjoint and each $\mathbf{p}_j$ is either a singleton or a connected word). By Lemma 6.9, the identities (6.1) can be used to convert any connected $\mathbf{p}_j$ into some sandwich $\mathbf{s}_j$ with $\mathsf{sim}(\mathbf{p}_j) = \mathsf{sim}(\mathbf{s}_j)$ and $\mathsf{non}(\mathbf{p}_j) = \mathsf{non}(\mathbf{s}_j)$. Therefore, it can be assumed that each $\mathbf{p}_j$ is either a singleton or a sandwich, so that $x\mathbf{u}_i x$ satisfies (R1). If either $k = 1$ or some $\mathbf{p}_j$ is a singleton, then $x\mathbf{u}_i x$ satisfies (R2) vacuously and so is refined. Hence, suppose that $k \geq 2$ and $\mathbf{p}_1, \mathbf{p}_2, \ldots, \mathbf{p}_k$ are all sandwiches with $\min(\mathbf{p}_1) \not\ll \min(\mathbf{p}_k)$. Then $\min(\mathbf{p}_k) \ll \min(\mathbf{p}_1)$ because $\mathbf{p}_1$ and $\mathbf{p}_k$ are disjoint. Consider any $j \in \{1, 2, \ldots, k\}$. Since $\mathbf{p}_j$ is a sandwich, it is of the form

$$\mathbf{p}_j = y_1^{c_1} y_2^{c_2} \cdots y_s^{c_s} \mathbf{w} y_1 y_2 \cdots y_s$$

for some $\mathbf{w} \in \mathscr{A}_\varnothing^+$ and $\mathscr{Y} = \{y_1 \ll y_2 \ll \cdots \ll y_s\} \subset \mathscr{A}$. If $s = 1$, then the deduction $\mathbf{p}_j \overset{(6.1a)}{\approx} y_1^n \mathbf{p}_j y_1^n$ clearly holds; if $s \geq 2$, then

$$\begin{aligned}
\mathbf{p}_j &\overset{(6.1m)}{\approx} y_1^{c_1} y_2^{c_2} \cdots y_s^{c_s} \mathbf{w} y_1 y_2 \cdots y_s \cdot y_1^n \\
&\overset{(6.1a)}{\approx} y_1^n \cdot y_1^{c_1} y_2^{c_2} \cdots y_s^{c_s} \mathbf{w} y_1 y_2 \cdots y_s \cdot y_1^n \\
&= y_1^n \mathbf{p}_j y_1^n.
\end{aligned}$$

Therefore, in any case, the deduction $\mathbf{p}_j \overset{(6.1)}{\approx} h_j^n \mathbf{p}_j h_j^n$ holds with $h_j = \mathsf{h}(\mathbf{p}_j)$. Hence,

$$\begin{aligned}
x\mathbf{u}_i x &\overset{(6.1)}{\approx} x(h_1^n \mathbf{p}_1 h_1^n)(h_2^n \mathbf{p}_2 h_2^n) \cdots (h_{k-1}^n \mathbf{p}_{k-1} h_{k-1}^n)(h_k^n \mathbf{p}_k h_k^n)x \\
&\overset{(6.1n)}{\approx} x(h_k^n \mathbf{p}_k h_k^n)(h_{k-1}^n \mathbf{p}_{k-1} h_{k-1}^n) \cdots (h_2^n \mathbf{p}_2 h_2^n)(h_1^n \mathbf{p}_1 h_1^n)x \\
&\overset{(6.1)}{\approx} x\mathbf{p}_k \mathbf{p}_{k-1} \cdots \mathbf{p}_2 \mathbf{p}_1 x,
\end{aligned}$$

where $\min(\mathbf{p}_k) \ll \min(\mathbf{p}_1)$. Therefore, the identities (6.1) can be used to convert $x\mathbf{u}_i x$ into some refined sandwich $\mathbf{r}_i = x\mathbf{p}_k \mathbf{p}_{k-1} \cdots \mathbf{p}_2 \mathbf{p}_1 x$ that satisfies (C). □

Lemma 6.14 *Let* $\mathbf{s}, \mathbf{s}' \in \mathscr{A}^+$ *be any sandwiches and* $x \in \mathscr{A}$ *be such that* $x \notin \mathsf{con}(\mathbf{s}) = \mathsf{con}(\mathbf{s}')$. *Suppose that* $x\mathbf{s}x \approx x\mathbf{s}'x \in \mathsf{id}\{\mathcal{L}_{3,n}\}$. *Then* $\mathbf{s} \approx \mathbf{s}' \in \mathsf{id}\{\mathcal{L}_{3,n}\}$.

Proof By assumption, $x\mathbf{s}x \approx x\mathbf{s}'x \in \mathsf{id}\{\mathcal{Z}_n\}$ and $x\mathbf{s}x \approx x\mathbf{s}'x \in \mathsf{id}\{\mathcal{L}_3\}$. Since the group $\mathcal{Z}_n$ possesses a unit element, it satisfies the identity $\mathbf{s} \approx \mathbf{s}'$. Hence, it suffices to show that $\mathbf{s} \approx \mathbf{s}' \in \mathsf{id}\{\mathcal{L}_3\}$. Seeking a contradiction, suppose that $\mathbf{s} \approx \mathbf{s}' \notin \mathsf{id}\{\mathcal{L}_3\}$. Then there exists some substitution $\varphi : \mathsf{con}(\mathbf{s}) \to \mathcal{L}_3$ such that $\mathbf{s}\varphi \neq \mathbf{s}'\varphi$. Since $\mathsf{con}(\mathbf{s}) = \mathsf{con}(\mathbf{s}')$, it is easily seen that $\mathbf{s}\varphi = \mathsf{e}$ if and only if $\mathbf{s}'\varphi = \mathsf{e}$. But $\mathbf{s}\varphi = \mathsf{e} = \mathbf{s}'\varphi$ contradicts the choice of φ, so that $\mathbf{s}\varphi, \mathbf{s}'\varphi \neq \mathsf{e}$. By the same argument, $\mathbf{s}\varphi, \mathbf{s}'\varphi \neq \mathsf{f}$. Further, since the words $\mathbf{s}$ and $\mathbf{s}'$ are connected, it is easily shown that $\mathbf{s}\varphi, \mathbf{s}'\varphi \notin \{\mathsf{ef}, \mathsf{fe}\}$. Hence, $\mathbf{s}\varphi, \mathbf{s}'\varphi \in \{0, \mathsf{fef}\}$. By symmetry, it suffices to assume that $\mathbf{s}\varphi = 0$ and $\mathbf{s}'\varphi = \mathsf{fef}$. Let $\widetilde{\varphi} : \mathscr{A} \to \mathcal{L}_3$ denote the extension of the substitution φ given by

$$z \mapsto \begin{cases} z\varphi & \text{if } z \in \mathsf{con}(\mathbf{s}) = \mathsf{con}(\mathbf{s}'), \\ \mathsf{f} & \text{otherwise.} \end{cases}$$

Then $(x\mathbf{s}x)\widetilde{\varphi} = 0$ and $(x\mathbf{s}'x)\widetilde{\varphi} = \mathsf{fef}$, so that $x\mathbf{s}x \approx x\mathbf{s}'x \notin \mathsf{id}\{\mathcal{L}_3\}$. □

Lemma 6.15 *Suppose that* $\mathbf{r} \approx \mathbf{r}' \in \mathsf{id}_{\mathrm{REF}}\{\mathcal{L}_{3,n}\}$, *where*

$$\mathbf{r} = x\left(\prod_{i=1}^{k} \mathbf{p}_i\right)x$$

is the refined sandwich in (6.4). *Then*

$$\mathbf{r}' = x\left(\prod_{i=1}^{k} \mathbf{p}_i'\right)x$$

for some $\mathbf{p}_1', \mathbf{p}_2', \ldots, \mathbf{p}_k' \in \mathscr{A}^+$ *such that each* $\mathbf{p}_i'$ *is either a singleton or a sandwich,* $x, \mathbf{p}_1', \mathbf{p}_2', \ldots, \mathbf{p}_k'$ *are pairwise disjoint, and* $\mathsf{con}(\mathbf{p}_i) = \mathsf{con}(\mathbf{p}_i')$ *for all* i. *Further, for any* i,

(i) $\mathbf{p}_i$ *and* $\mathbf{p}_i'$ *are either both singletons or both sandwiches*;
(ii) *if* $\mathbf{p}_i$ *and* $\mathbf{p}_i'$ *are both singletons, then* $\mathbf{p}_i = \mathbf{p}_i'$;
(iii) *if* $\mathbf{p}_i$ *and* $\mathbf{p}_i'$ *are both sandwiches, then* $\mathbf{p}_i \approx \mathbf{p}_i' \in \mathsf{id}_{\mathrm{SAN}}\{\mathcal{L}_{3,n}\}$.

Proof It follows from Lemma 6.11 that $\mathbf{r}' = x\mathbf{u}'x$ for some $\mathbf{u}' \in \mathscr{A}^+$ such that

(A) $x \notin \mathsf{con}(\mathbf{u}') = \mathsf{con}(\mathbf{p}_1\mathbf{p}_2 \cdots \mathbf{p}_k)$.

By the definition of a refined sandwich,

$$\mathbf{r}' = x\mathbf{u}'x = x\left(\prod_{i=1}^{k'} \mathbf{p}_i'\right)x,$$

where

(B) each $\mathbf{p}_i'$ is either a singleton or a sandwich and $x, \mathbf{p}_1', \mathbf{p}_2', \ldots, \mathbf{p}_{k'}'$ are pairwise disjoint.

It is first shown that $k = k'$ and $\mathsf{con}(\mathbf{p}_i) = \mathsf{con}(\mathbf{p}'_i)$ for all i. Let $\varphi : \mathscr{A} \to \mathscr{A}^+$ denote the substitution

$$z \mapsto \begin{cases} y_i^{2n} & \text{if } z \in \mathsf{con}(\mathbf{p}_i), \\ x^n & \text{otherwise.} \end{cases}$$

Then

$$\mathbf{r}\varphi \overset{(6.1\text{a})}{\approx} x^n\bigg(\prod_{i=1}^{k} y_i^{2n}\bigg)x^n \text{ and } \mathbf{r}'\varphi \overset{(6.1\text{a})}{\approx} x^n\bigg(\prod_{i=1}^{k'} (\mathbf{p}'_i\varphi)\bigg)x^n,$$

so that $x^n(\prod_{i=1}^{k} y_i^{2n})x^n \approx x^n(\prod_{i=1}^{k'}(\mathbf{p}'_i\varphi))x^n \in \mathsf{id}\{\mathcal{L}_3\}$. Since $x^n(\prod_{i=1}^{k} y_i^{2n})x^n \in \mathscr{P}_k^{\uparrow}$, it follows from Lemma 5.5(i) that

(C) $x^n(\prod_{i=1}^{k'}(\mathbf{p}'_i\varphi))x^n \in \mathscr{P}_k^{\uparrow} \cup \mathscr{P}_k^{\downarrow}$.

Further, since

$$\begin{aligned} \mathsf{con}(\mathbf{p}'_i\varphi) &\subseteq \mathsf{con}(\mathbf{u}'\varphi) \\ &\overset{(\text{A})}{=} \mathsf{con}((\mathbf{p}_1\mathbf{p}_2\cdots\mathbf{p}_k)\varphi) \\ &= \{y_1, y_2, \ldots, y_k\}, \end{aligned}$$

it follows that

(D) $\mathsf{con}(\mathbf{p}'_i\varphi) \subseteq \{y_1, y_2, \ldots, y_k\}$ for each $i \in \{1, 2, \ldots, k'\}$.

By (B), each $\mathbf{p}'_i$ is either a singleton or a sandwich. If $\mathbf{p}'_i$ is a singleton, then clearly $\mathsf{con}(\mathbf{p}'_i\varphi) = \{y_j\}$ for some j. Suppose that $\mathbf{p}'_i$ is a sandwich. Then $\mathbf{p}'_i$ is connected, so that by (C), the word $\mathbf{p}'_i\varphi$ is a connected factor of some word in $\mathscr{P}_k^{\uparrow} \cup \mathscr{P}_k^{\downarrow}$. The connected factors of words in $\mathscr{P}_k^{\uparrow} \cup \mathscr{P}_k^{\downarrow}$ are

$$\begin{aligned} &x^{s_1} y_1^{t_1} y_2^{t_2} \cdots y_k^{t_k} x^{s_2}, && s_1, s_2 \geq 1,\ t_1, t_2, \ldots, t_k \geq 2; \\ &x^{s_1} y_k^{t_k} y_{k-1}^{t_{k-1}} \cdots y_1^{t_1} x^{s_2}, && s_1, s_2 \geq 1,\ t_1, t_2, \ldots, t_k \geq 2; \\ &x^t,\ y_1^t,\ y_2^t,\ \ldots,\ y_k^t, && t \geq 2. \end{aligned}$$

But since $x \notin \mathsf{con}(\mathbf{p}'_i\varphi)$ by (D), the word $\mathbf{p}'_i\varphi$ is one of $y_1^t, y_2^t, \ldots, y_k^t$, so that $\mathsf{con}(\mathbf{p}'_i\varphi) = \{y_j\}$ for some j. Hence, regardless of whether $\mathbf{p}'_i$ is a singleton or a sandwich,

(E) $\mathsf{con}(\mathbf{p}'_i\varphi) = \{y_j\}$ for some j.

It follows that the inclusion $\mathsf{con}(\mathbf{p}_i') \subseteq \mathsf{con}(\mathbf{p}_j)$ holds for some j. By a symmetrical argument, the inclusion $\mathsf{con}(\mathbf{p}_j) \subseteq \mathsf{con}(\mathbf{p}_m')$ holds for some m, so that $\mathsf{con}(\mathbf{p}_i') \subseteq \mathsf{con}(\mathbf{p}_j) \subseteq \mathsf{con}(\mathbf{p}_m')$. Now by (B), the words $\mathbf{p}_i'$ and $\mathbf{p}_m'$ are either equal or disjoint. Therefore, $\mathbf{p}_i' = \mathbf{p}_m'$ is the only possibility, whence $\mathsf{con}(\mathbf{p}_i') = \mathsf{con}(\mathbf{p}_j)$. It has just been shown that for each $i \in \{1, 2, \ldots, k'\}$, there exists some $j \in \{1, 2, \ldots, k\}$ such that $\mathsf{con}(\mathbf{p}_i') = \mathsf{con}(\mathbf{p}_j)$. Since $\mathsf{con}(\mathbf{p}_1\mathbf{p}_2\cdots\mathbf{p}_k) = \mathsf{con}(\mathbf{p}_1'\mathbf{p}_2'\cdots\mathbf{p}_{k'}')$ by (A), it follows that

(F) $k = k'$;

(G) there exists a one-to-one correspondence between the sets

$$\mathsf{con}(\mathbf{p}_1), \mathsf{con}(\mathbf{p}_2), \ldots, \mathsf{con}(\mathbf{p}_k) \text{ and } \mathsf{con}(\mathbf{p}_1'), \mathsf{con}(\mathbf{p}_2'), \ldots, \mathsf{con}(\mathbf{p}_k').$$

Further, by (C) and (F), either $x^n(\prod_{i=1}^k(\mathbf{p}_i'\varphi))x^n \in \mathscr{P}_k^{\uparrow}$ or $x^n(\prod_{i=1}^k(\mathbf{p}_i'\varphi))x^n \in \mathscr{P}_k^{\downarrow}$. Hence, by (E),

$$(\mathbf{p}_1'\varphi, \mathbf{p}_2'\varphi, \ldots, \mathbf{p}_k'\varphi) \in \{(y_1^{t_1}, y_2^{t_2}, \ldots, y_k^{t_k}), (y_k^{t_k}, y_{k-1}^{t_{k-1}}, \ldots, y_1^{t_1}) \mid t_1, t_2, \ldots, t_k \geq 2\}.$$

It then follows from (G) that either

(H) $(\mathsf{con}(\mathbf{p}_1'), \mathsf{con}(\mathbf{p}_2'), \ldots, \mathsf{con}(\mathbf{p}_k')) = (\mathsf{con}(\mathbf{p}_1), \mathsf{con}(\mathbf{p}_2), \ldots, \mathsf{con}(\mathbf{p}_k))$ or

(H′) $(\mathsf{con}(\mathbf{p}_1'), \mathsf{con}(\mathbf{p}_2'), \ldots, \mathsf{con}(\mathbf{p}_k')) = (\mathsf{con}(\mathbf{p}_k), \mathsf{con}(\mathbf{p}_{k-1}), \ldots, \mathsf{con}(\mathbf{p}_1))$.

Now Lemma 5.4(i) implies that

(I) $\mathsf{sim}(\mathbf{r}) = \mathsf{sim}(\mathbf{r}')$ and $\mathsf{non}(\mathbf{r}) = \mathsf{non}(\mathbf{r}')$.

If $k = 1$, then (H) clearly holds, so assume that $k \geq 2$. There are two cases to consider.

CASE 1: $\mathbf{p}_1, \mathbf{p}_2, \ldots, \mathbf{p}_k$ are all sandwiches. Then in view of (B) and (I), either (H) or (H′) implies that $\mathbf{p}_1', \mathbf{p}_2', \ldots, \mathbf{p}_k'$ are also sandwiches. Hence, (H) must hold by (R2).

CASE 2: $\mathbf{p}_i = y$ is a singleton for some i. Then either (H) or (H′) implies that $\mathsf{con}(\mathbf{p}_j') = \{y\}$ for some j, whence $\mathbf{p}_j' = y$ by (I). Since $k \geq 2$, either $1 < i$ or $i < k$. By symmetry, it suffices to assume that $1 < i$. Let $\chi : \mathscr{A} \to \mathcal{L}_3$ denote the substitution

$$z \mapsto \begin{cases} \mathsf{e} & \text{if } z \in \mathsf{con}(\mathbf{p}_1\mathbf{p}_2\cdots\mathbf{p}_{i-1}), \\ \mathsf{ef} & \text{if } z = y, \\ \mathsf{f} & \text{otherwise.} \end{cases}$$

Then

$$\begin{aligned}\mathbf{r}\chi &= (x \cdot \mathbf{p}_1 \cdot \mathbf{p}_2 \cdots \mathbf{p}_{i-1} \cdot y \cdot \mathbf{p}_{i+1} \cdots \mathbf{p}_k \cdot x)\chi \\ &= \mathsf{f} \cdot \mathsf{e} \cdot \mathsf{e} \cdots \mathsf{e} \cdot \mathsf{ef} \cdot \mathsf{f} \cdots \mathsf{f} \cdot \mathsf{f} \\ &= \mathsf{fef}.\end{aligned} \tag{6.5}$$

If (H′) holds, then $\mathbf{r}'\chi$ equals the product (6.5) in reverse, that is,

$$\begin{aligned}\mathbf{r}'\chi &= \mathsf{f} \cdot \mathsf{f} \cdots \mathsf{f} \cdot \mathsf{ef} \cdot \mathsf{e} \cdot \mathsf{e} \cdots \mathsf{e} \cdot \mathsf{f} \\ &= 0.\end{aligned}$$

But this is impossible, so that (H′) cannot hold. Therefore, (H) must hold.

Hence, (H) holds in any case. It then follows from (B) and (I) that (i) and (ii) hold. It remains to verify that (iii) also holds. Suppose that $\mathbf{p}_i$ and $\mathbf{p}_i'$ are sandwiches. Let $\psi : \mathscr{A} \to \mathscr{A}^+$ denote the substitution

$$z \mapsto \begin{cases} z & \text{if } z \in \mathsf{con}(\mathbf{p}_i) = \mathsf{con}(\mathbf{p}_i'), \\ x^n & \text{otherwise.}\end{cases}$$

Then since $x(\mathbf{r}\psi)x \overset{(6.1a)}{\approx} x\mathbf{p}_i x$ and $x(\mathbf{r}'\psi)x \overset{(6.1a)}{\approx} x\mathbf{p}_i'x$, the deduction $\{(6.1a), \mathbf{r} \approx \mathbf{r}'\} \vdash x\mathbf{p}_i x \approx x\mathbf{p}_i'x$ holds. Therefore, $x\mathbf{p}_i x \approx x\mathbf{p}_i'x \in \mathsf{id}\{\mathcal{L}_{3,n}\}$ by Lemma 6.4, so that $\mathbf{p}_i \approx \mathbf{p}_i' \in \mathsf{id}\{\mathcal{L}_{3,n}\}$ by Lemma 6.14. □

6.5 An Explicit Identity Basis for $\mathcal{L}_{3,n}$

Theorem 6.16 *The identities* (6.1) *constitute an identity basis for the semigroup* $\mathcal{L}_{3,n}$.

By Lemma 6.10, the identities $\{(6.1)\} \cup \mathsf{id}_{\mathrm{SAN}}\{\mathcal{L}_{3,n}\}$ form an identity basis for the semigroup $\mathcal{L}_{3,n}$. Therefore, to prove Theorem 6.16, it suffices to establish the deduction $(6.1) \vdash \mathsf{id}_{\mathrm{SAN}}\{\mathcal{L}_{3,n}\}$. This is achieved by verifying the following statement for each $m \geq 1$:

(✠$_m$) If $\mathbf{s} \approx \mathbf{s}' \in \mathsf{id}_{\mathrm{SAN}}\{\mathcal{L}_{3,n}\}$ with $|\mathsf{non}(\mathbf{s})| \leq m$, then $(6.1) \vdash \mathbf{s} \approx \mathbf{s}'$.

Lemma 6.17 *The statement* (✠$_1$) *holds.*

Proof Suppose that $\mathbf{s} \approx \mathbf{s}' \in \mathsf{id}_{\mathrm{SAN}}\{\mathcal{L}_{3,n}\}$ with $|\mathsf{non}(\mathbf{s})| = 1$. Then by Lemma 6.13, there exists some finite $\Sigma \subseteq \mathsf{id}_{\mathrm{REF}}\{\mathcal{L}_{3,n}\}$ such that

(A) $\{(6.1), \mathbf{s} \approx \mathbf{s}'\} \sim \{(6.1)\} \cup \Sigma$;
(B) each $\mathbf{r} \approx \mathbf{r}' \in \Sigma$ satisfies $|\,\mathsf{sim}(\mathbf{r})| \leq |\,\mathsf{sim}(\mathbf{s})|$ and $|\,\mathsf{non}(\mathbf{r})| \leq |\,\mathsf{non}(\mathbf{s})| = 1$.

Consider any $\mathbf{r} \approx \mathbf{r}' \in \Sigma$. Generality is not lost by assuming that $\mathbf{r} = x(\prod_{i=1}^{k} \mathbf{p}_i)x$ is the refined sandwich in (6.4). Then Lemma 6.15 implies that $\mathbf{r}' = x(\prod_{i=1}^{k} \mathbf{p}_i')x$ for some $\mathbf{p}_1', \mathbf{p}_2', \ldots, \mathbf{p}_k' \in \mathscr{A}^+$ such that $\mathsf{con}(\mathbf{p}_i) = \mathsf{con}(\mathbf{p}_i')$ for all i. Since $|\,\mathsf{non}(\mathbf{r})| \leq 1$ by (B), it follows that $\mathsf{non}(\mathbf{r}) = \{x\}$. Therefore, every $\mathbf{p}_i$ is a singleton, whence $\mathbf{p}_i = \mathbf{p}_i'$ by Lemma 6.15. The identity $\mathbf{r} \approx \mathbf{r}'$ is thus trivial. Since the identity $\mathbf{r} \approx \mathbf{r}'$ is arbitrary in Σ, every identity in Σ is trivial. Consequently, the deduction $(6.1) \vdash \mathbf{s} \approx \mathbf{s}'$ follows from (A). □

Lemma 6.18 *Suppose that the statement* $(\maltese_m)$ *holds. Then the statement* $(\maltese_{m+1})$ *also holds.*

Proof Suppose that $\mathbf{s} \approx \mathbf{s}' \in \mathsf{id}_{\mathrm{SAN}}\{\mathscr{L}_{3,n}\}$ with $|\,\mathsf{non}(\mathbf{s})| = m + 1$. Then by Lemma 6.13, there exists some finite $\Sigma \subseteq \mathsf{id}_{\mathrm{REF}}\{\mathscr{L}_{3,n}\}$ such that

(A) $\{(6.1), \mathbf{s} \approx \mathbf{s}'\} \sim \{(6.1)\} \cup \Sigma$;
(B) each $\mathbf{r} \approx \mathbf{r}' \in \Sigma$ satisfies $|\,\mathsf{sim}(\mathbf{r})| \leq |\,\mathsf{sim}(\mathbf{s})|$ and $|\,\mathsf{non}(\mathbf{r})| \leq |\,\mathsf{non}(\mathbf{s})| = m + 1$.

Consider any $\mathbf{r} \approx \mathbf{r}' \in \Sigma$. Generality is not lost by assuming that $\mathbf{r} = x(\prod_{i=1}^{k} \mathbf{p}_i)x$ is the refined sandwich in (6.4). Then Lemma 6.15 implies that $\mathbf{r}' = x(\prod_{i=1}^{k} \mathbf{p}_i')x$ for some $\mathbf{p}_1', \mathbf{p}_2', \ldots, \mathbf{p}_k' \in \mathscr{A}^+$ such that $\mathsf{con}(\mathbf{p}_i) = \mathsf{con}(\mathbf{p}_i')$ for all i. By (i) of Lemma 6.15, the words $\mathbf{p}_i$ and $\mathbf{p}_i'$ are both singletons or both sandwiches.

CASE 1: $\mathbf{p}_i$ and $\mathbf{p}_i'$ are both singletons. Then $\mathbf{p}_i = \mathbf{p}_i'$ by (ii) of Lemma 6.15, so that the deduction $(6.1) \vdash \mathbf{p}_i \approx \mathbf{p}_i'$ holds vacuously.
CASE 2: $\mathbf{p}_i$ and $\mathbf{p}_i'$ are both sandwiches. Then $\mathbf{p}_i \approx \mathbf{p}_i' \in \mathsf{id}_{\mathrm{SAN}}\{\mathscr{L}_{3,n}\}$ by (iii) of Lemma 6.15. Since $|\,\mathsf{non}(\mathbf{p}_i)| < |\,\mathsf{non}(\mathbf{r})| \leq m + 1$ by (B), the deduction $(6.1) \vdash \mathbf{p}_i \approx \mathbf{p}_i'$ follows from $(\maltese_m)$.

Therefore, in any case, the deduction $(6.1) \vdash \mathbf{p}_i \approx \mathbf{p}_i'$ holds. Since

$$\mathbf{r} = x\left(\prod_{i=1}^{k} \mathbf{p}_i\right)x \overset{(6.1)}{\approx} x\left(\prod_{i=1}^{k} \mathbf{p}_i'\right)x = \mathbf{r}',$$

the deduction $(6.1) \vdash \mathbf{r} \approx \mathbf{r}'$ also holds. The identity $\mathbf{r} \approx \mathbf{r}'$ is arbitrary in Σ, so that $(6.1) \vdash \Sigma$. Consequently, the deduction $(6.1) \vdash \mathbf{s} \approx \mathbf{s}'$ follows from (A). □

The identity basis (6.1) for the semigroup $\mathscr{L}_{3,n}$ is quite involved. The following is a simpler identity basis for $\mathscr{L}_{3,n}$ that was first stated in Theorem 1.27.

Corollary 6.19 *The identities*

$$x^{n+2} \approx x^2, \tag{6.6a}$$

$$x^{n+1} y x^{n+1} \approx xyx, \tag{6.6b}$$

$$xhykxty \approx yhxkytx, \tag{6.6c}$$

$$x\bigg(\prod_{i=1}^{m}(y_i h_i y_i)\bigg)x \approx x\bigg(\prod_{i=m}^{1}(y_i h_i y_i)\bigg)x, \tag{6.6d}$$

where $m \in \{2, 3, 4, \ldots\}$, *constitute an identity basis for the semigroup* $\mathcal{L}_{3,n}$.

Proof By Theorem 6.16, it suffices to show that (6.1) and (6.6) are equivalent sets of identities. The deduction (6.1) $\vdash$ (6.6) is obvious, while verification of the deduction (6.6) $\vdash$ (6.1) is divided into six cases.

CASE 1: (6.6) $\vdash$ {(6.1a), (6.1n)}. This is obvious.
CASE 2: (6.6) $\vdash$ (6.1b). This deduction holds because

$$\begin{aligned} x^2yx &\overset{(6.6b)}{\approx} x^n(x^{2n}x)(xyx)x^{3n} \\ &\overset{(6.6d)}{\approx} x^n(xyx)(x^{2n}x)x^{3n} \\ &\overset{(6.6b)}{\approx} xyx^{4n+2} \\ &\overset{(6.6a)}{\approx} xyx^2. \end{aligned}$$

CASE 3: (6.6) $\vdash$ (6.1c). This deduction holds because

$$\begin{aligned} xyxzx &\overset{(6.6b)}{\approx} x^{n+1}(x^n yx^{n+1})(x^n zx^{n+1})x^n \\ &\overset{(6.6d)}{\approx} x^{n+1}(x^n zx^{n+1})(x^n yx^{n+1})x^n \\ &\overset{(6.6b)}{\approx} xzxyx. \end{aligned}$$

CASE 4: (6.6) $\vdash$ (6.1e). It is routinely shown that the identities

(A) $x\mathbf{h}_1 y\mathbf{h}_2 x\mathbf{h}_3 y \approx y\mathbf{h}_1 x\mathbf{h}_2 y\mathbf{h}_3 x$,

where $\mathbf{h}_i \in \{\varnothing, h_i\}$, are deducible from {(6.6c), (6.1a)}. For any $p \geq 2$, since

$$\begin{aligned}
x^p\mathbf{h}_1 y\mathbf{h}_2 x\mathbf{h}_3 y &\overset{\text{(A)}}{\approx} x^{p-1} y\mathbf{h}_1 x\mathbf{h}_2 y\mathbf{h}_3 x \\
&\overset{(6.1b)}{\approx} xy\mathbf{h}_1 x^{p-1}\mathbf{h}_2 y\mathbf{h}_3 x \\
&\overset{(6.1a)}{\approx} xy\mathbf{h}_1 x^{p-1+2n}\mathbf{h}_2 y\mathbf{h}_3 x \\
&\overset{\text{(A)}}{\approx} x^2\mathbf{h}_1 yx^{p-2+2n}\mathbf{h}_2 x\mathbf{h}_3 y \\
&\overset{(6.1b)}{\approx} x\mathbf{h}_1 yx^{p-1+2n}\mathbf{h}_2 x\mathbf{h}_3 y \\
&\overset{\text{(A)}}{\approx} y\mathbf{h}_1 x^{p+2n}\mathbf{h}_2 y\mathbf{h}_3 x \\
&\overset{(6.1a)}{\approx} y\mathbf{h}_1 x^p\mathbf{h}_2 y\mathbf{h}_3 x,
\end{aligned}$$

the identities

(B) $x^p\mathbf{h}_1 y\mathbf{h}_2 x\mathbf{h}_3 y \approx y\mathbf{h}_1 x^p\mathbf{h}_2 y\mathbf{h}_3 x$,

where $\mathbf{h}_i \in \{\varnothing, h_i\}$ and $p \geq 1$, are deducible from {(6.6c), (6.1a), (6.1b)}. Further, for any $p \geq 1$ and $q \geq 2$, since

$$\begin{aligned}
x^p\mathbf{h}_1 y^q\mathbf{h}_2 x\mathbf{h}_3 y &\overset{\text{(B)}}{\approx} y\mathbf{h}_1 x^p y^{q-1}\mathbf{h}_2 y\mathbf{h}_3 x \\
&\overset{(6.1b)}{\approx} y^{q-1}\mathbf{h}_1 x^p y\mathbf{h}_2 y\mathbf{h}_3 x \\
&\overset{(6.1a)}{\approx} y^{q-1+2n}\mathbf{h}_1 x^p y\mathbf{h}_2 y\mathbf{h}_3 x \\
&\overset{\text{(B)}}{\approx} y^{q-2+2n} x^p\mathbf{h}_1 y^2\mathbf{h}_2 x\mathbf{h}_3 y \\
&\overset{(6.1b)}{\approx} y^{q-1+2n} x^p\mathbf{h}_1 y\mathbf{h}_2 x\mathbf{h}_3 y \\
&\overset{\text{(B)}}{\approx} y^{q+2n}\mathbf{h}_1 x^p\mathbf{h}_2 y\mathbf{h}_3 x \\
&\overset{(6.1a)}{\approx} y^q\mathbf{h}_1 x^p\mathbf{h}_2 y\mathbf{h}_3 x,
\end{aligned}$$

the identities

(C) $x^p\mathbf{h}_1 y^q\mathbf{h}_2 x\mathbf{h}_3 y \approx y^q\mathbf{h}_1 x^p\mathbf{h}_2 y\mathbf{h}_3 x$,

where $\mathbf{h}_i \in \{\varnothing, h_i\}$, $p \geq 1$, and $q \geq 2$, are deducible from {(6.6c), (6.1a), (6.1b)}. Therefore,

$$\begin{aligned}
(6.6) &\vdash \{(6.6c), (6.1a), (6.1b)\} \qquad \text{by Cases 1–3} \\
&\vdash \{\text{(B)}, \text{(C)}\} \\
&\vdash (6.1e).
\end{aligned}$$

Case 5: (6.6) $\vdash$ (6.1d). The deduction (6.1e) $\vdash (\mathbf{h}_1 xy\mathbf{h}_2)^2 \approx (\mathbf{h}_1 yx\mathbf{h}_2)^2$ clearly holds. Since

$$\begin{aligned}(\mathbf{h}_1 xy\mathbf{h}_2)^3 &= \mathbf{h}_1(xy\mathbf{h}_2\mathbf{h}_1 xy)\mathbf{h}_2\mathbf{h}_1 xy\mathbf{h}_2\\ &\overset{(6.1e)}{\approx} \mathbf{h}_1(yx\mathbf{h}_2\mathbf{h}_1 yx\mathbf{h}_2\mathbf{h}_1 x)y\mathbf{h}_2\\ &\overset{(6.1e)}{\approx} \mathbf{h}_1(xy\mathbf{h}_2\mathbf{h}_1 xx\mathbf{h}_2\mathbf{h}_1 yy)\mathbf{h}_2\\ &\overset{(6.1e)}{\approx} \mathbf{h}_1 yx\mathbf{h}_2\mathbf{h}_1 yx\mathbf{h}_2\mathbf{h}_1 yx\mathbf{h}_2\\ &= (\mathbf{h}_1 yx\mathbf{h}_2)^3,\end{aligned}$$

the deduction (6.1e) $\vdash (\mathbf{h}_1 xy\mathbf{h}_2)^3 \approx (\mathbf{h}_1 yx\mathbf{h}_2)^3$ also holds. Therefore,

$$\begin{aligned}(6.6) &\vdash (6.1e) \qquad \text{by Case 4}\\ &\vdash \{(\mathbf{h}_1 xy\mathbf{h}_2)^2 \approx (\mathbf{h}_1 yx\mathbf{h}_2)^2, (\mathbf{h}_1 xy\mathbf{h}_2)^3 \approx (\mathbf{h}_1 yx\mathbf{h}_2)^3\}\\ &\vdash (6.1d).\end{aligned}$$

Case 6: (6.6) $\vdash$ {(6.1f), (6.1g), ..., (6.1m)}. The deduction {(6.1a), (6.1e)} $\vdash$ (6.1f) holds because

$$\begin{aligned}x\mathbf{h}_1 y\mathbf{h}_2 x\mathbf{h}_3 y &\overset{(6.1e)}{\approx} y\mathbf{h}_1 x\mathbf{h}_2 y\mathbf{h}_3 x\\ &\overset{(6.1a)}{\approx} yy^n\mathbf{h}_1 x\mathbf{h}_2 y\mathbf{h}_3 x\\ &\overset{(6.1e)}{\approx} xy^n\mathbf{h}_1 y\mathbf{h}_2 x\mathbf{h}_3 y.\end{aligned}$$

Since the deduction (6.6) $\vdash$ {(6.1a), (6.1e)} holds by Cases 1 and 4, the deduction (6.6) $\vdash$ (6.1f) also holds. The identities (6.1g)–(6.1m) are deducible from (6.6) in a similar manner. □

6.6 Nonexistence of Irredundant Identity Bases for $\mathcal{L}_{3,n}$

Lemma 6.20 *The identity* $\mathfrak{q}_m^{\uparrow} \approx \mathfrak{q}_m^{\downarrow}$ *is deducible from* $\{(6.1a), \mathfrak{q}_{m+1}^{\uparrow} \approx \mathfrak{q}_{m+1}^{\downarrow}\}$.

Proof Since

$$\mathfrak{q}_m^{\uparrow} \overset{(6.1a)}{\approx} x\left(\prod_{i=1}^{m}(y_i h_i y_i)\right)(x^n x^n x^n)x$$

$$\approx \ x(x^n x^n x^n)\Bigg(\prod_{i=m}^{1}(y_i h_i y_i)\Bigg)x \qquad \text{by } \mathsf{q}_{m+1}^{\uparrow} \approx \mathsf{q}_{m+1}^{\downarrow}$$

$$\overset{(6.1\text{a})}{\approx} \ \mathsf{q}_m^{\downarrow},$$

the lemma follows. □

Theorem 6.21 *The semigroup $\mathcal{L}_{3,n}$ has no irredundant identity bases.*

Proof Let $\Lambda_0 = \{(6.1\text{a}), (6.1\text{b}), (6.6\text{a}), (6.6\text{b}), (6.6\text{c})\}$, and let λ_k denote the identity $\mathsf{q}_{k+1}^{\uparrow} \approx \mathsf{q}_{k+1}^{\downarrow}$. Then by Theorem 6.16 and Corollary 6.19, the set $\Lambda = \Lambda_0 \cup \{\lambda_1, \lambda_2, \lambda_3, \ldots\}$ is an identity basis for $\mathcal{L}_{3,n}$. By Lemma 6.20, the deduction $\Lambda_0 \cup \{\lambda_{k+1}\} \vdash \lambda_k$ holds for each $k \geq 1$. Hence, assumption (a) of Theorem 6.3 holds for Λ, and it remains to verify that assumption (b) of the same theorem also holds for Λ. For this purpose, first note that by Lemma 5.4(i), the following are easily established for any identity $\mathbf{w} \approx \mathbf{w}' \in \mathsf{id}\{\mathcal{Z}_n\}$:

(A) if $\mathbf{w}, \mathbf{w}' \in \mathcal{Q}_m^{\uparrow}$, then $\{(6.1\text{a}), (6.1\text{b})\} \vdash \mathbf{w} \approx \mathbf{w}'$;
(B) if $\mathbf{w}, \mathbf{w}' \in \mathcal{Q}_m^{\downarrow}$, then $\{(6.1\text{a}), (6.1\text{b})\} \vdash \mathbf{w} \approx \mathbf{w}'$.

Let Σ be any identity basis for $\mathcal{L}_{3,n}$. Since $\lambda_i \in \mathsf{id}\{\mathcal{L}_{3,n}\}$, some sequence

$$\mathsf{q}_{k+1}^{\uparrow} = \mathbf{w}_1, \mathbf{w}_2, \ldots, \mathbf{w}_r = \mathsf{q}_{k+1}^{\downarrow}$$

of words exist, where

(C) each identity $\mathbf{w}_j \approx \mathbf{w}_{j+1}$ is directly deducible from some identity $\sigma_j \in \Sigma$.

Further, since $\mathsf{q}_{k+1}^{\uparrow} \in \mathcal{Q}_{k+1}^{\uparrow}$ and $\mathsf{q}_{k+1}^{\uparrow} \approx \mathbf{w}_j \in \mathsf{id}\{\mathcal{L}_{3,n}\}$ for all j, it follows from Lemma 5.5(ii) that $\mathbf{w}_j \in \mathcal{Q}_{k+1}^{\uparrow} \cup \mathcal{Q}_{k+1}^{\downarrow}$ for all j. Now $\mathbf{w}_r = \mathsf{q}_{k+1}^{\downarrow} \in \mathcal{Q}_{k+1}^{\downarrow}$ implies the existence of the least integer ℓ such that $\mathbf{w}_\ell \in \mathcal{Q}_{k+1}^{\downarrow}$. Hence, $\mathbf{w}_1, \mathbf{w}_2, \ldots, \mathbf{w}_{\ell-1} \in \mathcal{Q}_{k+1}^{\uparrow}$. Since

$$\begin{aligned}
\mathsf{q}_{k+1}^{\uparrow} &= \mathbf{w}_1 \\
&\overset{\Lambda_0}{\approx} \mathbf{w}_{\ell-1} && \text{by (A)} \\
&\overset{\sigma_{\ell-1}}{\approx} \mathbf{w}_\ell && \text{by (C)} \\
&\overset{\Lambda_0}{\approx} \mathbf{w}_r && \text{by (B)} \\
&= \mathsf{q}_{k+1}^{\downarrow},
\end{aligned}$$

the deduction $\Lambda_0 \cup \{\sigma_{\ell-1}\} \vdash \lambda_k$ holds. Therefore, assumption (b) of Theorem 6.3 holds with θ_k being $\sigma_{\ell-1}$. □

6.7 Summary

The sufficient condition in Theorem 6.3 for the nonexistence of irredundant identity bases was published in Lee (2018a). The explicit identity basis (6.1) for the semigroup $\mathcal{L}_{3,n}$ and the nonexistence of irredundant identity bases for $\mathcal{L}_{3,n}$ constitute the main results of Lee (2015).

Part II

Involution Semigroups

Involution Semigroups with Infinite Irredundant Identity Bases 7

This chapter is a detailed study of the identities satisfied by the involution semigroup

$$\langle \mathcal{L}_{3,n}, {}^{\star\mathfrak{R}} \rangle = \langle \mathcal{L}_3, {}^{\star} \rangle \times \langle \mathcal{Z}_n, {}^{\mathfrak{R}} \rangle.$$

Refer to Sect. 1.5.1 for the roles $\langle \mathcal{L}_{3,n}, {}^{\star\mathfrak{R}} \rangle$ plays in solving several open problems.

In Sects. 7.1–7.3, identities satisfied by the involution semigroup $\langle \mathcal{L}_{3,n}, {}^{\star\mathfrak{R}} \rangle$ are examined. In Sect. 7.4, an explicit identity basis is exhibited for $\langle \mathcal{L}_{3,n}, {}^{\star\mathfrak{R}} \rangle$ with $\mathfrak{R} > 1$, and the existence of an infinite irredundant identity basis for it is established in Sect. 7.5. In Sect. 7.6, it is shown that $\langle \mathcal{L}_{3,n}, {}^{\star\mathfrak{R}} \rangle$ satisfies the same identities as some proper involution subsemigroup. It turns out that an involution subsemigroup of $\langle \mathcal{L}_{3,3}, {}^{\star 2} \rangle$ of order eight has an infinite irredundant identity basis.

Throughout this chapter, unless otherwise specified, n represents a fixed arbitrary positive integer and $\mathfrak{R}$ represents any number in

$$\mathsf{sq}(n) = \{\mathfrak{R} \in \{1, 2, \ldots, n\} \mid \mathfrak{R}^2 \equiv 1 \pmod{n}\}.$$

Note that $\mathsf{sq}(1) = \mathsf{sq}(2) = \{1\}$. Therefore, whenever the condition $\mathfrak{R} > 1$ is assumed, then it is understood that $n \geq 3$.

7.1 Identities Satisfied by $\langle \mathcal{L}_{3,n}, {}^{\star\mathfrak{R}} \rangle$

In this section, some identities satisfied by $\langle \mathcal{L}_{3,n}, {}^{\star\mathfrak{R}} \rangle$ are presented. These identities are shown in Sect. 7.4 to form an identity basis for the involution semigroup $\langle \mathcal{L}_{3,n}, {}^{\star\mathfrak{R}} \rangle$ with $\mathfrak{R} > 1$.

Lemma 7.1 *The involution semigroup* $\langle \mathcal{L}_{3,n}, {}^{\star\mathfrak{R}} \rangle$ *satisfies the identities*

E. W. H. Lee, *Advances in the Theory of Varieties of Semigroups*, Frontiers in Mathematics, https://doi.org/10.1007/978-3-031-16497-2_7

$$x^{n+2} \approx x^2, \quad x^{n+1}yx \approx xyx, \quad xyx^{n+1} \approx xyx, \tag{7.1a}$$

$$x^2yx \approx xyx^2, \tag{7.1b}$$

$$xyxzx \approx xzxyx, \tag{7.1c}$$

$$(\mathbf{h}_1 xy\mathbf{h}_2)^{n+1} \approx (\mathbf{h}_1 yx\mathbf{h}_2)^{n+1}, \tag{7.1d}$$

$$x^p\mathbf{h}_1 y^q\mathbf{h}_2 x\mathbf{h}_3 y \approx y^q\mathbf{h}_1 x^p\mathbf{h}_2 y\mathbf{h}_3 x, \tag{7.1e}$$

$$x\mathbf{h}_1 y\mathbf{h}_2 x\mathbf{h}_3 y \approx xy^n\mathbf{h}_1 y\mathbf{h}_2 x\mathbf{h}_3 y, \tag{7.1f}$$

$$x\mathbf{h}_1 y\mathbf{h}_2 x\mathbf{h}_3 y \approx y^n x\mathbf{h}_1 y\mathbf{h}_2 x\mathbf{h}_3 y, \tag{7.1g}$$

$$x\mathbf{h}_1 y\mathbf{h}_2 x\mathbf{h}_3 y \approx x\mathbf{h}_1 x^n y\mathbf{h}_2 x\mathbf{h}_3 y, \tag{7.1h}$$

$$x\mathbf{h}_1 y\mathbf{h}_2 x\mathbf{h}_3 y \approx x\mathbf{h}_1 yx^n\mathbf{h}_2 x\mathbf{h}_3 y, \tag{7.1i}$$

$$x\mathbf{h}_1 y\mathbf{h}_2 x\mathbf{h}_3 y \approx x\mathbf{h}_1 y\mathbf{h}_2 xy^n\mathbf{h}_3 y, \tag{7.1j}$$

$$x\mathbf{h}_1 y\mathbf{h}_2 x\mathbf{h}_3 y \approx x\mathbf{h}_1 y\mathbf{h}_2 y^n x\mathbf{h}_3 y, \tag{7.1k}$$

$$x\mathbf{h}_1 y\mathbf{h}_2 x\mathbf{h}_3 y \approx x\mathbf{h}_1 y\mathbf{h}_2 x\mathbf{h}_3 x^n y, \tag{7.1l}$$

$$x\mathbf{h}_1 y\mathbf{h}_2 x\mathbf{h}_3 y \approx x\mathbf{h}_1 y\mathbf{h}_2 x\mathbf{h}_3 yx^n, \tag{7.1m}$$

$$\left(\prod_{i=1}^{r} x_i^{c_i}\right)\mathbf{h}_1\left(\prod_{i=1}^{r} x_i\right)^2 \approx \left(\prod_{i=1}^{r} x_i^{c_i+1}\right)\mathbf{h}_1\left(\prod_{i=1}^{r} x_i\right), \tag{7.1n}$$

$$\left(\prod_{i=1}^{r} x_i^{c_i}\right)\left(\prod_{i=1}^{r} x_i\right)\mathbf{h}_1\left(\prod_{i=1}^{r} x_i\right) \approx \left(\prod_{i=1}^{r} x_i^{c_i+1}\right)\mathbf{h}_1\left(\prod_{i=1}^{r} x_i\right), \tag{7.1o}$$

$$(x^{2n})^\star \approx x^{2n}, \tag{7.1p}$$

$$x^\star\mathbf{h}_1 x^\star \approx x^{\mathfrak{R}}\mathbf{h}_1 x^{\mathfrak{R}}, \tag{7.1q}$$

$$x^\star\mathbf{h}_1 x\mathbf{h}_2 x \approx x^{\mathfrak{R}}\mathbf{h}_1 x\mathbf{h}_2 x, \tag{7.1r}$$

$$x\mathbf{h}_1 x^\star\mathbf{h}_2 x \approx x\mathbf{h}_1 x^{\mathfrak{R}}\mathbf{h}_2 x, \tag{7.1s}$$

$$x\mathbf{h}_1 x\mathbf{h}_2 x^\star \approx x\mathbf{h}_1 x\mathbf{h}_2 x^{\mathfrak{R}}, \tag{7.1t}$$

$$x^\star\mathbf{h}_1 y^{\circledast_1}\mathbf{h}_2 x^{\circledast_2}\mathbf{h}_3 y^{\circledast_3} \approx x^{\mathfrak{R}}\mathbf{h}_1 y^{\circledast_1}\mathbf{h}_2 x^{\circledast_2}\mathbf{h}_3 y^{\circledast_3}, \tag{7.1u}$$

$$x^{\circledast_1}\mathbf{h}_1 y^\star\mathbf{h}_2 x^{\circledast_2}\mathbf{h}_3 y^{\circledast_3} \approx x^{\circledast_1}\mathbf{h}_1 y^{\mathfrak{R}}\mathbf{h}_2 x^{\circledast_2}\mathbf{h}_3 y^{\circledast_3}, \tag{7.1v}$$

$$x^{\circledast_1}\mathbf{h}_1 y^{\circledast_2}\mathbf{h}_2 x^\star\mathbf{h}_3 y^{\circledast_3} \approx x^{\circledast_1}\mathbf{h}_1 y^{\circledast_2}\mathbf{h}_2 x^{\mathfrak{R}}\mathbf{h}_3 y^{\circledast_3}, \tag{7.1w}$$

$$x^{\circledast_1}\mathbf{h}_1 y^{\circledast_2}\mathbf{h}_2 x^{\circledast_3}\mathbf{h}_3 y^\star \approx x^{\circledast_1}\mathbf{h}_1 y^{\circledast_2}\mathbf{h}_2 x^{\circledast_3}\mathbf{h}_3 y^{\mathfrak{R}}, \tag{7.1x}$$

$$x\left(\prod_{i=1}^{m}(y_i\mathbf{h}_i y_i^\star)\right)x^\star \approx x\left(\prod_{i=m}^{1}(y_i\mathbf{h}_i y_i^\star)\right)x^\star, \tag{7.1y}$$

where $\mathbf{h}_i \in \{\varnothing, h_i\}$, $\circledast_i \in \{1, \star\}$, $p, q, c_i, r \in \{1, 2, 3, \ldots\}$, *and* $m \in \{2, 3, 4, \ldots\}$.

Proof Since the group $\langle \mathbb{Z}_n, {}^{\Re} \rangle$ satisfies the identities $\{x^n y \approx y, xy \approx yx, x^\star \approx x^{\Re}\}$, it is easily seen to satisfy the identities (7.1). Hence, it remains to verify that $\langle \mathcal{L}_3, {}^\star \rangle \models (7.1)$.

CASE 1: $\langle \mathcal{L}_3, {}^\star \rangle \models \{(7.1\text{a}), (7.1\text{b}), \ldots, (7.1\text{o})\}$. This follows from Lemmas 6.4 and 6.5 because the plain identities (7.1a)–(7.1o) are satisfied by $\mathcal{L}_3$.

CASE 2: $\langle \mathcal{L}_3, {}^\star \rangle \models \{(7.1\text{p}), (7.1\text{q}), \ldots, (7.1\text{x})\}$. This is routine.

CASE 3: $\langle \mathcal{L}_3, {}^\star \rangle \models (7.1\text{y})$. It is easily seen that $\langle \mathcal{L}_3, {}^\star \rangle$ satisfies the identity $\sigma : xy^\star x^\star \approx xyx^\star$. Since

$$
\begin{aligned}
x\bigg(\prod_{i=1}^{m}(y_i \mathbf{h}_i y_i^\star)\bigg)x^\star &\overset{\sigma}{\approx} x\bigg(\prod_{i=1}^{m}(y_i \mathbf{h}_i y_i^\star)\bigg)^{\star} x^\star \\
&\overset{(\text{inv})}{\approx} x\bigg(\prod_{i=m}^{1}(y_i \mathbf{h}_i^\star y_i^\star)\bigg)x^\star \\
&\overset{\sigma}{\approx} x\bigg(\prod_{i=m}^{1}(y_i \mathbf{h}_i y_i^\star)\bigg)x^\star,
\end{aligned}
$$

the present case holds. □

7.2 Connected Identities and ⋆-Sandwich Identities

In this section, a specific type of identities, called *⋆-sandwich identities*, is introduced. It is shown in Lemma 7.8 that the involution semigroup $\langle \mathcal{L}_{3,n}, {}^{\star\Re} \rangle$ has an identity basis that consists of $\{(\text{inv}), (7.1)\}$ and ⋆-sandwich identities.

7.2.1 Connected Identities

A word $\mathbf{w} \in (\mathscr{A} \cup \mathscr{A}^\star)^+_\varnothing$ is *simple* if $\mathsf{occ}(x, \overline{\mathbf{w}}) \leq 1$ for all $x \in \mathscr{A}$; it is *non-simple* otherwise. A simple word that contains only one variable is called a *singleton*.

Lemma 7.2 *Let* $\mathbf{w} \approx \mathbf{w}' \in \mathsf{id}_\mathsf{W}\{\langle \mathcal{L}_3, {}^\star \rangle\}$. *Then* $\mathsf{sim}(\overline{\mathbf{w}}) = \mathsf{sim}(\overline{\mathbf{w}'})$ *and* $\mathsf{non}(\overline{\mathbf{w}}) = \mathsf{non}(\overline{\mathbf{w}'})$.

Proof For each $x \in \mathscr{A}$, let $\varphi_x : \mathscr{A} \to \mathcal{L}_3$ denote the substitution

$$z \mapsto \begin{cases} \mathsf{fef} & \text{if } z = x, \\ \mathsf{f} & \text{otherwise.} \end{cases}$$

Then for any $\mathbf{w} \in (\mathscr{A} \cup \mathscr{A}^\star)^+$, it is routinely shown that

$$\mathbf{w}\varphi_x = \begin{cases} \mathsf{fef} & \text{if } x \in \mathsf{sim}(\overline{\mathbf{w}}), \\ 0 & \text{if } x \in \mathsf{non}(\overline{\mathbf{w}}), \\ \mathsf{f} & \text{if } x \notin \mathsf{con}(\overline{\mathbf{w}}). \end{cases}$$

It is then easily shown that either $\mathsf{sim}(\overline{\mathbf{w}}) \neq \mathsf{sim}(\overline{\mathbf{w}}')$ or $\mathsf{non}(\overline{\mathbf{w}}) \neq \mathsf{non}(\overline{\mathbf{w}}')$ implies that $\mathbf{w}\varphi_x \neq \mathbf{w}'\varphi_x$ for some $x \in \mathscr{A}$. Therefore, $\mathbf{w} \approx \mathbf{w}' \notin \mathsf{id}_\mathsf{W}\{\langle \mathcal{L}_3, {}^\star \rangle\}$. □

Two words $\mathbf{w}_1, \mathbf{w}_2 \in (\mathscr{A} \cup \mathscr{A}^\star)^+_\varnothing$ are *disjoint* if $\mathsf{con}(\overline{\mathbf{w}}_1) \cap \mathsf{con}(\overline{\mathbf{w}}_2) = \varnothing$. A non-simple word $\mathbf{w}$ is *connected* if it cannot be decomposed as a product of two disjoint nonempty words, that is, if $\mathbf{w} = \mathbf{w}_1\mathbf{w}_2$ for some $\mathbf{w}_1, \mathbf{w}_2 \in (\mathscr{A} \cup \mathscr{A}^\star)^+$, then $\mathsf{con}(\overline{\mathbf{w}}_1) \cap \mathsf{con}(\overline{\mathbf{w}}_2) \neq \varnothing$. An identity is *connected* if it is formed by a pair of connected words.

Lemma 7.3 *Let* $\mathbf{w} \approx \mathbf{w}' \in \mathsf{id}_\mathsf{W}\{\langle \mathcal{L}_{3,n}, {}^{\star\mathfrak{R}} \rangle\}$. *Suppose that* $\mathbf{w} = \mathbf{w}_1\mathbf{w}_2$ *for some disjoint words* $\mathbf{w}_1, \mathbf{w}_2 \in (\mathscr{A} \cup \mathscr{A}^\star)^+$. *Then* $\mathbf{w}' = \mathbf{w}_1'\mathbf{w}_2'$ *for some disjoint words* $\mathbf{w}_1', \mathbf{w}_2' \in (\mathscr{A} \cup \mathscr{A}^\star)^+$ *such that* $\mathbf{w}_1 \approx \mathbf{w}_1', \mathbf{w}_2 \approx \mathbf{w}_2' \in \mathsf{id}_\mathsf{W}\{\langle \mathcal{L}_{3,n}, {}^{\star\mathfrak{R}} \rangle\}$. *In particular,* $\mathbf{w}$ *is connected if and only if* $\mathbf{w}'$ *is connected.*

Proof It follows from Lemma 7.2 that

(A) $\mathsf{con}(\overline{\mathbf{w}}) = \mathsf{con}(\overline{\mathbf{w}}') = \mathscr{C}$.

Let $\varphi : \mathscr{C} \to \mathcal{L}_3$ denote the substitution

$$z \mapsto \begin{cases} \mathsf{e} & \text{if } z \in \mathsf{con}(\overline{\mathbf{w}}_1), \\ \mathsf{f} & \text{if } z \in \mathsf{con}(\overline{\mathbf{w}}_2). \end{cases}$$

Then $\mathbf{w}'\varphi = \mathbf{w}\varphi = \mathsf{ef}$, so that $\mathbf{w}' = \mathbf{w}_1'\mathbf{w}_2'$ for some $\mathbf{w}_1', \mathbf{w}_2' \in (\mathscr{A} \cup \mathscr{A}^\star)^+$ with $\mathsf{con}(\overline{\mathbf{w}}_1') \subseteq \mathsf{con}(\overline{\mathbf{w}}_1)$ and $\mathsf{con}(\overline{\mathbf{w}}_2') \subseteq \mathsf{con}(\overline{\mathbf{w}}_2)$. Since

$$\mathsf{con}(\overline{\mathbf{w}}_1') \cup \mathsf{con}(\overline{\mathbf{w}}_2') = \mathsf{con}(\overline{\mathbf{w}}_1) \cup \mathsf{con}(\overline{\mathbf{w}}_2)$$

by (A) and $\mathsf{con}(\overline{\mathbf{w}}_1) \cap \mathsf{con}(\overline{\mathbf{w}}_2) = \varnothing$ by assumption,

(B) $\mathsf{con}(\overline{\mathbf{w}}_1) = \mathsf{con}(\overline{\mathbf{w}}_1') = \mathscr{C}_1$ and $\mathsf{con}(\overline{\mathbf{w}}_2) = \mathsf{con}(\overline{\mathbf{w}}_2') = \mathscr{C}_2$ with $\mathscr{C}_1 \cap \mathscr{C}_2 = \varnothing$.

Now $\mathbf{w} \approx \mathbf{w}' \in \mathsf{id}\{\langle \mathbb{Z}_n, {}^{\Re}\rangle\}$ with $\mathbf{w} = \mathbf{w}_1\mathbf{w}_2$ and $\mathbf{w}' = \mathbf{w}_1'\mathbf{w}_2'$. Since the group $\mathbb{Z}_n$ possesses a unit element, it follows from (B) that $\mathbf{w}_1 \approx \mathbf{w}_1', \mathbf{w}_2 \approx \mathbf{w}_2' \in \mathsf{id}\{\langle \mathbb{Z}_n, {}^{\Re}\rangle\}$. Therefore, it remains to verify that $\mathbf{w}_1 \approx \mathbf{w}_1', \mathbf{w}_2 \approx \mathbf{w}_2' \in \mathsf{id}\{\langle \mathcal{L}_3, {}^\star\rangle\}$.

Suppose that $\mathbf{w}_1 \approx \mathbf{w}_1' \notin \mathsf{id}\{\langle \mathcal{L}_3, {}^\star\rangle\}$. Then there exists some substitution $\chi : \mathscr{C}_1 \to \mathcal{L}_3$ such that $\mathbf{w}_1\chi \neq \mathbf{w}_1'\chi$. As observed in Lemma 2.2(i), the semigroup $\mathcal{L}_3$ is idempotent-separable, so that $(\mathbf{w}_1\chi)a \neq (\mathbf{w}_1'\chi)a$ for some $a \in \{\mathsf{e}, \mathsf{f}\}$. Let $\widetilde{\chi} : \mathscr{C} \to \mathcal{L}_3$ denote the extension of the substitution χ given by

$$z \mapsto \begin{cases} z\chi & \text{if } z \in \mathscr{C}_1, \\ a & \text{if } z \in \mathscr{C}_2. \end{cases}$$

Then $\mathbf{w}_2\widetilde{\chi} = \mathbf{w}_2'\widetilde{\chi} = a$, so that $\mathbf{w}\widetilde{\chi} = (\mathbf{w}_1\chi)a$ and $\mathbf{w}'\widetilde{\chi} = (\mathbf{w}_1'\chi)a$. Therefore, $\mathbf{w}\widetilde{\chi} \neq \mathbf{w}'\widetilde{\chi}$, whence the contradiction $\mathbf{w} \approx \mathbf{w}' \notin \mathsf{id}\{\langle \mathcal{L}_3, {}^\star\rangle\}$ is obtained. Thus, $\mathbf{w}_1 \approx \mathbf{w}_1' \in \mathsf{id}\{\langle \mathcal{L}_3, {}^\star\rangle\}$. By a symmetrical argument, $\mathbf{w}_2 \approx \mathbf{w}_2' \in \mathsf{id}\{\langle \mathcal{L}_3, {}^\star\rangle\}$. □

Any word $\mathbf{u} \in (\mathscr{A} \cup \mathscr{A}^\star)^+$ can be uniquely decomposed into a product

$$\mathbf{u} = \mathbf{p}_1\mathbf{p}_2 \cdots \mathbf{p}_m,$$

where $\mathbf{p}_1, \mathbf{p}_2, \ldots, \mathbf{p}_m$ are pairwise disjoint words, each of which is either a singleton or a connected word; this is called the *natural decomposition* of $\mathbf{u}$.

Lemma 7.4 *The involution semigroup $\langle \mathcal{L}_{3,n}, {}^{\star\Re}\rangle$ possesses an identity basis that consists of* (inv) *and connected identities.*

Proof In view of Remark 2.10(ii), generality is not lost by assuming that $\{(\text{inv})\} \cup \Sigma$ is an identity basis for $\langle \mathcal{L}_{3,n}, {}^{\star\Re}\rangle$ for some $\Sigma \subseteq \mathsf{id}_\mathsf{W}\{\langle \mathcal{L}_{3,n}, {}^{\star\Re}\rangle\}$. Let $\mathbf{u} \approx \mathbf{u}'$ be any identity in Σ. Suppose that $\mathbf{u} = \prod_{i=1}^m \mathbf{p}_i$ is the natural decomposition of $\mathbf{u}$. Then it follows from Lemma 7.3 that $\mathbf{u}'$ can be decomposed as $\mathbf{u}' = \prod_{i=1}^m \mathbf{p}_i'$ with $\mathbf{p}_i \approx \mathbf{p}_i' \in \mathsf{id}_\mathsf{W}\{\langle \mathcal{L}_{3,n}, {}^{\star\Re}\rangle\}$ for all i.

CASE 1: $\mathbf{p}_i$ is a singleton, say $\mathbf{p}_i \in \{x, x^\star\}$ for some $x \in \mathscr{A}$. Then $\mathbf{p}_i' \in \{x, x^\star\}$ by Lemma 7.2. But $x \approx x^\star \notin \mathsf{id}_\mathsf{W}\{\langle \mathcal{L}_3, {}^\star\rangle\}$ because $\mathsf{ef} \neq (\mathsf{ef})^\star$ in $\langle \mathcal{L}_3, {}^\star\rangle$. Therefore, $\mathbf{p}_i = \mathbf{p}_i'$.

CASE 2: $\mathbf{p}_i$ is a connected word. Then $\mathbf{p}_i'$ is connected by Lemma 7.3.

Therefore, for each i, the identity $\mathbf{p}_i \approx \mathbf{p}_i'$ is either trivial or connected. Consequently, when the identity $\mathbf{u} \approx \mathbf{u}'$ in $\{(\text{inv})\} \cup \Sigma$ is replaced by the connected identities from $\{\mathbf{p}_i \approx \mathbf{p}_i' \mid 1 \leq i \leq m\}$, the resulting set remains an identity basis for $\langle \mathcal{L}_{3,n}, {}^{\star\Re}\rangle$. □

7.2.2 ⋆-Sandwich Identities

Recall from Chap. 6 that for any finite nonempty subset $\mathscr{X} = \{x_1 \ll x_2 \ll \cdots \ll x_r\}$ of $\mathscr{A}$, the shortest word in the set $\mathscr{X}^{\boxplus} = \{x_1^{c_1} x_2^{c_2} \cdots x_r^{c_r} \mid 1 \leq c_1, c_2, \ldots, c_r \leq n\}$ is $\vec{\mathscr{X}} = x_1 x_2 \cdots x_r$. A connected word $\mathbf{s} \in (\mathscr{A} \cup \mathscr{A}^\star)^+$ is called a *⋆-sandwich* if it is of one of the following forms:

(⋆S1) $\mathbf{s} = x\mathbf{u}x^\star$ for some $x \in \mathscr{A}$ and $\mathbf{u} \in (\mathscr{A} \cup \mathscr{A}^\star)^+_\varnothing$ such that $x \notin \mathsf{con}(\overline{\mathbf{u}})$;
(⋆S2) $\mathbf{s} = x^\star\mathbf{u}x$ for some $x \in \mathscr{A}$ and $\mathbf{u} \in (\mathscr{A} \cup \mathscr{A}^\star)^+_\varnothing$ such that $x \notin \mathsf{con}(\overline{\mathbf{u}})$;
(⋆S3) $\mathbf{s} = \mathbf{x} \prod_{i=1}^{\ell} (\mathbf{u}_i \vec{\mathscr{X}})$ for some $\ell \geq 1$, finite nonempty $\mathscr{X} \subseteq \mathscr{A}$, $\mathbf{x} \in \mathscr{X}^{\boxplus}$, and $\mathbf{u}_i \in (\mathscr{A} \cup \mathscr{A}^\star)^+_\varnothing$ such that

(a) $\vec{\mathscr{X}}, \mathbf{u}_1, \mathbf{u}_2, \ldots, \mathbf{u}_\ell$ are pairwise disjoint;
(b) if $\ell \geq 2$, then $\mathbf{u}_i \neq \varnothing$ for all i and $\min(\overline{\mathbf{u}}_1) \ll \min(\overline{\mathbf{u}}_2) \ll \cdots \ll \min(\overline{\mathbf{u}}_\ell)$.

Specifically, for any $k \in \{1, 2, 3\}$, a ⋆-sandwich from (⋆Sk) is said to be of *type* (⋆Sk). The *level* of the ⋆-sandwich in (⋆S3) is the number ℓ, and the *level* of any ⋆-sandwich in (⋆S1) and (⋆S2) is defined to be one.

Remark 7.5 In (⋆S3), due to (b), the only scenario in which some of the factors $\mathbf{u}_1, \mathbf{u}_2, \ldots, \mathbf{u}_\ell$ of $\mathbf{s}$ can be empty is when $\ell = 1$ and $\mathbf{u}_1 = \varnothing$, that is, $\mathbf{s} = \mathbf{x}\vec{\mathscr{X}}$.

An identity $\mathbf{s} \approx \mathbf{s}'$ is a *⋆-sandwich identity* if $\mathbf{s}$ and $\mathbf{s}'$ are ⋆-sandwiches. Denote by

$$\mathsf{id}_{\mathrm{SAN}}\{\langle \mathcal{L}_{3,n}, {}^{\star\mathfrak{R}} \rangle\}$$

the set of all ⋆-sandwich identities satisfied by the involution semigroup $\langle \mathcal{L}_{3,n}, {}^{\star\mathfrak{R}} \rangle$.

Lemma 7.6 *Suppose that* $\mathbf{w}$ *is any connected word. Then the identities* (7.1) *can be used to convert* $\mathbf{w}$ *into some word* $\mathbf{w}'$ *such that* $\mathsf{h}(\overline{\mathbf{w}'}) = \mathsf{t}(\overline{\mathbf{w}'})$.

Proof Suppose that $\mathsf{h}(\overline{\mathbf{w}}) \neq \mathsf{t}(\overline{\mathbf{w}})$. Then since the word $\mathbf{w}$ is connected, there exist distinct variables $x_1, x_2, \ldots, x_r \in \mathscr{A}$ occurring in $\mathbf{w}$ in the following overlapping manner:

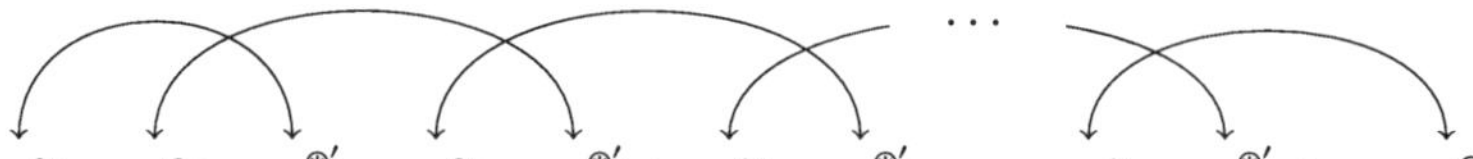

$$\mathbf{w} = x_1^{\circledast_1}\, \mathbf{a}_1\, x_2^{\circledast_2}\, \mathbf{a}_2\, x_1^{\circledast_1'}\, \mathbf{b}_1\, x_3^{\circledast_3}\, \mathbf{a}_3\, x_2^{\circledast_2'}\, \mathbf{b}_2\, x_4^{\circledast_4}\, \mathbf{a}_4\, x_3^{\circledast_3'}\, \mathbf{b}_3 \cdots x_r^{\circledast_r}\, \mathbf{a}_r\, x_{r-1}^{\circledast_{r-1}'}\, \mathbf{b}_{r-1}\, x_r^{\circledast_r'},$$

where $\mathbf{a}_i, \mathbf{b}_i \in (\mathscr{A} \cup \mathscr{A}^\star)^+_\varnothing$ and $\circledast_i, \circledast_i' \in \{1, \star\}$. (Note that $\mathbf{a}_i$ follows $x_i^{\circledast_i}$ and $\mathbf{b}_i$ follows $x_i^{\circledast_i'}$.) Then the identities (7.1b) and (7.1u)–(7.1x) can be used to convert $\mathbf{w}$ into

$$\mathbf{u} = x_1^{p_1}\mathbf{a}_1 x_2^{p_2}\mathbf{a}_2 x_1 \mathbf{b}_1 x_3^{p_3}\mathbf{a}_3 x_2 \mathbf{b}_2 x_4^{p_4}\mathbf{a}_4 x_3 \mathbf{b}_3 \cdots x_r^{p_r}\mathbf{a}_r x_{r-1}\mathbf{b}_{r-1}x_r$$

for some $p_1, p_2, \ldots, p_r \geq 1$. Hence,

$$\begin{aligned}
\mathbf{u} &\overset{(7.1\text{m})}{\approx} x_1^{p_1}\mathbf{a}_1 x_2^{p_2}\mathbf{a}_2 x_1 \mathbf{b}_1 x_3^{p_3}\mathbf{a}_3 x_2 x_1^n \mathbf{b}_2 x_4^{p_4}\mathbf{a}_4 x_3 \mathbf{b}_3 \cdots x_r^{p_r}\mathbf{a}_r x_{r-1}\mathbf{b}_{r-1}x_r \\
&\overset{(7.1\text{m})}{\approx} x_1^{p_1}\mathbf{a}_1 x_2^{p_2}\mathbf{a}_2 x_1 \mathbf{b}_1 x_3^{p_3}\mathbf{a}_3 x_2 x_1^n \mathbf{b}_2 x_4^{p_4}\mathbf{a}_4 x_3 x_1^n \mathbf{b}_3 \cdots x_r^{p_r}\mathbf{a}_r x_{r-1}\mathbf{b}_{r-1}x_r \\
&\;\;\vdots \\
&\overset{(7.1\text{m})}{\approx} x_1^{p_1}\mathbf{a}_1 x_2^{p_2}\mathbf{a}_2 x_1 \mathbf{b}_1 x_3^{p_3}\mathbf{a}_3 x_2 x_1^n \mathbf{b}_2 x_4^{p_4}\mathbf{a}_4 x_3 x_1^n \mathbf{b}_3 \cdots x_r^{p_r}\mathbf{a}_r x_{r-1} x_1^n \mathbf{b}_{r-1}x_r \\
&\overset{(7.1\text{m})}{\approx} x_1^{p_1}\mathbf{a}_1 x_2^{p_2}\mathbf{a}_2 x_1 \mathbf{b}_1 x_3^{p_3}\mathbf{a}_3 x_2 x_1^n \mathbf{b}_2 x_4^{p_4}\mathbf{a}_4 x_3 x_1^n \mathbf{b}_3 \cdots x_r^{p_r}\mathbf{a}_r x_{r-1} x_1^n \mathbf{b}_{r-1}x_r x_1^n \\
&= \mathbf{w}',
\end{aligned}$$

where $\mathsf{h}(\overline{\mathbf{w}}') = x_1 = \mathsf{t}(\overline{\mathbf{w}}')$. □

Lemma 7.7 *Suppose that* $\mathbf{w}$ *is any connected word. Then the identities* (7.1) *can be used to convert* $\mathbf{w}$ *into some* ⋆*-sandwich* $\mathbf{s}$ *with* $\mathsf{sim}(\overline{\mathbf{w}}) = \mathsf{sim}(\overline{\mathbf{s}})$ *and* $\mathsf{non}(\overline{\mathbf{w}}) = \mathsf{non}(\overline{\mathbf{s}})$.

Proof By Lemma 7.6, one can assume that $x_1 = \mathsf{h}(\overline{\mathbf{w}}) = \mathsf{t}(\overline{\mathbf{w}})$. Then $\mathbf{w}$ can be written as

$$\mathbf{w} = x_1^{\circledast_0}\prod_{i=1}^{m_1}(\mathbf{w}_{1,i}\, x_1^{\circledast_i}),$$

where $m_1 \geq 1$, $\circledast_i \in \{1, \star\}$, and $\mathbf{w}_{1,i} \in (\mathscr{A} \cup \mathscr{A}^\star)^+_\varnothing$ with $x_1 \notin \mathsf{con}(\overline{\mathbf{w}}_{1,i})$ for all i. If $m_1 = 1$, then $\mathbf{w}$ is either $x_1\mathbf{w}_{1,1}x_1$, $x_1\mathbf{w}_{1,1}x_1^\star$, $x_1^\star\mathbf{w}_{1,1}x_1$, or $x_1^\star\mathbf{w}_{1,1}x_1^\star$; the first three words are ⋆-sandwiches, while the identities (7.1) can be used to convert the fourth word into a ⋆-sandwich of type (⋆S3):

$$\begin{aligned}
x_1^\star\mathbf{w}_{1,1}x_1^\star &\overset{(7.1\text{q})}{\approx} x_1^{\mathfrak{R}}\mathbf{w}_{1,1}x_1^{\mathfrak{R}} \\
&\overset{(7.1\text{b})}{\approx} x_1^{2\mathfrak{R}-1}\mathbf{w}_{1,1}x_1 \\
&\overset{(7.1\text{a})}{\approx} x_1^{c}\mathbf{w}_{1,1}x_1,
\end{aligned}$$

where $c \in \{1, 2, \ldots, n\}$ is such that $c \equiv 2\mathfrak{R} - 1 \pmod n$. Therefore, assume that $m_1 \geq 2$, so that the identities (7.1q)–(7.1t) can be used to replace any $x_1^\star$ by $x_1^{\mathfrak{R}}$.

Suppose that $\mathbf{w}_{1,i}$ and $\mathbf{w}_{1,j}$ are not disjoint with $i \neq j$, say $x_2 \in \mathsf{con}(\overline{\mathbf{w}}_{1,i}) \cap \mathsf{con}(\overline{\mathbf{w}}_{1,j})$. Then $\mathbf{w} = x_1 \mathbf{h}_1 x_2^{\circledast'_1} \mathbf{h}_2 x_1 \mathbf{h}_3 x_2^{\circledast'_2} \mathbf{h}_4 x_1$ for some $\mathbf{h}_i \in (\mathscr{A} \cup \mathscr{A}^\star)^+_\varnothing$ and $\circledast'_i \in \{1, \star\}$. The identities (7.1u)–(7.1x) can first be used to replace any $x_2^\star$ by $x_2^{\Re}$, and the identities (7.1f)–(7.1m) can then be used to perform the replacement $(x_1, x_2) \mapsto (x_1 x_2^n, x_1^n x_2)$. The resulting word has the form

$$x_1^{c_{0,1}} x_2^{c_{0,2}} \prod_{i=1}^{m_2} (\mathbf{w}_{2,i}\, x_1^{c_{i,1}} x_2^{c_{i,2}}),$$

where $m_2 \geq 2$, $c_{i,j} \geq 1$, and $\mathbf{w}_{2,i} \in (\mathscr{A} \cup \mathscr{A}^\star)^+_\varnothing$ with $x_1, x_2 \notin \mathsf{con}(\overline{\mathbf{w}}_{2,i})$ for all i. Similarly, if $\mathbf{w}_{2,i}$ and $\mathbf{w}_{2,j}$ are not disjoint with $i \neq j$, say $x_3 \in \mathsf{con}(\overline{\mathbf{w}}_{2,i}) \cap \mathsf{con}(\overline{\mathbf{w}}_{2,j})$, then the identities (7.1u)–(7.1x) can first be used to replace any $x_3^\star$ by $x_3^{\Re}$, and the identities (7.1f)–(7.1m) can then be used to perform the replacement $(x_2^{c_{i,2}}, x_3) \mapsto (x_2^{c_{i,2}} x_3^n, x_1^n x_2^n x_3)$. The resulting word has the form

$$x_1^{c_{0,1}} x_2^{c_{0,2}} x_3^{c_{0,3}} \prod_{i=1}^{m_3} (\mathbf{w}_{3,i}\, x_1^{c_{i,1}} x_2^{c_{i,2}} x_3^{c_{i,3}}),$$

where $m_3 \geq 2$, $c_{i,j} \geq 1$, and $\mathbf{w}_{3,i} \in (\mathscr{A} \cup \mathscr{A}^\star)^+_\varnothing$ with $x_1, x_2, x_3 \notin \mathsf{con}(\overline{\mathbf{w}}_{3,i})$ for all i. This can be repeated until a word of the form

$$x_1^{c_{0,1}} x_2^{c_{0,2}} \cdots x_r^{c_{0,r}} \prod_{i=1}^{m_r} (\mathbf{w}_{r,i}\, x_1^{c_{i,1}} x_2^{c_{i,2}} \cdots x_r^{c_{i,r}})$$

is obtained, where $m_r \geq 2$, $c_{i,j} \geq 1$, and $\mathbf{w}_{r,1}, \mathbf{w}_{r,2}, \ldots, \mathbf{w}_{r,m_r} \in (\mathscr{A} \cup \mathscr{A}^\star)^+_\varnothing$ are pairwise disjoint words with $x_1, x_2, \ldots, x_r \notin \mathsf{con}(\overline{\mathbf{w}}_{r,i})$ for all i. It is easily seen that the identities (7.1a) and (7.1b) can be used to convert this word into

$$\mathbf{w}' = x_1^{c_1} x_2^{c_2} \cdots x_r^{c_r} \prod_{i=1}^{m_r} (\mathbf{w}_{r,i}\, x_1 x_2 \cdots x_r),$$

where $1 \leq c_i \leq n$. Let π be the permutation of $\{1, 2, \ldots, r\}$ such that $x_{1\pi} \ll x_{2\pi} \ll \cdots \ll x_{r\pi}$. Then

$$\begin{aligned} \mathbf{w}' &\overset{(7.1a)}{\approx} x_1^{c_1} x_2^{c_2} \cdots x_r^{c_r} \mathbf{w}_{r,1} x_1 x_2 \cdots x_r \prod_{i=2}^{m_r} (\mathbf{w}_{r,i} (x_1 x_2 \cdots x_r)^{n+1}) \\ &\overset{(7.1e)}{\approx} x_{1\pi}^{c_{1\pi}} x_{2\pi}^{c_{2\pi}} \cdots x_{r\pi}^{c_{r\pi}} \mathbf{w}_{r,1} x_{1\pi} x_{2\pi} \cdots x_{r\pi} \prod_{i=2}^{m_r} (\mathbf{w}_{r,i} (x_1 x_2 \cdots x_r)^{n+1}) \end{aligned}$$

$$\overset{(7.1d)}{\approx} x_{1\pi}^{c_{1\pi}} x_{2\pi}^{c_{2\pi}} \cdots x_{r\pi}^{c_{r\pi}} \mathbf{w}_{r,1} x_{1\pi} x_{2\pi} \cdots x_{r\pi} \prod_{i=2}^{m_r} (\mathbf{w}_{r,i} (x_{1\pi} x_{2\pi} \cdots x_{r\pi})^{n+1})$$

$$\overset{(7.1a)}{\approx} x_{1\pi}^{c_{1\pi}} x_{2\pi}^{c_{2\pi}} \cdots x_{r\pi}^{c_{r\pi}} \prod_{i=1}^{m_r} (\mathbf{w}_{r,i}\, x_{1\pi} x_{2\pi} \cdots x_{r\pi}).$$

In summary, the identities (7.1) can be used to convert $\mathbf{w}$ into a word of the form

$$\mathbf{s} = \mathbf{x} \prod_{i=1}^{\ell} (\mathbf{u}_i \vec{\mathscr{X}}),$$

where $\ell \geq 1$, $\mathscr{X} = \{x_1 \ll x_2 \ll \cdots \ll x_r\} \subseteq \mathscr{A}$, $\mathbf{x} = x_1^{c_1} x_2^{c_2} \cdots x_r^{c_r} \in \mathscr{X}^{\boxplus}$, and $\mathbf{u}_i \in (\mathscr{A} \cup \mathscr{A}^\star)^+_\varnothing$ are such that $\vec{\mathscr{X}}, \mathbf{u}_1, \mathbf{u}_2, \ldots, \mathbf{u}_\ell$ are pairwise disjoint. Let c_i' be the number in $\{1, 2, \ldots, n\}$ such that $c_i + 1 \equiv c_i' \pmod n$. If $\mathbf{u}_1 = \varnothing$, then

$$\begin{aligned} \mathbf{s} &= x_1^{c_1} x_2^{c_2} \cdots x_r^{c_r} \vec{\mathscr{X}} \mathbf{u}_2 \vec{\mathscr{X}} \prod_{i=3}^{\ell} (\mathbf{u}_i \vec{\mathscr{X}}) \\ &\overset{(7.1o)}{\approx} x_1^{c_1+1} x_2^{c_2+1} \cdots x_r^{c_r+1} \mathbf{u}_2 \vec{\mathscr{X}} \prod_{i=3}^{\ell} (\mathbf{u}_i \vec{\mathscr{X}}) \\ &\overset{(7.1a)}{\approx} x_1^{c_1'} x_2^{c_2'} \cdots x_r^{c_r'} \mathbf{u}_2 \vec{\mathscr{X}} \prod_{i=3}^{\ell} (\mathbf{u}_i \vec{\mathscr{X}}); \end{aligned}$$

if $\mathbf{u}_k = \varnothing$ for some $k \in \{2, 3, \ldots, \ell\}$, then

$$\begin{aligned} \mathbf{s} &= x_1^{c_1} x_2^{c_2} \cdots x_r^{c_r} \left(\prod_{i=1}^{k-1} (\mathbf{u}_i \vec{\mathscr{X}}) \right) \vec{\mathscr{X}} \left(\prod_{i=k+1}^{\ell} (\mathbf{u}_i \vec{\mathscr{X}}) \right) \\ &\overset{(7.1n)}{\approx} x_1^{c_1+1} x_2^{c_2+1} \cdots x_r^{c_r+1} \left(\prod_{i=1}^{k-1} (\mathbf{u}_i \vec{\mathscr{X}}) \right) \left(\prod_{i=k+1}^{\ell} (\mathbf{u}_i \vec{\mathscr{X}}) \right) \\ &\overset{(7.1a)}{\approx} x_1^{c_1'} x_2^{c_2'} \cdots x_r^{c_r'} \left(\prod_{i=1}^{k-1} (\mathbf{u}_i \vec{\mathscr{X}}) \right) \left(\prod_{i=k+1}^{\ell} (\mathbf{u}_i \vec{\mathscr{X}}) \right). \end{aligned}$$

Hence, for any i, if the factor $\mathbf{u}_i$ is empty, then the $\vec{\mathscr{X}}$ that follows it can be "combined" with the prefix $\mathbf{x}$. Therefore, it can further be assumed that either $\mathbf{s} = \mathbf{x}\vec{\mathscr{X}}$ or $\mathbf{u}_i \neq \varnothing$ for all i. If $\mathbf{s} = \mathbf{x}\vec{\mathscr{X}}$, then the word $\mathbf{s}$ is a ⋆-sandwich of type (⋆S3). Hence, assume that $\mathbf{u}_i \neq \varnothing$ for all i; in this case, it remains to show that if $\ell \geq 2$, then the identities (7.1) can be used to rearrange the factors $\mathbf{u}_1, \mathbf{u}_2, \ldots, \mathbf{u}_\ell$ until $\mathbf{s}$ satisfies (⋆S3b) and so is a ⋆-sandwich of type

($\star$S3). To interchange $\mathbf{u}_i$ and $\mathbf{u}_{i+1}$ when $i \geq 2$, the identity (7.1c) can clearly be used. To interchange $\mathbf{u}_1$ and $\mathbf{u}_2$,

$$\begin{aligned}
\mathbf{s} &\overset{(7.1a)}{\approx} \mathbf{x}\mathbf{u}_1\vec{\mathscr{X}}\mathbf{u}_2\vec{\mathscr{X}}\vec{\mathscr{X}}\vec{\mathscr{X}}^{n-1}\prod_{i=3}^{\ell}(\mathbf{u}_i\vec{\mathscr{X}}) \\
&\overset{(7.1b)}{\approx} \vec{\mathscr{X}}\mathbf{u}_1\vec{\mathscr{X}}\mathbf{u}_2\vec{\mathscr{X}}\mathbf{x}\vec{\mathscr{X}}^{n-1}\prod_{i=3}^{\ell}(\mathbf{u}_i\vec{\mathscr{X}}) \\
&\overset{(7.1c)}{\approx} \vec{\mathscr{X}}\mathbf{u}_2\vec{\mathscr{X}}\mathbf{u}_1\vec{\mathscr{X}}\mathbf{x}\vec{\mathscr{X}}^{n-1}\prod_{i=3}^{\ell}(\mathbf{u}_i\vec{\mathscr{X}}) \\
&\overset{(7.1b)}{\approx} \mathbf{x}\mathbf{u}_2\vec{\mathscr{X}}\mathbf{u}_1\vec{\mathscr{X}}\vec{\mathscr{X}}\vec{\mathscr{X}}^{n-1}\prod_{i=3}^{\ell}(\mathbf{u}_i\vec{\mathscr{X}}) \\
&\overset{(7.1a)}{\approx} \mathbf{x}\mathbf{u}_2\vec{\mathscr{X}}\mathbf{u}_1\vec{\mathscr{X}}\prod_{i=3}^{\ell}(\mathbf{u}_i\vec{\mathscr{X}}).
\end{aligned}$$

Throughout this proof, the identities (7.1) have been used to convert $\mathbf{w}$ into some $\star$-sandwich $\mathbf{s}$. Therefore, $\mathbf{w} \approx \mathbf{s} \in \mathsf{id}\{\langle \mathcal{L}_3, {}^\star\rangle\}$ by Lemma 7.1, so that $\mathsf{sim}(\overline{\mathbf{w}}) = \mathsf{sim}(\overline{\mathbf{s}})$ and $\mathsf{non}(\overline{\mathbf{w}}) = \mathsf{non}(\overline{\mathbf{s}})$ by Lemma 7.2. □

Lemma 7.8 *The identities* $\{(\text{inv}), (7.1)\} \cup \mathsf{id}_{\text{SAN}}\{\langle \mathcal{L}_{3,n}, {}^{\star\Re}\rangle\}$ *constitute an identity basis for the involution semigroup* $\langle \mathcal{L}_{3,n}, {}^{\star\Re}\rangle$.

Proof It follows from Lemmas 7.1 and 7.4 that $\{(\text{inv}), (7.1)\}$ and the connected identities in $\mathsf{id}_{\mathsf{W}}\{\langle \mathcal{L}_{3,n}, {}^{\star\Re}\rangle\}$ form an identity basis for $\langle \mathcal{L}_{3,n}, {}^{\star\Re}\rangle$. The result then follows from Lemma 7.7. □

7.3 Restrictions on $\star$-Sandwich Identities

The present section establishes some properties of $\star$-sandwich identities satisfied by the involution semigroup $\langle \mathcal{L}_{3,n}, {}^{\star\Re}\rangle$. In Sect. 7.3.1, it is shown that any two $\star$-sandwiches that form such an identity must share the same type and level. In Sect. 7.3.2, *refined $\star$-sandwich identities* are introduced; these are identities formed by certain $\star$-sandwiches of level one. It is shown in Lemma 7.15 that refined $\star$-sandwich identities satisfied by the involution semigroup $\langle \mathcal{L}_{3,n}, {}^{\star\Re}\rangle$ are of a very specific form.

7.3.1 Type of ⋆-Sandwiches Forming ⋆-Sandwich Identities

Lemma 7.9 *Suppose that* $\mathbf{s} \approx \mathbf{s}' \in \mathsf{id}_{\mathrm{SAN}}\{\langle \mathcal{L}_{3,n}, {}^{\star\Re} \rangle\}$, *where* $\mathbf{s} = x\mathbf{u}x^\star$ *is the* ⋆*-sandwich in* (⋆S1). *Then* $\mathbf{s}' = x\mathbf{u}'x^\star$ *for some* $\mathbf{u}' \in (\mathscr{A} \cup \mathscr{A}^\star)^+_\varnothing$ *such that* $x \notin \mathsf{con}(\overline{\mathbf{u}}) = \mathsf{con}(\overline{\mathbf{u}}')$. *Consequently,* $\mathbf{s}$ *is of type* (⋆S1) *if and only if* $\mathbf{s}'$ *is of type* (⋆S1).

Proof Let $\varphi : \mathscr{A} \to \mathcal{L}_3$ denote the substitution

$$z \mapsto \begin{cases} \mathsf{fe} & \text{if } z = x, \\ \mathsf{e} & \text{otherwise.} \end{cases}$$

Then $\mathbf{s}'\varphi = \mathbf{s}\varphi = \mathsf{fef}$, so that $\mathbf{s}'\varphi = \mathsf{fe} \cdot \mathsf{e} \cdot \mathsf{e} \cdots \mathsf{e} \cdot \mathsf{ef}$. Therefore, $\mathbf{s} = x\mathbf{u}'x^\star$ for some $\mathbf{u}' \in (\mathscr{A} \cup \mathscr{A}^\star)^+_\varnothing$ such that $x \notin \mathsf{con}(\overline{\mathbf{u}}')$. Further, $\mathsf{con}(\overline{\mathbf{s}}) = \mathsf{con}(\overline{\mathbf{s}}')$ by Lemma 7.2, so that $x \notin \mathsf{con}(\overline{\mathbf{u}}) = \mathsf{con}(\overline{\mathbf{u}}')$. □

Lemma 7.10 *Suppose that* $\mathbf{s} \approx \mathbf{s}' \in \mathsf{id}_{\mathrm{SAN}}\{\langle \mathcal{L}_{3,n}, {}^{\star\Re} \rangle\}$, *where* $\mathbf{s} = x^\star\mathbf{u}x$ *is the* ⋆*-sandwich in* (⋆S2). *Then* $\mathbf{s}' = x^\star\mathbf{u}'x$ *for some* $\mathbf{u}' \in (\mathscr{A} \cup \mathscr{A}^\star)^+_\varnothing$ *such that* $x \notin \mathsf{con}(\overline{\mathbf{u}}) = \mathsf{con}(\overline{\mathbf{u}}')$. *Consequently,* $\mathbf{s}$ *is of type* (⋆S2) *if and only if* $\mathbf{s}'$ *is of type* (⋆S2).

Proof This is symmetrical to Lemma 7.9. □

Lemma 7.11 *Suppose that* $\mathbf{s} \approx \mathbf{s}' \in \mathsf{id}_{\mathrm{SAN}}\{\langle \mathcal{L}_{3,n}, {}^{\star\Re} \rangle\}$, *where*

$$\mathbf{s} = \mathbf{x} \prod_{i=1}^{\ell} (\mathbf{u}_i \vec{\mathscr{X}})$$

is the ⋆*-sandwich in* (⋆S3). *Then*

$$\mathbf{s}' = \mathbf{x} \prod_{i=1}^{\ell} (\mathbf{u}'_i \vec{\mathscr{X}})$$

for some $\mathbf{u}'_1, \mathbf{u}'_2, \ldots, \mathbf{u}'_\ell \in (\mathscr{A} \cup \mathscr{A}^\star)^+_\varnothing$ *such that* $\mathsf{con}(\overline{\mathbf{u}}_i) = \mathsf{con}(\overline{\mathbf{u}}'_i)$ *for all* i. *Consequently,* $\mathbf{s}$ *is of type* (⋆S3) *if and only if* $\mathbf{s}'$ *is of type* (⋆S3).

Proof By Lemma 7.2,

(A) $\mathsf{sim}(\overline{\mathbf{s}}) = \mathsf{sim}(\overline{\mathbf{s}}')$ and $\mathsf{non}(\overline{\mathbf{s}}) = \mathsf{non}(\overline{\mathbf{s}}')$.

By Lemmas 7.9 and 7.10, the ⋆-sandwich $\mathbf{s}'$ is of type (⋆S3). Hence,

(B) $\mathbf{s}' = \mathbf{y}\prod_{i=1}^{\ell'}(\mathbf{u}_i'\vec{\mathscr{Y}})$ for some $\ell' \geq 1$, finite nonempty $\mathscr{Y} \subseteq \mathscr{A}$, $\mathbf{y} \in \mathscr{Y}^{\boxplus}$, and $\mathbf{u}_i' \in (\mathscr{A} \cup \mathscr{A}^\star)^+_\varnothing$ such that
- (a) $\vec{\mathscr{Y}}, \mathbf{u}_1', \mathbf{u}_2', \ldots, \mathbf{u}_{\ell'}'$ are pairwise disjoint;
- (b) if $\ell' \geq 2$, then $\mathbf{u}_i' \neq \varnothing$ for all i and $\min(\overline{\mathbf{u}}_1') \ll \min(\overline{\mathbf{u}}_2') \ll \cdots \ll \min(\overline{\mathbf{u}}_{\ell'}')$.

Suppose that $\mathscr{X} \neq \mathscr{Y}$, say $x \in \mathscr{X}\backslash\mathscr{Y}$. Then $x \in \mathscr{X} \subseteq \mathsf{non}(\overline{\mathbf{s}}) = \mathsf{non}(\overline{\mathbf{s}}')$ by (A), so that (Ba) implies that $x \in \mathsf{con}(\overline{\mathbf{u}}_k')$ for some $k \in \{1, 2, \ldots, \ell'\}$. Let $\varphi : \mathscr{A} \to \mathcal{L}_3$ denote the substitution

$$z \mapsto \begin{cases} \mathsf{e} & \text{if } z \in \mathsf{con}(\overline{\mathbf{u}}_k'), \\ \mathsf{f} & \text{otherwise.} \end{cases}$$

Then $\mathbf{s}'\varphi = \mathsf{fef}$ by (Ba). On the other hand, $\mathbf{x}\varphi = \vec{\mathscr{X}}\varphi \neq \mathsf{f}$ due to $x\varphi = \mathsf{e}$, so that the contradiction $\mathbf{s}\varphi \neq \mathsf{fef}$ is deduced. Therefore, $\mathscr{X} = \mathscr{Y}$. It then follows from (A) and (B) that

(C) $\mathbf{s}' = \mathbf{y}\prod_{i=1}^{\ell'}(\mathbf{u}_i'\vec{\mathscr{X}})$ with $\mathbf{y} \in \mathscr{X}^{\boxplus}$;
(D) $\bigcup_{i=1}^{\ell} \mathsf{con}(\overline{\mathbf{u}}_i) = \bigcup_{i=1}^{\ell'} \mathsf{con}(\overline{\mathbf{u}}_i')$.

Suppose that i is such that $\mathsf{con}(\overline{\mathbf{u}}_i) \nsubseteq \mathsf{con}(\overline{\mathbf{u}}_j')$ for any j. Then in view of (D), there exist distinct $x, y \in \mathsf{con}(\overline{\mathbf{u}}_i)$ such that $x \in \mathsf{con}(\overline{\mathbf{u}}_j')$ and $y \in \mathsf{con}(\overline{\mathbf{u}}_{j'}')$ with $j < j'$. Let $\chi : \mathscr{A} \to \mathcal{L}_3$ denote the substitution

$$z \mapsto \begin{cases} \mathsf{e} & \text{if } z \in \mathsf{con}(\overline{\mathbf{u}}_i), \\ \mathsf{f} & \text{otherwise.} \end{cases}$$

Then $\mathbf{s}\chi = \mathsf{fef}$ by (⋆S3a). Since $\mathscr{X} \cap \mathsf{con}(\overline{\mathbf{u}}_i) = \varnothing$ by (⋆S3a), it follows that $z\chi = \mathsf{f}$ for all $z \in \mathscr{X}$. Hence,

$$\begin{aligned} \mathbf{s}'\chi &= \cdots \mathbf{u}_j'\chi \cdots \vec{\mathscr{X}}\chi \cdots \mathbf{u}_{j'}'\chi \cdots \\ &= \cdots \mathsf{e} \cdots \mathsf{f} \cdots \mathsf{e} \cdots \\ &= 0 \neq \mathbf{s}\chi \end{aligned}$$

is a contradiction. Therefore, the inclusion $\mathsf{con}(\overline{\mathbf{u}}_i) \subseteq \mathsf{con}(\overline{\mathbf{u}}_j')$ holds for some j. By a symmetrical argument, the inclusion $\mathsf{con}(\overline{\mathbf{u}}_j') \subseteq \mathsf{con}(\overline{\mathbf{u}}_k)$ holds for some k, so that $\mathsf{con}(\overline{\mathbf{u}}_i) \subseteq \mathsf{con}(\overline{\mathbf{u}}_j') \subseteq \mathsf{con}(\overline{\mathbf{u}}_k)$. By (⋆S3a), the words $\mathbf{u}_i$ and $\mathbf{u}_k$ are either equal or disjoint. Hence, $\mathbf{u}_i = \mathbf{u}_k$ is the only possibility, so that $\mathsf{con}(\overline{\mathbf{u}}_i) = \mathsf{con}(\overline{\mathbf{u}}_j')$. It has just been shown that for each $i \in \{1, 2, \ldots, \ell\}$, there exists some $j \in \{1, 2, \ldots, \ell'\}$ such that $\mathsf{con}(\overline{\mathbf{u}}_i) = \mathsf{con}(\overline{\mathbf{u}}'_j)$. Therefore, $\ell = \ell'$ by (Ba) and (D), and it follows from (C) that

$$\mathbf{s}' = \mathbf{y} \prod_{i=1}^{\ell} (\mathbf{u}'_i \vec{\mathscr{X}})$$

with $\{\mathsf{con}(\overline{\mathbf{u}}_i) \mid 1 \leq i \leq \ell\} = \{\mathsf{con}(\overline{\mathbf{u}}'_i) \mid 1 \leq i \leq \ell\}$. If $\ell \geq 2$, then (⋆S3b) and (Bb) imply that

(E) $(\mathsf{con}(\overline{\mathbf{u}}_1), \mathsf{con}(\overline{\mathbf{u}}_2), \ldots, \mathsf{con}(\overline{\mathbf{u}}_\ell)) = (\mathsf{con}(\overline{\mathbf{u}}'_1), \mathsf{con}(\overline{\mathbf{u}}'_2), \ldots, \mathsf{con}(\overline{\mathbf{u}}'_\ell))$;

if $\ell = 1$, then (E) holds vacuously.

It remains to show that $\mathbf{x} = \mathbf{y}$. Suppose that $\mathscr{X} = \{x_1 \ll x_2 \ll \cdots \ll x_r\}$. Then by definition, $\vec{\mathscr{X}} = x_1 x_2 \cdots x_r$. Since $\mathbf{x}, \mathbf{y} \in \mathscr{X}^{\boxplus}$, there exist

(F) $c_1, c_2, \ldots, c_r, d_1, d_2, \ldots, d_r \in \{1, 2, \ldots, n\}$

such that $\mathbf{x} = x_1^{c_1} x_2^{c_2} \cdots x_r^{c_r}$ and $\mathbf{y} = x_1^{d_1} x_2^{d_2} \cdots x_r^{d_r}$. For any $j \in \{1, 2, \ldots, r\}$,

$$\begin{aligned} \mathsf{occ}(x_j, \overline{\mathbf{s}}) &= \mathsf{occ}(x_j, \overline{\mathbf{x}}) + \sum_{i=1}^{\ell} (\mathsf{occ}(x_j, \overline{\mathbf{u}}_i) + \mathsf{occ}(x_j, \vec{\mathscr{X}})) \\ &= c_j + \sum_{i=1}^{\ell} (0 + 1) \\ &= c_j + \ell; \end{aligned}$$

similarly, $\mathsf{occ}(x_j, \overline{\mathbf{s}}') = d_j + \ell$. Since $\mathsf{occ}(x_j, \overline{\mathbf{s}}) \equiv \mathsf{occ}(x_j, \overline{\mathbf{s}}') \pmod{n}$ due to the assumption $\mathcal{Z}_n \models \overline{\mathbf{s}} \approx \overline{\mathbf{s}}'$, it follows that $c_j \equiv d_j \pmod{n}$, whence $c_j = d_j$ by (F). Consequently, $\mathbf{x} = \mathbf{y}$. □

Lemma 7.12 *Suppose that* $\mathbf{s} \approx \mathbf{s}' \in \mathsf{id}_{\mathrm{SAN}}\{\langle \mathcal{L}_{3,n}, {}^{\star\Re} \rangle\}$, *where*

$$\mathbf{s} = \mathbf{x} \prod_{i=1}^{\ell} (\mathbf{u}_i \vec{\mathscr{X}}) \quad \text{and} \quad \mathbf{s}' = \mathbf{x} \prod_{i=1}^{\ell} (\mathbf{u}'_i \vec{\mathscr{X}})$$

are the ⋆-sandwiches of type (⋆S3) in Lemma 7.11. Then

$$\{(7.1), \mathbf{s} \approx \mathbf{s}'\} \sim \{(7.1)\} \cup \{x\mathbf{u}_i x \approx x\mathbf{u}'_i x \mid 1 \leq i \leq \ell\}.$$

Proof For each $i \in \{1, 2, \ldots, \ell\}$, let $\varphi_i : \mathscr{A} \to \mathscr{A}^+$ denote the substitution

$$z \mapsto \begin{cases} z & \text{if } z \in \mathsf{con}(\overline{\mathbf{u}}_i) = \mathsf{con}(\overline{\mathbf{u}}_i'), \\ x^{2n} & \text{otherwise.} \end{cases}$$

Note that for any $\mathbf{w} \in (\mathscr{A} \cup \mathscr{A}^\star)^+$ such that $\mathbf{w}$ and $\mathbf{u}_i$ are disjoint, the image $\mathbf{w}\varphi_i$ belongs to $\{x^{2n}, (x^{2n})^\star\}^+$. Therefore, the identity (7.1p) can be used to convert $\mathbf{w}\varphi_i$ into the plain word $\overline{\mathbf{w}\varphi_i}$ in $\{x^{2n}\}^+$. Hence, by ($\star$S3a),

$$\begin{aligned} x(\mathbf{s}\varphi_i)x &= x((\mathbf{x}\mathbf{u}_1\vec{\mathscr{X}} \cdots \mathbf{u}_{i-1}\vec{\mathscr{X}})\varphi_i)\mathbf{u}_i((\vec{\mathscr{X}}\mathbf{u}_{i+1}\vec{\mathscr{X}} \cdots \mathbf{u}_\ell\vec{\mathscr{X}})\varphi_i)x \\ &\overset{(7.1\text{p})}{\approx} x\overline{((\mathbf{x}\mathbf{u}_1\vec{\mathscr{X}} \cdots \mathbf{u}_{i-1}\vec{\mathscr{X}})\varphi_i)}\mathbf{u}_i\overline{((\vec{\mathscr{X}}\mathbf{u}_{i+1}\vec{\mathscr{X}} \cdots \mathbf{u}_\ell\vec{\mathscr{X}})\varphi_i)}x \\ &\overset{(7.1\text{a})}{\approx} x\mathbf{u}_i x, \end{aligned}$$

so that $(7.1) \vdash x(\mathbf{s}\varphi_i)x \approx x\mathbf{u}_i x$. Similarly, the deduction $(7.1) \vdash x(\mathbf{s}'\varphi_i)x \approx x\mathbf{u}_i'x$ holds. Hence, the deduction $\{(7.1), \mathbf{s} \approx \mathbf{s}'\} \vdash x\mathbf{u}_i x \approx x\mathbf{u}_i' x$ holds. Conversely,

$$\begin{aligned} \mathbf{s} &= \overbrace{x_1^{c_1}x_2^{c_2}\cdots x_r^{c_r}}^{\mathbf{x}}\,\mathbf{u}_1\,\overbrace{x_1x_2\cdots x_r}^{\vec{\mathscr{X}}}\prod_{i=2}^{\ell}(\mathbf{u}_i\vec{\mathscr{X}}) \\ &\overset{(7.1\text{k})}{\approx} x_1^{c_1}x_2^{c_2}\cdots x_r^{c_r}\mathbf{u}_1x_r^n x_1x_2\cdots x_r\prod_{i=2}^{\ell}(\mathbf{u}_i\vec{\mathscr{X}}) \\ &\approx x_1^{c_1}x_2^{c_2}\cdots x_r^{c_r}\mathbf{u}_1'x_r^n x_1x_2\cdots x_r\prod_{i=2}^{\ell}(\mathbf{u}_i'\vec{\mathscr{X}}) \qquad \text{by } x\mathbf{u}_i x \approx x\mathbf{u}_i'x \\ &\overset{(7.1\text{k})}{\approx} x_1^{c_1}x_2^{c_2}\cdots x_r^{c_r}\mathbf{u}_1'x_1x_2\cdots x_r\prod_{i=2}^{\ell}(\mathbf{u}_i'\vec{\mathscr{X}}) \\ &= \mathbf{s}'. \end{aligned}$$

Therefore, the deduction $\{(7.1)\} \cup \{x\mathbf{u}_i x \approx x\mathbf{u}_i'x \mid 1 \le i \le \ell\} \vdash \mathbf{s} \approx \mathbf{s}'$ holds. □

7.3.2 Refined $\star$-Sandwich Identities

By Lemmas 7.9–7.11, any identity in $\mathsf{id}_{\mathrm{SAN}}\{\langle \mathcal{L}_{3,n}, {}^{\star\mathfrak{R}}\rangle\}$ is formed by a pair of $\star$-sandwiches that share the same type and level. Hence, it is unambiguous to define the *type* and *level* of a $\star$-sandwich identity $\mathbf{s} \approx \mathbf{s}'$ in $\mathsf{id}_{\mathrm{SAN}}\{\langle \mathcal{L}_{3,n}, {}^{\star\mathfrak{R}}\rangle\}$ to be, respectively, the type and level shared by the ⋆-sandwiches $\mathbf{s}$ and $\mathbf{s}'$. The present subsection examines identities in $\mathsf{id}_{\mathrm{SAN}}\{\langle \mathcal{L}_{3,n}, {}^{\star\Re}\rangle\}$ of level one.

Consider a word

$$\mathbf{r} = x\left(\prod_{i=1}^{k} \mathbf{p}_i\right)x^{\circledast}, \tag{7.2}$$

where $k \geq 1$, $\circledast \in \{1, \star\}$, $x \in \mathscr{A}$, and $\mathbf{p}_1, \mathbf{p}_2, \ldots, \mathbf{p}_k \in (\mathscr{A} \cup \mathscr{A}^\star)^+$ are such that $x, \mathbf{p}_1, \mathbf{p}_2, \ldots, \mathbf{p}_k$ are pairwise disjoint. Note that depending on $\circledast$, the word $\mathbf{r}$ is a level one ⋆-sandwich of type (⋆S1) or (⋆S3). This ⋆-sandwich is *refined* if it satisfies all of the following conditions:

(⋆R1) each $\mathbf{p}_i$ is either a singleton or a ⋆-sandwich;
(⋆R2) if $k \geq 2$ and $\mathbf{p}_1, \mathbf{p}_2, \ldots, \mathbf{p}_k$ are all ⋆-sandwiches, then $\min(\overline{\mathbf{p}}_1) \ll \min(\overline{\mathbf{p}}_k)$.

An identity $\mathbf{r} \approx \mathbf{r}'$ is a *refined ⋆-sandwich identity* if $\mathbf{r}$ and $\mathbf{r}'$ are refined ⋆-sandwiches. Denote by

$$\mathsf{id}_{\mathrm{REF}}\{\langle \mathcal{L}_{3,n}, {}^{\star\Re}\rangle\}$$

the set of all refined ⋆-sandwich identities satisfied by the involution semigroup $\langle \mathcal{L}_{3,n}, {}^{\star\Re}\rangle$.

Lemma 7.13 *Suppose that* $\mathbf{s} \approx \mathbf{s}' \in \mathsf{id}_{\mathrm{SAN}}\{\langle \mathcal{L}_{3,n}, {}^{\star\Re}\rangle\}$. *Then there exists some finite subset* Σ *of* $\mathsf{id}_{\mathrm{REF}}\{\langle \mathcal{L}_{3,n}, {}^{\star\Re}\rangle\}$ *such that*

$$\{(\mathrm{inv}), (7.1), \mathbf{s} \approx \mathbf{s}'\} \sim \{(\mathrm{inv}), (7.1)\} \cup \Sigma.$$

Further, each identity $\mathbf{r} \approx \mathbf{r}'$ *in* Σ *can be chosen so that* $|\,\mathsf{sim}(\overline{\mathbf{r}})| \leq |\,\mathsf{sim}(\overline{\mathbf{s}})|$ *and* $|\,\mathsf{non}(\overline{\mathbf{r}})| \leq |\,\mathsf{non}(\overline{\mathbf{s}})|$.

Proof There are three cases depending on the type of the ⋆-sandwich identity $\mathbf{s} \approx \mathbf{s}'$.

CASE 1: $\mathbf{s} \approx \mathbf{s}'$ is of type (⋆S1). Then by Lemma 7.9,

$$\mathbf{s} = x\mathbf{u}x^\star \;\text{ and }\; \mathbf{s}' = x\mathbf{u}'x^\star$$

for some $x \in \mathscr{A}$ and $\mathbf{u}, \mathbf{u}' \in (\mathscr{A} \cup \mathscr{A}^\star)^+_\varnothing$ such that $x \notin \mathsf{con}(\overline{\mathbf{u}}) = \mathsf{con}(\overline{\mathbf{u}}')$. If $\mathsf{con}(\overline{\mathbf{u}}) = \mathsf{con}(\overline{\mathbf{u}}') = \varnothing$, then the identity $\mathbf{s} \approx \mathbf{s}'$ is trivial, so that the result holds with $\Sigma = \varnothing$. Therefore, assume that $\mathsf{con}(\overline{\mathbf{u}}) = \mathsf{con}(\overline{\mathbf{u}'}) \neq \varnothing$. In what follows, it is shown that the identities $\{(\mathrm{inv}), (7.1)\}$ can be used to convert $\mathbf{s}$ into some refined ⋆-sandwich $\mathbf{r}$. Similarly, the identities $\{(\mathrm{inv}), (7.1)\}$ can be used to convert $\mathbf{s}'$ into some refined $\star$-sandwich $\mathbf{r}'$. Hence, the equivalence

$$\{(\text{inv}), (7.1), \mathbf{s} \approx \mathbf{s}'\} \sim \{(\text{inv}), (7.1), \mathbf{r} \approx \mathbf{r}'\}$$

holds. In particular, the deduction $\{(\text{inv}), (7.1)\} \vdash \mathbf{s} \approx \mathbf{r}$ also holds. Hence, $\mathbf{s} \approx \mathbf{r} \in \mathsf{id}_{\mathsf{W}}\{\langle \mathcal{L}_3, {}^\star\rangle\}$ by Lemma 7.1, so that $|\,\mathsf{sim}(\overline{\mathbf{r}})| = |\,\mathsf{sim}(\overline{\mathbf{s}})|$ and $|\,\mathsf{non}(\overline{\mathbf{r}})| = |\,\mathsf{non}(\overline{\mathbf{s}})|$ by Lemma 7.2.

Let $\mathbf{u} = \prod_{i=1}^{k} \mathbf{p}_i$ be the natural decomposition of $\mathbf{u}$, so that $x, \mathbf{p}_1, \mathbf{p}_2, \ldots, \mathbf{p}_k$ are pairwise disjoint and each $\mathbf{p}_i$ is either a singleton or a connected word. By Lemma 7.7, the identities (7.1) can be used to convert any connected $\mathbf{p}_i$ into some $\star$-sandwich $\mathbf{s}_i$ with $\mathsf{sim}(\overline{\mathbf{p}}_i) = \mathsf{sim}(\overline{\mathbf{s}}_i)$ and $\mathsf{non}(\overline{\mathbf{p}}_i) = \mathsf{non}(\overline{\mathbf{s}}_i)$. Therefore, it can be assumed that $\mathbf{s}$ satisfies ($\star$R1). If $\mathbf{s}$ also satisfies ($\star$R2), then $\mathbf{s}$ is already refined. Hence, suppose that $\mathbf{s}$ does not satisfy ($\star$R2), that is, $\mathbf{p}_1, \mathbf{p}_2, \ldots, \mathbf{p}_k$ are all $\star$-sandwiches with $k \geq 2$, but $\min(\overline{\mathbf{p}}_1) \not\ll \min(\overline{\mathbf{p}}_k)$. Then $\min(\overline{\mathbf{p}}_k) \ll \min(\overline{\mathbf{p}}_1)$ because $\mathbf{p}_1$ and $\mathbf{p}_k$ are disjoint. Let x_1 be the first variable of $\overline{\mathbf{p}}_i$. If $\mathbf{p}_i$ is of type ($\star$S1), then $\mathbf{p}_i = x_1 \mathbf{w} x_1^\star$ for some $\mathbf{w} \in (\mathscr{A} \cup \mathscr{A}^\star)^+_\varnothing$. If $\mathbf{p}_i$ is of type ($\star$S2), then $\mathbf{p}_i = x_1^\star \mathbf{w} x_1$ for some $\mathbf{w} \in (\mathscr{A} \cup \mathscr{A}^\star)^+_\varnothing$, so that $\mathbf{p}_i \overset{(\text{inv})}{\approx} x_1^\star \mathbf{w} (x_1^\star)^\star$. If $\mathbf{p}_i$ is of type ($\star$S3), then

$$\mathbf{p}_i = x_1^{c_1} x_2^{c_2} \cdots x_r^{c_r} \mathbf{w} x_1 x_2 \cdots x_r$$

for some $\mathbf{w} \in (\mathscr{A} \cup \mathscr{A}^\star)^+_\varnothing$ and $\mathscr{X} = \{x_1 \ll x_2 \ll \cdots \ll x_r\} \subseteq \mathscr{A}$, so that

$$\begin{aligned} \mathbf{p}_i &\overset{(7.1\text{m})}{\approx} x_1^{c_1} x_2^{c_2} \cdots x_r^{c_r} \mathbf{w} x_1 x_2 \cdots x_r x_1^{2n} \\ &\overset{(7.1\text{t})}{\approx} x_1^{c_1} x_2^{c_2} \cdots x_r^{c_r} \mathbf{w} x_1 x_2 \cdots x_r x_1^{2n-\Re} x_1^\star. \end{aligned}$$

Hence, regardless of type, there exist $h_i \in \mathscr{A} \cup \mathscr{A}^\star$ and $\mathbf{w}_i \in (\mathscr{A} \cup \mathscr{A}^\star)^+_\varnothing$ such that $\mathbf{p}_i \overset{(7.1)}{\approx} h_i \mathbf{w}_i h_i^\star$. Therefore,

$$\begin{aligned} \mathbf{s} &= x\left(\prod_{i=1}^{k} \mathbf{p}_i\right)x^\star \\ &\overset{(7.1)}{\approx} x\left(\prod_{i=1}^{k} (h_i \mathbf{w}_i h_i^\star)\right)x^\star \\ &\overset{(7.1\text{y})}{\approx} x\left(\prod_{i=k}^{1} (h_i \mathbf{w}_i h_i^\star)\right)x^\star \\ &\overset{(7.1)}{\approx} x\left(\prod_{i=k}^{1} \mathbf{p}_i\right)x^\star. \end{aligned}$$

Since $\min(\overline{\mathbf{p}}_k) \ll \min(\overline{\mathbf{p}}_1)$, the word $\mathbf{r} = x(\prod_{i=k}^{1} \mathbf{p}_i)x^\star$ is the required refined ⋆-sandwich.

CASE 2: $\mathbf{s} \approx \mathbf{s}'$ is of type (⋆S2). Then by Lemma 7.10,

$$\mathbf{s} = x^\star \mathbf{u} x \quad \text{and} \quad \mathbf{s}' = x^\star \mathbf{u}' x$$

for some $x \in \mathscr{A}$ and $\mathbf{u}, \mathbf{u}' \in (\mathscr{A} \cup \mathscr{A}^\star)^+_\varnothing$ such that $x \notin \mathsf{con}(\overline{\mathbf{u}}) = \mathsf{con}(\overline{\mathbf{u}}')$. It is clear that the equivalence $\{(\text{inv}), \mathbf{s} \approx \mathbf{s}'\} \sim \{(\text{inv}), x\mathbf{u}x^\star \approx x\mathbf{u}'x^\star\}$ holds. Since $x\mathbf{u}x^\star \approx x\mathbf{u}'x^\star$ is a ⋆-sandwich identity of type (⋆S1), the result follows from Case 1.

CASE 3: $\mathbf{s} \approx \mathbf{s}'$ is of type (⋆S3). Then by Lemma 7.11,

$$\mathbf{s} = \mathbf{x} \prod_{i=1}^{\ell} (\mathbf{u}_i \vec{\mathscr{X}}) \quad \text{and} \quad \mathbf{s}' = \mathbf{x} \prod_{i=1}^{\ell} (\mathbf{u}'_i \vec{\mathscr{X}})$$

for some $\ell \geq 1$, nonempty $\mathscr{X} \subseteq \mathscr{A}$, $\mathbf{x} \in \mathscr{X}^{\boxplus}$, and $\mathbf{u}_i, \mathbf{u}'_i \in (\mathscr{A} \cup \mathscr{A}^\star)^+_\varnothing$ such that $\mathsf{con}(\overline{\mathbf{u}}_i) = \mathsf{con}(\overline{\mathbf{u}}'_i)$ for each i and $\vec{\mathscr{X}}, \mathbf{u}_1, \mathbf{u}_2, \ldots, \mathbf{u}_\ell$ are pairwise disjoint. By Lemma 7.12, the equivalence

(A) $\{(7.1), \mathbf{s} \approx \mathbf{s}'\} \sim \{(7.1)\} \cup \{x\mathbf{u}_i x \approx x\mathbf{u}'_i x \mid 1 \leq i \leq \ell\}$

holds. It is easily seen that

(B) $|\mathsf{sim}(\overline{x\mathbf{u}_i x})| \leq |\mathsf{sim}(\overline{\mathbf{s}})|$ and $|\mathsf{non}(\overline{x\mathbf{u}_i x})| \leq |\mathsf{non}(\overline{\mathbf{s}})|$ for all i.

For each i, the argument in Case 1 can be repeated to show that the equivalence

(C) $\{(\text{inv}), (7.1), x\mathbf{u}_i x \approx x\mathbf{u}'_i x\} \sim \{(\text{inv}), (7.1), \mathbf{r}_i \approx \mathbf{r}'_i\}$

holds for some $\mathbf{r}_i \approx \mathbf{r}'_i \in \mathsf{id}_{\text{REF}}\{\langle \mathcal{L}_{3,n}, {}^{\star\Re} \rangle\}$ such that

(D) $|\mathsf{sim}(\overline{\mathbf{r}}_i)| = |\mathsf{sim}(\overline{x\mathbf{u}_i x})|$ and $|\mathsf{non}(\overline{\mathbf{r}}_i)| = |\mathsf{non}(\overline{x\mathbf{u}_i x})|$.

Hence, the equivalence

$$\{(\text{inv}), (7.1), \mathbf{s} \approx \mathbf{s}'\} \sim \{(\text{inv}), (7.1)\} \cup \{\mathbf{r}_i \approx \mathbf{r}'_i \mid 1 \leq i \leq \ell\}$$

holds by (A) and (C), where $|\mathsf{sim}(\overline{\mathbf{r}}_i)| \leq |\mathsf{sim}(\overline{\mathbf{s}})|$ and $|\mathsf{non}(\overline{\mathbf{r}}_i)| \leq |\mathsf{non}(\overline{\mathbf{s}})|$ by (B) and (D). □

Lemma 7.14 *Let* $\mathbf{s}, \mathbf{s}' \in (\mathscr{A} \cup \mathscr{A}^\star)^+$ *be any ⋆-sandwiches and* $x \in \mathscr{A}$ *be such that* $x \notin \mathsf{con}(\overline{\mathbf{s}}) = \mathsf{con}(\overline{\mathbf{s}}')$. *Suppose that* $x\mathbf{s}x \approx x\mathbf{s}'x \in \mathsf{id}_{\mathsf{W}}\{\langle \mathcal{L}_{3,n}, {}^{\star\Re} \rangle\}$. *Then* $\mathbf{s} \approx \mathbf{s}' \in \mathsf{id}_{\mathsf{W}}\{\langle \mathcal{L}_{3,n}, {}^{\star\Re} \rangle\}$.

Proof This is established in the same manner as Lemma 6.14. □

Lemma 7.15 *Let* $\Re > 1$. *Suppose that* $\mathbf{r} \approx \mathbf{r}' \in \mathsf{id}_{\mathrm{REF}}\{\langle \mathcal{L}_{3,n}, {}^{\star\Re}\rangle\}$, *where*

$$\mathbf{r} = x\left(\prod_{i=1}^{k} \mathbf{p}_i\right)x^{\circledast}$$

is the refined $\star$*-sandwich in* (7.2). *Then*

$$\mathbf{r}' = x\left(\prod_{i=1}^{k} \mathbf{p}_i'\right)x^{\circledast}$$

for some $\mathbf{p}_1', \mathbf{p}_2', \ldots, \mathbf{p}_k' \in (\mathscr{A} \cup \mathscr{A}^\star)^+$ *such that each* $\mathbf{p}_i'$ *is either a singleton or a* $\star$*-sandwich,* $x, \mathbf{p}_1', \mathbf{p}_2', \ldots, \mathbf{p}_k'$ *are pairwise disjoint, and* $\mathsf{con}(\overline{\mathbf{p}}_i) = \mathsf{con}(\overline{\mathbf{p}}_i')$ *for all* i. *Further, for each* i,

(i) $\mathbf{p}_i$ *and* $\mathbf{p}_i'$ *are either both singletons or both* $\star$*-sandwiches;*
(ii) *if* $\mathbf{p}_i$ *and* $\mathbf{p}_i'$ *are both singletons, then* $\mathbf{p}_i = \mathbf{p}_i'$;
(iii) *if* $\mathbf{p}_i$ *and* $\mathbf{p}_i'$ *are both* $\star$*-sandwiches, then* $\mathbf{p}_i \approx \mathbf{p}_i' \in \mathsf{id}_{\mathrm{SAN}}\{\langle \mathcal{L}_{3,n}, {}^{\star\Re}\rangle\}$.

Proof It follows from Lemmas 7.9 and 7.11 that $\mathbf{r}' = x\mathbf{u}'x^{\circledast}$ for some $\mathbf{u}' \in (\mathscr{A} \cup \mathscr{A}^\star)^+$ with

(A) $x \notin \mathsf{con}(\overline{\mathbf{u}}') = \mathsf{con}(\overline{\mathbf{p}}_1\overline{\mathbf{p}}_2 \cdots \overline{\mathbf{p}}_k)$.

Let $\mathbf{u}' = \prod_{i=1}^{k'} \mathbf{p}_i'$ be the natural decomposition of $\mathbf{u}'$, so that

$$\mathbf{r}' = x\left(\prod_{i=1}^{k'} \mathbf{p}_i'\right)x^{\circledast},$$

where

(B) each $\mathbf{p}_i'$ is either a singleton or a $\star$-sandwich and $x, \mathbf{p}_1', \mathbf{p}_2', \ldots, \mathbf{p}_k'$ are pairwise disjoint.

It is first shown that $k = k'$ and $\mathsf{con}(\overline{\mathbf{p}}_i) = \mathsf{con}(\overline{\mathbf{p}}_i')$ for all i. Let $\varphi : \mathscr{A} \to \mathscr{A}^+$ denote the substitution

$$z \mapsto \begin{cases} y_i^{2n} & \text{if } z \in \mathsf{con}(\overline{\mathbf{p}}_i), \\ x^{2n} & \text{otherwise.} \end{cases}$$

Note that for any $\mathbf{w} \in (\mathscr{A} \cup \mathscr{A}^\star)^+$, the image $\mathbf{w}\varphi$ belongs to

$$(\mathscr{A}\varphi \cup (\mathscr{A}\varphi)^\star)^+ = \{x^{2n}, y_1^{2n}, y_2^{2n}, \ldots, y_k^{2n}, (x^{2n})^\star, (y_1^{2n})^\star, (y_2^{2n})^\star, \ldots, (y_k^{2n})^\star\}^+.$$

Therefore, the identity (7.1p) can be used to convert $\mathbf{w}\varphi$ into the plain word $\overline{\mathbf{w}\varphi}$ in

$$\{x^{2n}, y_1^{2n}, y_2^{2n}, \ldots, y_k^{2n}\}^+.$$

Hence,

$$\mathbf{r}\varphi \overset{(7.1\text{p})}{\approx} \overline{\mathbf{r}\varphi} = x^{2n}\left(\prod_{i=1}^{k} y_i^{2n|\mathbf{p}_i|}\right)x^{2n} \text{ and } \mathbf{r}'\varphi \overset{(7.1\text{p})}{\approx} \overline{\mathbf{r}'\varphi} = x^{2n}\left(\prod_{i=1}^{k'} \overline{\mathbf{p}_i'\varphi}\right)x^{2n}.$$

Since $\overline{\mathbf{r}\varphi} \approx \overline{\mathbf{r}'\varphi} \in \mathsf{id}\{\mathcal{L}_3\}$ and $\overline{\mathbf{r}\varphi} \in \mathscr{P}_k^{\uparrow}$, it follows from Lemma 5.5(i) that $\overline{\mathbf{r}'\varphi} \in \mathscr{P}_k^{\uparrow} \cup \mathscr{P}_k^{\downarrow}$, that is,

(C) $x^{2n}(\prod_{i=1}^{k'} \overline{\mathbf{p}_i'\varphi})x^{2n} \in \mathscr{P}_k^{\uparrow} \cup \mathscr{P}_k^{\downarrow}$.

Further, since

$$\begin{aligned} \mathsf{con}(\overline{\mathbf{p}_i'\varphi}) &\subseteq \mathsf{con}(\overline{\mathbf{u}'\varphi}) \\ &\overset{(\mathrm{A})}{=} \mathsf{con}(\overline{(\mathbf{p}_1\mathbf{p}_2\cdots\mathbf{p}_k)\varphi}) \\ &= \{y_1, y_2, \ldots, y_k\}, \end{aligned}$$

it follows that

(D) $\mathsf{con}(\overline{\mathbf{p}_i'\varphi}) \subseteq \{y_1, y_2, \ldots, y_k\}$ for each $i \in \{1, 2, \ldots, k'\}$.

By (B), each $\mathbf{p}_i'$ is a singleton or a ⋆-sandwich. If $\mathbf{p}_i'$ is a singleton, then clearly $\mathsf{con}(\overline{\mathbf{p}_i'\varphi}) = \{y_j\}$ for some j. Suppose that $\mathbf{p}_i'$ is a ⋆-sandwich. Then $\mathbf{p}_i'$ is connected, so that by (C), the word $\overline{\mathbf{p}_i'\varphi}$ is a connected factor of $\overline{\mathbf{r}'\varphi} \in \mathscr{P}_k^{\uparrow} \cup \mathscr{P}_k^{\downarrow}$. The connected factors of words in $\mathscr{P}_k^{\uparrow} \cup \mathscr{P}_k^{\downarrow}$ are

$$\begin{aligned} &x^{s_1}y_1^{t_1}y_2^{t_2}\cdots y_k^{t_k}x^{s_2}, && s_1, s_2 \geq 1,\ t_1, t_2, \ldots, t_k \geq 2; \\ &x^{s_1}y_k^{t_k}y_{k-1}^{t_{k-1}}\cdots y_1^{t_1}x^{s_2}, && s_1, s_2 \geq 1,\ t_1, t_2, \ldots, t_k \geq 2; \\ &x^t,\ y_1^t,\ y_2^t,\ \ldots,\ y_k^t, && t \geq 2. \end{aligned}$$

But since $x \notin \mathsf{con}(\overline{\mathbf{p}_i'\varphi})$ by (D), the word $\overline{\mathbf{p}_i'\varphi}$ is one of $y_1^t, y_2^t, \ldots, y_k^t$, so that $\mathsf{con}(\overline{\mathbf{p}_i'\varphi}) = \{y_j\}$ for some j. Hence, regardless of whether $\mathbf{p}_i'$ is a singleton or a ⋆-sandwich,

(E) $\mathsf{con}(\overline{\mathbf{p}_i'\varphi}) = \{y_j\}$ for some j.

It follows that the inclusion $\mathsf{con}(\overline{\mathbf{p}}_i') \subseteq \mathsf{con}(\overline{\mathbf{p}}_j)$ holds for some j. By a symmetrical argument, the inclusion $\mathsf{con}(\overline{\mathbf{p}}_j) \subseteq \mathsf{con}(\overline{\mathbf{p}}_m')$ holds for some m, so that $\mathsf{con}(\overline{\mathbf{p}}_i') \subseteq \mathsf{con}(\overline{\mathbf{p}}_j) \subseteq \mathsf{con}(\overline{\mathbf{p}}_m')$. Now by (B), the words $\mathbf{p}_i'$ and $\mathbf{p}_m'$ are either equal or disjoint. Therefore, $\mathbf{p}_i' = \mathbf{p}_m'$ is the only possibility, whence $\mathsf{con}(\overline{\mathbf{p}}_i') = \mathsf{con}(\overline{\mathbf{p}}_j)$. It has just been shown that for each $i \in \{1, 2, \ldots, k'\}$, there exists some $j \in \{1, 2, \ldots, k\}$ such that $\mathsf{con}(\overline{\mathbf{p}}_i') = \mathsf{con}(\overline{\mathbf{p}}_j)$. Since $\mathsf{con}(\overline{\mathbf{p}}_1\overline{\mathbf{p}}_2\cdots\overline{\mathbf{p}}_k) = \mathsf{con}(\overline{\mathbf{p}}_1'\overline{\mathbf{p}}_2'\cdots\overline{\mathbf{p}}_{k'}')$ by (A), it follows that

(F) $k = k'$;
(G) there exists a one-to-one correspondence between the sets

$$\mathsf{con}(\overline{\mathbf{p}}_1), \mathsf{con}(\overline{\mathbf{p}}_2), \ldots, \mathsf{con}(\overline{\mathbf{p}}_k) \text{ and } \mathsf{con}(\overline{\mathbf{p}}_1'), \mathsf{con}(\overline{\mathbf{p}}_2'), \ldots, \mathsf{con}(\overline{\mathbf{p}}_k').$$

Further, by (C) and (F), either $x^{2n}(\prod_{i=1}^{k}\overline{\mathbf{p}_i'\varphi})x^{2n} \in \mathscr{P}_k^{\uparrow}$ or $x^{2n}(\prod_{i=1}^{k}\overline{\mathbf{p}_i'\varphi})x^{2n} \in \mathscr{P}_k^{\downarrow}$. Hence, by (E),

$$(\overline{\mathbf{p}_1'\varphi}, \overline{\mathbf{p}_2'\varphi}, \ldots, \overline{\mathbf{p}_k'\varphi}) \in \{(y_1^{t_1}, y_2^{t_2}, \ldots, y_k^{t_k}), (y_k^{t_k}, y_{k-1}^{t_{k-1}}, \ldots, y_1^{t_1}) \mid t_1, t_2, \ldots, t_k \geq 2\}.$$

It then follows from (G) that either

(H) $(\mathsf{con}(\overline{\mathbf{p}}_1'), \mathsf{con}(\overline{\mathbf{p}}_2'), \ldots, \mathsf{con}(\overline{\mathbf{p}}_k')) = (\mathsf{con}(\overline{\mathbf{p}}_1), \mathsf{con}(\overline{\mathbf{p}}_2), \ldots, \mathsf{con}(\overline{\mathbf{p}}_k))$ or
(H′) $(\mathsf{con}(\overline{\mathbf{p}}_1'), \mathsf{con}(\overline{\mathbf{p}}_2'), \ldots, \mathsf{con}(\overline{\mathbf{p}}_k')) = (\mathsf{con}(\overline{\mathbf{p}}_k), \mathsf{con}(\overline{\mathbf{p}}_{k-1}), \ldots, \mathsf{con}(\overline{\mathbf{p}}_1))$.

Now Lemma 7.2 implies that

(I) $\mathsf{sim}(\overline{\mathbf{r}}) = \mathsf{sim}(\overline{\mathbf{r}}')$ and $\mathsf{non}(\overline{\mathbf{r}}) = \mathsf{non}(\overline{\mathbf{r}}')$.

If $k = 1$, then (H) clearly holds, so assume that $k \geq 2$. There are two cases to consider.

CASE 1: $\mathbf{p}_1, \mathbf{p}_2, \ldots, \mathbf{p}_k$ are all $\star$-sandwiches. Then in view of (B) and (I), either (H) or (H′) implies that $\mathbf{p}_1', \mathbf{p}_2', \ldots, \mathbf{p}_k'$ are also $\star$-sandwiches. Hence, (H) must hold by ($\star$R2).

CASE 2: $\mathbf{p}_i$ is a singleton for some i. Then $\mathbf{p}_i \in \{y, y^\star\}$ for some $y \in \mathsf{sim}(\overline{\mathbf{r}})$. It follows from either (H) or (H′) that $\mathsf{con}(\overline{\mathbf{p}}_j') = \{y\}$ for some j, whence $\mathbf{p}_j' \in \{y, y^\star\}$ by

(I). If $\mathbf{p}_i \neq \mathbf{p}'_j$, so that $(\mathbf{p}_i, \mathbf{p}'_j) \in \{(y, y^\star), (y^\star, y)\}$, then since the group $\langle \mathbb{Z}_n, {}^{\Re} \rangle$ satisfies $\mathbf{r} \approx \mathbf{r}'$ and possesses a unit element, it also satisfies the identity $y^\star \approx y$; this is impossible because $\mathsf{g}^\star \neq \mathsf{g}$ under the assumption $\Re > 1$. Therefore, $\mathbf{p}_i = \mathbf{p}'_j \in \{y, y^\star\}$. Now since $k \geq 2$, either $1 < i$ or $i < k$. By symmetry, assume that $1 < i$. Let $\chi : \mathscr{A} \to \mathcal{L}_3$ denote the substitution

$$z \mapsto \begin{cases} \mathsf{e} & \text{if } z \in \mathsf{con}(\overline{\mathbf{p}}_1 \overline{\mathbf{p}}_2 \cdots \overline{\mathbf{p}}_{i-1}), \\ \mathsf{ef} & \text{if } z = y \text{ and } \mathbf{p}_i = y, \\ \mathsf{fe} & \text{if } z = y \text{ and } \mathbf{p}_i = y^\star, \\ \mathsf{f} & \text{otherwise.} \end{cases}$$

Then

$$\begin{aligned} \mathbf{r}\chi &= (x \cdot \mathbf{p}_1 \cdot \mathbf{p}_2 \cdots \mathbf{p}_{i-1} \cdot \mathbf{p}_i \cdot \mathbf{p}_{i+1} \cdots \mathbf{p}_k \cdot x)\chi \\ &= \mathsf{f} \cdot \mathsf{e} \cdot \mathsf{e} \cdots \mathsf{e} \cdot \mathsf{ef} \cdot \mathsf{f} \cdots \mathsf{f} \cdot \mathsf{f} \qquad (7.3) \\ &= \mathsf{fef}. \end{aligned}$$

If (H′) holds, then $\mathbf{r}'\chi$ equals the product (7.3) in reverse, that is,

$$\begin{aligned} \mathbf{r}'\chi &= \mathsf{f} \cdot \mathsf{f} \cdots \mathsf{f} \cdot \mathsf{ef} \cdot \mathsf{e} \cdot \mathsf{e} \cdots \mathsf{e} \cdot \mathsf{f} \\ &= 0. \end{aligned}$$

But this is impossible, so that (H′) cannot hold. Therefore, (H) must hold.

Hence, (H) holds in any case. Then (i) and (ii) hold by (B), (I), and the assumption that $\Re > 1$. It remains to verify that (iii) also holds. Suppose that $\mathbf{p}_i$ and $\mathbf{p}'_i$ are ⋆-sandwiches. Let $\psi : \mathscr{A} \to \mathscr{A}^+$ denote the substitution

$$z \mapsto \begin{cases} z & \text{if } z \in \mathsf{con}(\overline{\mathbf{p}}_i) = \mathsf{con}(\overline{\mathbf{p}}'_i), \\ x^{2n} & \text{otherwise.} \end{cases}$$

Then the deduction $\{(\text{inv}), (7.1\text{a}), (7.1\text{p})\} \vdash \{x(\mathbf{r}\psi)x \approx x\mathbf{p}_i x, x(\mathbf{r}'\psi)x \approx x\mathbf{p}'_i x\}$ holds, so the deduction $\{(\text{inv}), (7.1\text{a}), (7.1\text{p}), \mathbf{r} \approx \mathbf{r}'\} \vdash x\mathbf{p}_i x \approx x\mathbf{p}'_i x$ follows. Therefore, $x\mathbf{p}_i x \approx x\mathbf{p}'_i x \in \mathsf{id}_{\mathsf{W}}\{\langle \mathcal{L}_{3,n}, {}^{\star\Re} \rangle\}$ by Lemma 7.1, whence $\mathbf{p}_i \approx \mathbf{p}'_i \in \mathsf{id}_{\mathsf{W}}\{\langle \mathcal{L}_{3,n}, {}^{\star\Re} \rangle\}$ by Lemma 7.14. □

7.4 An Explicit Identity Basis for $\langle \mathcal{L}_{3,n}, {}^{\star\mathfrak{R}} \rangle$ with $\mathfrak{R} > 1$

Recall that the main goal of the present chapter is to show that for each $\mathfrak{R} > 1$, the involution semigroup $\langle \mathcal{L}_{3,n}, {}^{\star\mathfrak{R}} \rangle$ possesses an infinite irredundant identity basis. The proof of this result, given in Sect. 7.5, requires an explicit identity basis. In Sect. 7.4.1, the identities $\{(\text{inv}), (7.1)\}$ are shown to form an identity basis for $\langle \mathcal{L}_{3,n}, {}^{\star\mathfrak{R}} \rangle$ with $\mathfrak{R} > 1$. In Sect. 7.4.2, a simpler identity basis for this involution semigroup is deduced. Although the identity basis $\{(\text{inv}), (7.1)\}$ serves the purpose of Sect. 7.5, a simpler identity basis vastly reduces the number of deductions required.

7.4.1 An Identity Basis from (7.1)

Theorem 7.16 *Let* $\mathfrak{R} > 1$. *Then the identities* $\{(\text{inv}), (7.1)\}$ *constitute an identity basis for the involution semigroup* $\langle \mathcal{L}_{3,n}, {}^{\star\mathfrak{R}} \rangle$.

By Lemma 7.8, the identities $\{(\text{inv}), (7.1)\} \cup \mathsf{id}_{\text{SAN}}\{\langle \mathcal{L}_{3,n}, {}^{\star\mathfrak{R}} \rangle\}$ form an identity basis for the involution semigroup $\langle \mathcal{L}_{3,n}, {}^{\star\mathfrak{R}} \rangle$. In the remainder of this subsection, the following statement is established for each $m \geq 1$:

($\clubsuit_m$) If $\mathbf{s} \approx \mathbf{s}' \in \mathsf{id}_{\text{SAN}}\{\langle \mathcal{L}_{3,n}, {}^{\star\mathfrak{R}} \rangle\}$ with $|\,\mathsf{non}(\overline{\mathbf{s}})| \leq m$, then $\{(\text{inv}), (7.1)\} \vdash \mathbf{s} \approx \mathbf{s}'$.

The deduction $\{(\text{inv}), (7.1)\} \vdash \mathsf{id}_{\text{SAN}}\{\langle \mathcal{L}_{3,n}, {}^{\star\mathfrak{R}} \rangle\}$ thus follows, so that Theorem 7.16 is proved.

Lemma 7.17 *The statement* ($\clubsuit_1$) *holds.*

Proof Suppose that $\mathbf{s} \approx \mathbf{s}' \in \mathsf{id}_{\text{SAN}}\{\langle \mathcal{L}_{3,n}, {}^{\star\mathfrak{R}} \rangle\}$ with $|\,\mathsf{non}(\overline{\mathbf{s}})| = 1$. Then by Lemma 7.13, there exists some finite $\Sigma \subseteq \mathsf{id}_{\text{REF}}\{\langle \mathcal{L}_{3,n}, {}^{\star\mathfrak{R}} \rangle\}$ such that

(A) $\{(\text{inv}), (7.1), \mathbf{s} \approx \mathbf{s}'\} \sim \{(\text{inv}), (7.1)\} \cup \Sigma$;
(B) each $\mathbf{r} \approx \mathbf{r}' \in \Sigma$ satisfies $|\,\mathsf{sim}(\overline{\mathbf{r}})| \leq |\,\mathsf{sim}(\overline{\mathbf{s}})|$ and $|\,\mathsf{non}(\overline{\mathbf{r}})| \leq |\,\mathsf{non}(\overline{\mathbf{s}})| = 1$.

Consider any $\mathbf{r} \approx \mathbf{r}' \in \Sigma$. Generality is not lost by assuming that $\mathbf{r} = x(\prod_{i=1}^{k} \mathbf{p}_i)x^{\circledast}$ is the refined $\star$-sandwich in (7.2). Then it follows from Lemma 7.15 that $\mathbf{r}' = x(\prod_{i=1}^{k} \mathbf{p}_i')x^{\circledast}$ for some $\mathbf{p}_1', \mathbf{p}_2', \ldots, \mathbf{p}_k' \in (\mathscr{A} \cup \mathscr{A}^\star)^+$ such that $\mathsf{con}(\overline{\mathbf{p}}_i) = \mathsf{con}(\overline{\mathbf{p}}_i')$ for all i. Since $1 = |\,\mathsf{non}(\overline{\mathbf{r}})| = |\,\mathsf{non}(\overline{\mathbf{r}}')|$ by (B) and Lemma 7.2, it follows from (i) and (ii) of Lemma 7.15 that $\mathbf{p}_i$ and $\mathbf{p}_i'$ are the same singleton. Hence, the identity $\mathbf{r} \approx \mathbf{r}'$ is trivial. Since the identity $\mathbf{r} \approx \mathbf{r}'$ is arbitrary in Σ, every identity in Σ is trivial. Consequently, the deduction $\{(\text{inv}), (7.1)\} \vdash \mathbf{s} \approx \mathbf{s}'$ follows from (A). □

Lemma 7.18 *Suppose that the statement* ($\clubsuit_m$) *holds. Then the statement* ($\clubsuit_{m+1}$) *also holds.*

Proof Suppose that $\mathbf{s} \approx \mathbf{s}' \in \mathsf{id}_{\mathrm{SAN}}\{\langle \mathcal{L}_{3,n}, {}^{\star\mathfrak{R}} \rangle\}$ with $|\,\mathsf{non}(\overline{\mathbf{s}})| = m + 1$. Then by Lemma 7.13, there exists some finite $\Sigma \subseteq \mathsf{id}_{\mathrm{REF}}\{\langle \mathcal{L}_{3,n}, {}^{\star\mathfrak{R}} \rangle\}$ such that

(A) $\{(\mathrm{inv}), (7.1), \mathbf{s} \approx \mathbf{s}'\} \sim \{(\mathrm{inv}), (7.1)\} \cup \Sigma$;
(B) each $\mathbf{r} \approx \mathbf{r}' \in \Sigma$ satisfies $|\,\mathsf{sim}(\overline{\mathbf{r}})| \leq |\,\mathsf{sim}(\overline{\mathbf{s}})|$ and $|\,\mathsf{non}(\overline{\mathbf{r}})| \leq |\,\mathsf{non}(\overline{\mathbf{s}})| = m + 1$.

Consider any $\mathbf{r} \approx \mathbf{r}' \in \Sigma$. Generality is not lost by assuming that $\mathbf{r} = x(\prod_{i=1}^{k} \mathbf{p}_i)x^{\circledast}$ is the refined $\star$-sandwich in (7.2). Then it follows from Lemma 7.15 that $\mathbf{r}' = x(\prod_{i=1}^{k} \mathbf{p}_i')x^{\circledast}$ for some $\mathbf{p}_1', \mathbf{p}_2', \ldots, \mathbf{p}_k' \in (\mathscr{A} \cup \mathscr{A}^{\star})^+$ such that $\mathsf{con}(\overline{\mathbf{p}}_i) = \mathsf{con}(\overline{\mathbf{p}}_i')$ for all i. By (i) of Lemma 7.15, the words $\mathbf{p}_i$ and $\mathbf{p}_i'$ are both singletons or both $\star$-sandwiches.

CASE 1: $\mathbf{p}_i$ and $\mathbf{p}_i'$ are singletons. Then $\mathbf{p}_i = \mathbf{p}_i'$ by (ii) of Lemma 7.15, so that the deduction $\{(\mathrm{inv}), (7.1)\} \vdash \mathbf{p}_i \approx \mathbf{p}_i'$ holds vacuously.
CASE 2: $\mathbf{p}_i$ and $\mathbf{p}_i'$ are $\star$-sandwiches. Then $\mathbf{p}_i \approx \mathbf{p}_i' \in \mathsf{id}_{\mathrm{SAN}}\{\langle \mathcal{L}_{3,n}, {}^{\star\mathfrak{R}} \rangle\}$ by (iii) of Lemma 7.15. Since $|\,\mathsf{non}(\overline{\mathbf{p}}_i)| < |\,\mathsf{non}(\overline{\mathbf{r}})| \leq m + 1$ by (B), the deduction $\{(\mathrm{inv}), (7.1)\} \vdash \mathbf{p}_i \approx \mathbf{p}_i'$ holds by ($\clubsuit_m$).

Therefore, the deduction $\{(\mathrm{inv}), (7.1)\} \vdash \mathbf{p}_i \approx \mathbf{p}_i'$ holds in any case. Since

$$\mathbf{r} = x\left(\prod_{i=1}^{k} \mathbf{p}_i\right)x^{\circledast} \overset{(\mathrm{inv}),(7.1)}{\approx} x\left(\prod_{i=1}^{k} \mathbf{p}_i'\right)x^{\circledast} = \mathbf{r}',$$

the deduction $\{(\mathrm{inv}), (7.1)\} \vdash \mathbf{r} \approx \mathbf{r}'$ also holds. The identity $\mathbf{r} \approx \mathbf{r}'$ is arbitrary in Σ, so that $\{(\mathrm{inv}), (7.1)\} \vdash \Sigma$. Consequently, the deduction $\{(\mathrm{inv}), (7.1)\} \vdash \mathbf{s} \approx \mathbf{s}'$ follows from (A). □

7.4.2 A Simpler Identity Basis

Lemma 7.19 *The identities* (inv) *and*

$$x^{n+2} \approx x^2, \tag{7.4a}$$

$$x^{n+1}yx^{n+1} \approx xyx, \tag{7.4b}$$

$$xhykxty \approx yhxkytx, \tag{7.4c}$$

$$x^{\star}yx^{\star} \approx x^{\mathfrak{R}}yx^{\mathfrak{R}}, \tag{7.4d}$$

$$xhx^{\star}kx \approx xhx^{\mathfrak{R}}kx, \tag{7.4e}$$

$$x^\star \mathbf{h}_1 y^{\circledast_1} \mathbf{h}_2 x^{\circledast_2} \mathbf{h}_3 y^{\circledast_3} \approx x^{\Re} \mathbf{h}_1 y^{\circledast_1} \mathbf{h}_2 x^{\circledast_2} \mathbf{h}_3 y^{\circledast_3}, \tag{7.4f}$$

$$x^{\circledast_1} \mathbf{h}_1 y^\star \mathbf{h}_2 x^{\circledast_2} \mathbf{h}_3 y^{\circledast_3} \approx x^{\circledast_1} \mathbf{h}_1 y^{\Re} \mathbf{h}_2 x^{\circledast_2} \mathbf{h}_3 y^{\circledast_3}, \tag{7.4g}$$

$$x^{\circledast_1} \mathbf{h}_1 y^{\circledast_2} \mathbf{h}_2 x^\star \mathbf{h}_3 y^{\circledast_3} \approx x^{\circledast_1} \mathbf{h}_1 y^{\circledast_2} \mathbf{h}_2 x^{\Re} \mathbf{h}_3 y^{\circledast_3}, \tag{7.4h}$$

$$x^{\circledast_1} \mathbf{h}_1 y^{\circledast_2} \mathbf{h}_2 x^{\circledast_3} \mathbf{h}_3 y^\star \approx x^{\circledast_1} \mathbf{h}_1 y^{\circledast_2} \mathbf{h}_2 x^{\circledast_3} \mathbf{h}_3 y^{\Re}, \tag{7.4i}$$

$$x\left(\prod_{i=1}^{m}(y_i \mathbf{h}_i y_i^\star)\right)x^\star \approx x\left(\prod_{i=m}^{1}(y_i \mathbf{h}_i y_i^\star)\right)x^\star, \tag{7.4j}$$

where $\mathbf{h}_i \in \{\varnothing, h_i\}$, $\circledast_i \in \{1, \star\}$, *and* $m \in \{2, 3, 4, \ldots\}$, *are equivalent to* $\{(\text{inv}), (7.1)\}$.

Proof The deduction (7.1) $\vdash$ (7.4) clearly holds. Verification of the deduction $\{(\text{inv}), (7.4)\} \vdash (7.1)$ is divided into five cases.

CASE 1: (7.4) $\vdash$ $\{(7.1\text{a}), (7.4\text{b}), \ldots, (7.1\text{o})\}$. The identities (7.4a)–(7.4c) and (6.6a)–(6.6c) coincide, while the deduction (7.4) $\vdash$ (6.6d) holds because

$$\begin{aligned}
x\left(\prod_{i=1}^{m}(y_i h_i y_i)\right)x &\overset{(7.4\text{b})}{\approx} x^{\Re} x^{n+1-\Re}\left(\prod_{i=1}^{m}(y_i^{n+1-\Re} y_i^{\Re} h_i y_i^{n+1-\Re} y_i^{\Re})\right)x^{\Re} x^{n+1-\Re} \\
&\overset{(7.4\text{d})}{\approx} x^\star x^{n+1-\Re}\left(\prod_{i=1}^{m}(y_i^{n+1-\Re} y_i^\star h_i y_i^{n+1-\Re} y_i^\star)\right)x^\star x^{n+1-\Re} \\
&\overset{(7.4\text{j})}{\approx} x^\star x^{n+1-\Re}\left(\prod_{i=m}^{1}(y_i^{n+1-\Re} y_i^\star h_i y_i^{n+1-\Re} y_i^\star)\right)x^\star x^{n+1-\Re} \\
&\overset{(7.4\text{d})}{\approx} x^{\Re} x^{n+1-\Re}\left(\prod_{i=m}^{1}(y_i^{n+1-\Re} y_i^{\Re} h_i y_i^{n+1-\Re} y_i^{\Re})\right)x^{\Re} x^{n+1-\Re} \\
&\overset{(7.4\text{b})}{\approx} x\left(\prod_{i=m}^{1}(y_i h_i y_i)\right)x.
\end{aligned}$$

Therefore, the deduction (7.4) $\vdash$ (6.6) holds, whence by Corollary 6.19, all plain identities satisfied by $\langle \mathcal{L}_{3,n}, {}^{\star\Re} \rangle$ are deducible from (7.4). In particular, the present case follows.

CASE 2: $\{(\text{inv}), (7.4)\} \vdash (7.1\text{p})$. This deduction holds because

$$\begin{aligned}
(x^{2n})^\star &\overset{(\text{inv})}{\approx} (x^\star)^{2n} \\
&\overset{(7.4\text{a})}{\approx} (x^\star)^{2n}(x^\star)^n(x^\star)^{2n}
\end{aligned}$$

$$\begin{aligned}
&\overset{(7.4d)}{\approx} (x^{\mathfrak{R}})^{2n}(x^\star)^n(x^{\mathfrak{R}})^{2n} \\
&\overset{(7.4e)}{\approx} (x^{\mathfrak{R}})^{2n}(x^{\mathfrak{R}})^n(x^{\mathfrak{R}})^{2n} \\
&\overset{(7.4a)}{\approx} x^{2n}.
\end{aligned}$$

CASE 3: (7.4) $\vdash$ (7.1q). If $\mathbf{h}_1 = h_1$, then clearly (7.4d) $\vdash$ (7.1q). If $\mathbf{h}_1 = \varnothing$, then the identity (7.1q) reduces to $(x^\star)^2 \approx (x^{\mathfrak{R}})^2$, whence the deduction (7.4) $\vdash$ (7.1q) holds because

$$\begin{aligned}
(x^\star)^2 &\overset{(7.4a)}{\approx} (x^\star)^{2n}(x^\star)^2(x^\star)^{2n} \\
&\overset{(7.4d)}{\approx} (x^{\mathfrak{R}})^{2n}(x^\star)^2(x^{\mathfrak{R}})^{2n} \\
&\overset{(7.4e)}{\approx} (x^{\mathfrak{R}})^{2n}(x^{\mathfrak{R}})^2(x^{\mathfrak{R}})^{2n} \\
&\overset{(7.4a)}{\approx} (x^{\mathfrak{R}})^2.
\end{aligned}$$

CASE 4: (7.4) $\vdash$ {(7.1r), (7.1s), (7.1t)}. The deduction (7.4) $\vdash$ (7.1r) holds because

$$\begin{aligned}
x^\star\mathbf{h}_1x\mathbf{h}_2x &\overset{(7.4b)}{\approx} x^\star\mathbf{h}_1x^{n+1-\mathfrak{R}}x^{\mathfrak{R}}\mathbf{h}_2x^{n+1-\mathfrak{R}}x^{\mathfrak{R}} \\
&\overset{(7.4d)}{\approx} x^\star\mathbf{h}_1x^{n+1-\mathfrak{R}}x^\star\mathbf{h}_2x^{n+1-\mathfrak{R}}x^\star \\
&\overset{(7.4d)}{\approx} x^{\mathfrak{R}}\mathbf{h}_1x^{n+1-\mathfrak{R}}x^\star\mathbf{h}_2x^{n+1-\mathfrak{R}}x^{\mathfrak{R}} \\
&\overset{(7.4e)}{\approx} x^{\mathfrak{R}}\mathbf{h}_1x^{n+1-\mathfrak{R}}x^{\mathfrak{R}}\mathbf{h}_2x^{n+1-\mathfrak{R}}x^{\mathfrak{R}} \\
&\overset{(7.4b)}{\approx} x^{\mathfrak{R}}\mathbf{h}_1x\mathbf{h}_2x.
\end{aligned}$$

The deductions (7.4) $\vdash$ (7.1s) and (7.4) $\vdash$ (7.1t) are similarly established.

CASE 5: (7.4) $\vdash$ {(7.1u), (7.1v), ..., (7.1y)}. This deduction vacuously holds because the identities (7.4f)–(7.4j) and (7.1u)–(7.1y) coincide. □

Corollary 7.20 *Let* $\mathfrak{R} > 1$. *Then the identities* {(inv), (7.4)} *constitute an identity basis for the involution semigroup* $\langle \mathcal{L}_{3,n}, {}^{\star\mathfrak{R}} \rangle$.

Proof This follows from Theorem 7.16 and Lemma 7.19. □

7.5 An Infinite Irredundant Identity Basis for $\langle \mathcal{L}_{3,n}, {}^{\star\mathfrak{R}} \rangle$ with $\mathfrak{R} > 1$

Theorem 7.21 *Let $\mathfrak{R} > 1$. Then some subset of* (7.4) *constitutes an infinite irredundant identity basis for the involution semigroup $\langle \mathcal{L}_{3,n}, {}^{\star\mathfrak{R}} \rangle$.*

The identities in (7.4j) are far from being irredundant. In Sect. 7.5.1, an irredundant set Ω_∞ of identities from (7.4j) is chosen with the property that (7.4j) $\sim \Omega_\infty$. The proof of Theorem 7.21 is then given in Sect. 7.5.2.

7.5.1 The Identities (7.4j)

For each $m \geq 1$, let $\mathscr{B}^m$ denote the set of all binary vectors of dimension m, that is,

$$\mathscr{B}^m = \{(b_1, b_2, \ldots, b_m) \mid b_1, b_2, \ldots, b_m \in \{0, 1\}\}.$$

The vectors in $\mathscr{B}^m$ are lexicographically ordered by $<$ as follows:

$$(b_1, b_2, \ldots, b_m) < (b'_1, b'_2, \ldots, b'_m)$$

if there exists a least $\ell \in \{1, 2, \ldots, m\}$ such that $b_\ell < b'_\ell$ and $b_i = b'_i$ for any $i < \ell$. The *reversal* of a vector $\mathrm{v} = (b_1, b_2, \ldots, b_m) \in \mathscr{B}^m$ is the vector $\mathrm{v}^\triangleleft = (b_m, b_{m-1}, \ldots, b_1)$. If $\mathrm{v}^\triangleleft = \mathrm{v}$, then v is a *palindrome*. The set $\mathscr{B}^m$ can be partitioned into

$$\mathscr{B}^m_= = \{\mathrm{v} \in \mathscr{B}^m \mid \mathrm{v} = \mathrm{v}^\triangleleft\},$$

$$\mathscr{B}^m_< = \{\mathrm{v} \in \mathscr{B}^m \mid \mathrm{v} < \mathrm{v}^\triangleleft\},$$

$$\text{and } \mathscr{B}^m_> = \{\mathrm{v} \in \mathscr{B}^m \mid \mathrm{v} > \mathrm{v}^\triangleleft\}.$$

Each vector $\mathrm{v} = (b_1, b_2, \ldots, b_m) \in \mathscr{B}^m$ is associated with the words

$$\mathrm{v}^\uparrow = x\Bigg(\prod_{i=1}^{m}(y_i h_i^{b_i} y_i^\star)\Bigg)x^\star = x(y_1 h_1^{b_1} y_1^\star)(y_2 h_2^{b_2} y_2^\star)\cdots(y_m h_m^{b_m} y_m^\star)x^\star$$

$$\text{and } \mathrm{v}^\downarrow = x\Bigg(\prod_{i=m}^{1}(y_i h_i^{b_i} y_i^\star)\Bigg)x^\star = x(y_m h_m^{b_m} y_m^\star)(y_{m-1} h_{m-1}^{b_{m-1}} y_{m-1}^\star)\cdots(y_1 h_1^{b_1} y_1^\star)x^\star,$$

where $h_i^0 = \varnothing$ and $h_i^1 = h_i$, so that $\bigcup_{m\geq 2}\{\mathrm{v}^\uparrow \approx \mathrm{v}^\downarrow \mid \mathrm{v} \in \mathscr{B}^m\} = \{(7.4\mathrm{j})\}$. For each $m \geq 2$, the identities in $\{\mathrm{v}^\uparrow \approx \mathrm{v}^\downarrow \mid \mathrm{v} \in \mathscr{B}^m\}$ are not irredundant: if $\mathrm{v} \in \mathscr{B}^m$ is not a palindrome, then v and $\mathrm{v}^\triangleleft$ are distinct vectors such that the associated identities $\mathrm{v}^\uparrow \approx \mathrm{v}^\downarrow$ and $(\mathrm{v}^\triangleleft)^\uparrow \approx (\mathrm{v}^\triangleleft)^\downarrow$ are equivalent. This redundancy can be eliminated by choosing identities $\mathrm{V}^{\uparrow} \approx \mathrm{V}^{\downarrow}$ with V taken from only $\mathscr{B}^m_=$ or $\mathscr{B}^m_<$. Therefore, by letting

$$\Omega_\infty = \bigcup_{m \geq 2} \{\mathrm{V}^{\uparrow} \approx \mathrm{V}^{\downarrow} \mid \mathrm{V} \in \mathscr{B}^m_= \cup \mathscr{B}^m_<\},$$

the equivalence (7.4j) $\sim \Omega_\infty$ holds and no two identities in Ω_∞ are equivalent.

7.5.2 Proof of Theorem 7.21

Since the identity system (7.4a)–(7.4i) is finite, by Corollary 7.20, it contains some minimal subsystem Ω_0 such that $\Omega_0 \cup \Omega_\infty$ is an identity basis for $\langle \mathcal{L}_{3,n}, {}^{\star\mathfrak{R}} \rangle$. Let

$$\mathrm{V}_{\mathrm{fix}} = (b_1, b_2, \ldots, b_m)$$

be a fixed vector in $\mathscr{B}^m_= \cup \mathscr{B}^m_<$. In this subsection, it is shown that the identity $\mathrm{V}^{\uparrow}_{\mathrm{fix}} \approx \mathrm{V}^{\downarrow}_{\mathrm{fix}}$ in Ω_∞ is not deducible from

$$\{(\mathrm{inv})\} \cup \Omega_0 \cup (\Omega_\infty \backslash \{\mathrm{V}^{\uparrow}_{\mathrm{fix}} \approx \mathrm{V}^{\downarrow}_{\mathrm{fix}}\}). \tag{7.5}$$

Since Ω_∞ is infinite, the set $\{(\mathrm{inv})\} \cup \Omega_0 \cup \Omega_\infty$ is an infinite irredundant identity basis for the involution semigroup $\langle \mathcal{L}_{3,n}, {}^{\star\mathfrak{R}} \rangle$ with $\mathfrak{R} > 1$. This completes the proof of Theorem 7.21.

Seeking a contradiction, suppose that the identity $\mathrm{V}^{\uparrow}_{\mathrm{fix}} \approx \mathrm{V}^{\downarrow}_{\mathrm{fix}}$ is deducible from the identities (7.5). Then there exists a sequence

$$\mathrm{V}^{\uparrow}_{\mathrm{fix}} = \mathbf{t}_1, \mathbf{t}_2, \ldots, \mathbf{t}_r = \mathrm{V}^{\downarrow}_{\mathrm{fix}}$$

of terms, where each identity $\mathbf{t}_i \approx \mathbf{t}_{i+1}$ is directly deducible from some identity $\mathbf{u}_i \approx \mathbf{v}_i$ in (7.5). If every identity $\mathbf{u}_i \approx \mathbf{v}_i$ is from (inv), then it follows from Lemma 2.11 that $\mathrm{V}^{\uparrow}_{\mathrm{fix}} = \lfloor \mathbf{t}_i \rfloor$ for all i, whence the contradiction $\mathrm{V}^{\uparrow}_{\mathrm{fix}} = \lfloor \mathbf{t}_r \rfloor = \mathrm{V}^{\downarrow}_{\mathrm{fix}}$ is obtained. Therefore, some identity $\mathbf{u}_i \approx \mathbf{v}_i$ is not from (inv); let $\ell \geq 1$ be the least possible integer such that $\mathbf{u}_\ell \approx \mathbf{v}_\ell$ is not from (inv). Then $\mathbf{u}_\ell \approx \mathbf{v}_\ell$ is from $\Omega_0 \cup (\Omega_\infty \backslash \{\mathrm{V}^{\uparrow}_{\mathrm{fix}} \approx \mathrm{V}^{\downarrow}_{\mathrm{fix}}\})$, while $\mathbf{u}_1 \approx \mathbf{v}_1, \mathbf{u}_2 \approx \mathbf{v}_2, \ldots, \mathbf{u}_{\ell-1} \approx \mathbf{v}_{\ell-1}$ are from (inv), whence $\mathrm{V}^{\uparrow}_{\mathrm{fix}} \approx \mathbf{t}_\ell$ is deducible from (inv).

In the remainder of this subsection, it is shown that the identity $\mathbf{u}_\ell \approx \mathbf{v}_\ell$ belongs to neither Ω_0 nor $\Omega_\infty \backslash \{\mathrm{V}^{\uparrow}_{\mathrm{fix}} \approx \mathrm{V}^{\downarrow}_{\mathrm{fix}}\}$. It follows that $\mathbf{u}_\ell \approx \mathbf{v}_\ell$ is contradictorily not an identity from (7.5).

Lemma 7.22

(i) $\lfloor \mathbf{t}_\ell \rfloor = \mathrm{V}^{\uparrow}_{\mathrm{fix}}$.

(ii) *The word* $\lfloor \mathbf{t}_\ell \rfloor$ *does not contain any of the following factors:*

$$x^{\circledast_1}\mathbf{h}_1x^{\circledast_1}, \tag{7.6a}$$

$$x^{\circledast_1}\mathbf{h}_1x^{\circledast_2}\mathbf{h}_2x^{\circledast_3}, \tag{7.6b}$$

$$x^{\circledast_1}\mathbf{h}_1y^{\circledast_2}\mathbf{h}_2x^{\circledast_3}\mathbf{h}_3y^{\circledast_4}, \tag{7.6c}$$

$$x^{\circledast_1}\mathbf{h}_1y^{\circledast_2}\mathbf{h}_2z^{\circledast_3}\mathbf{h}_3z^{\circledast_4}\mathbf{h}_4y^{\circledast_5}\mathbf{h}_5x^{\circledast_6}, \tag{7.6d}$$

$$x^{\circledast_1}\mathbf{h}_1y^{\circledast_2}\mathbf{h}_2z^{\circledast_3}t^{\circledast_4}\mathbf{h}_3y^{\circledast_5}\mathbf{h}_4x^{\circledast_6}, \tag{7.6e}$$

where $x, y, z, t \in \mathscr{A}$, $\mathbf{h}_i \in (\mathscr{A} \cup \mathscr{A}^\star)^+_\varnothing$, *and* $\circledast_i \in \{1, \star\}$.

Proof

(i) Since the deductions

$$\begin{aligned}(\text{inv}) &\vdash \{\mathbf{u}_i \approx \mathbf{v}_i \mid 1 \le i \le \ell - 1\}\\ &\vdash \{\mathbf{t}_i \approx \mathbf{t}_{i+1} \mid 1 \le i \le \ell - 1\}\\ &\vdash \mathbf{t}_\ell \approx \mathbf{t}_1 = \mathsf{v}^{\uparrow}_{\text{fix}}\end{aligned}$$

hold, it follows from Lemma 2.11 that $\lfloor \mathbf{t}_\ell \rfloor = \mathsf{v}^{\uparrow}_{\text{fix}}$.

(ii) This is a consequence of part (i). □

Since the identity $\mathbf{t}_\ell \approx \mathbf{t}_{\ell+1}$ is directly deducible from $\mathbf{u}_\ell \approx \mathbf{v}_\ell$, there exists some substitution $\varphi : \mathscr{A} \to \mathsf{T}(\mathscr{A})$ such that $\mathbf{u}_\ell\varphi$ is a subterm of $\mathbf{t}_\ell$.

Lemma 7.23 *The identity* $\mathbf{u}_\ell \approx \mathbf{v}_\ell$ *is not from* Ω_0.

Proof Suppose that the identity $\mathbf{u}_\ell \approx \mathbf{v}_\ell$ is from either (7.4c) or (7.4f)–(7.4i). Then

$$\mathbf{u}_\ell = \mathbf{h}_0x^{\circledast_1}\mathbf{h}_1y^{\circledast_2}\mathbf{h}_2x^{\circledast_3}\mathbf{h}_3y^{\circledast_4}\mathbf{h}_4$$

for some $\mathbf{h}_i \in (\mathscr{A} \cup \mathscr{A}^\star)^+_\varnothing$ and $\circledast_i \in \{1, \star\}$. Let $z \in \mathsf{con}(\overline{\lfloor x\varphi \rfloor})$ and $t \in \mathsf{con}(\overline{\lfloor y\varphi \rfloor})$, so that

$$\begin{aligned}\lfloor \mathbf{u}_\ell\varphi \rfloor &= \lfloor \mathbf{h}_0\varphi \rfloor \lfloor (x\varphi)^{\circledast_1} \rfloor \lfloor \mathbf{h}_1\varphi \rfloor \lfloor (y\varphi)^{\circledast_2} \rfloor \lfloor \mathbf{h}_2\varphi \rfloor \lfloor (x\varphi)^{\circledast_3} \rfloor \lfloor \mathbf{h}_3\varphi \rfloor \lfloor (y\varphi)^{\circledast_4} \rfloor \lfloor \mathbf{h}_4\varphi \rfloor\\ &= \lfloor \mathbf{h}_0\varphi \rfloor \cdots z^{\circledast'_1} \cdots \lfloor \mathbf{h}_1\varphi \rfloor \cdots t^{\circledast'_2} \cdots \lfloor \mathbf{h}_2\varphi \rfloor \cdots z^{\circledast'_3} \cdots \lfloor \mathbf{h}_3\varphi \rfloor \cdots t^{\circledast'_4} \cdots \lfloor \mathbf{h}_4\varphi \rfloor\end{aligned}$$

for some $\circledast'_i \in \{1, \star\}$. Since $\mathbf{u}_\ell\varphi$ is a subterm of $\mathbf{t}_\ell$, as observed in Remark 2.10(iii), either $\lfloor \mathbf{u}_\ell\varphi \rfloor$ or $\lfloor (\mathbf{u}_\ell\varphi)^\star \rfloor$ is a factor of the word $\lfloor \mathbf{t}_\ell \rfloor$, whence $\lfloor \mathbf{t}_\ell \rfloor$ contains a factor of the form (7.6c). But this is impossible by Lemma 7.22(ii).

If the identity $\mathbf{u}_\ell \approx \mathbf{v}_\ell$ is from {(7.4a), (7.4b), (7.4d), (7.4e)}, then a similar argument shows that the word $\lfloor \mathbf{t}_\ell \rfloor$ contains a factor of the form (7.6a) or (7.6b), which again is impossible by Lemma 7.22(ii). □

Lemma 7.24 *The identity* $\mathbf{u}_\ell \approx \mathbf{v}_\ell$ *is not from* $\Omega_\infty \backslash \{\mathtt{V}^\uparrow_{\mathrm{fix}} \approx \mathtt{V}^\downarrow_{\mathrm{fix}}\}$.

Proof It suffices to show that if the identity $\mathbf{u}_\ell \approx \mathbf{v}_\ell$ belongs to Ω_∞, then it is precisely $\mathtt{V}^\uparrow_{\mathrm{fix}} \approx \mathtt{V}^\downarrow_{\mathrm{fix}}$. Suppose that $\mathbf{u}_\ell \approx \mathbf{v}_\ell$ is the identity $\mathtt{W}^\uparrow \approx \mathtt{W}^\downarrow$ in Ω_∞, where $\mathtt{W} = (c_1, c_2, \ldots, c_n) \in \mathscr{B}^n_= \cup \mathscr{B}^n_<$ for some $n \geq 2$. Then $\mathbf{u}_\ell\varphi = \mathtt{W}^\uparrow\varphi$ or $\mathbf{u}_\ell\varphi = \mathtt{W}^\downarrow\varphi$ is a subterm of $\mathbf{t}_\ell$; by symmetry, it suffices to assume that

$$\mathbf{u}_\ell\varphi = \mathtt{W}^\uparrow\varphi = (x\varphi)\left(\prod_{i=1}^{n}((y_i\varphi)(h_i^{c_i}\varphi)(y_i\varphi)^\star)\right)(x\varphi)^\star$$

is a subterm of $\mathbf{t}_\ell$.

Suppose that the word $\lfloor x\varphi \rfloor$ is not a singleton. Then $\overline{\lfloor x\varphi \rfloor} = \cdots x_1 x_2 \cdots$ for some $x_1, x_2 \in \mathscr{A}$. Choose any $q \in \mathsf{con}(\lfloor y_1\varphi \rfloor)$. Then

$$\begin{aligned}\lfloor \mathtt{W}^\uparrow\varphi \rfloor &= \lfloor x\varphi \rfloor \lfloor y_1\varphi \rfloor \lfloor h_1^{c_1}\varphi \rfloor \lfloor (y_1\varphi)^\star \rfloor \cdots \lfloor (x\varphi)^\star \rfloor \\ &= \underbrace{(\cdots x_1^{\circledast_1} x_2^{\circledast_2} \cdots)}_{\lfloor x\varphi \rfloor} \underbrace{(\cdots q^{\circledast_3} \cdots)}_{\lfloor y_1\varphi \rfloor} \lfloor h_1^{c_1}\varphi \rfloor \underbrace{(\cdots q^{\circledast_4} \cdots)}_{\lfloor (y_1\varphi)^\star \rfloor} \cdots \underbrace{(\cdots x_2^{\circledast_5} x_1^{\circledast_6} \cdots)}_{\lfloor (x\varphi)^\star \rfloor}\end{aligned}$$

for some $\circledast_i \in \{1, \star\}$. Since $\mathtt{W}^\uparrow\varphi$ is a subterm of $\mathbf{t}_\ell$, as observed in Remark 2.10(iii), either $\lfloor \mathtt{W}^\uparrow\varphi \rfloor$ or $\lfloor (\mathtt{W}^\uparrow\varphi)^\star \rfloor$ is a factor of the word $\lfloor \mathbf{t}_\ell \rfloor$, whence $\lfloor \mathbf{t}_\ell \rfloor$ contains a factor of the form (7.6b) or (7.6d), depending on whether or not the variables x_1, x_2, and q are all distinct. But this is impossible by Lemma 7.22(ii). Therefore,

(A) the word $\lfloor x\varphi \rfloor$ is a singleton.

Suppose that the word $\lfloor y_i\varphi \rfloor$ is not a singleton. Then $\overline{\lfloor y_i\varphi \rfloor} = \cdots x_1 x_2 \cdots$ for some $x_1, x_2 \in \mathscr{A}$. Choose any $q \in \mathsf{con}(\lfloor x\varphi \rfloor)$. Then

$$\begin{aligned}\lfloor \mathtt{W}^\uparrow\varphi \rfloor &= \lfloor x\varphi \rfloor \cdots \lfloor y_i\varphi \rfloor \lfloor h_i^{c_i}\varphi \rfloor \lfloor (y_i\varphi)^\star \rfloor \cdots \lfloor (x\varphi)^\star \rfloor \\ &= \underbrace{(\cdots q^{\circledast_1} \cdots)}_{\lfloor x\varphi \rfloor} \cdots \underbrace{(\cdots x_1^{\circledast_2} x_2^{\circledast_3} \cdots)}_{\lfloor y_i\varphi \rfloor} \lfloor h_i^{c_i}\varphi \rfloor \underbrace{(\cdots x_2^{\circledast_4} x_1^{\circledast_5} \cdots)}_{\lfloor (y_i\varphi)^\star \rfloor} \cdots \underbrace{(\cdots q^{\circledast_6} \cdots)}_{\lfloor (x\varphi)^\star \rfloor}\end{aligned}$$

for some $\circledast_j \in \{1, \star\}$. Since $\mathtt{W}^\uparrow\varphi$ is a subterm of $\mathbf{t}_\ell$, by Remark 2.10(iii), either $\lfloor \mathtt{W}^\uparrow\varphi \rfloor$ or $\lfloor (\mathtt{W}^\uparrow\varphi)^\star \rfloor$ is a factor of the word $\lfloor \mathbf{t}_\ell \rfloor$, whence $\lfloor \mathbf{t}_\ell \rfloor$ contains a factor of the form (7.6b) or (7.6d), depending on whether or not the variables x_1, x_2, and q are all distinct. But this is impossible by Lemma 7.22(ii). Therefore,

(B) the word $\lfloor y_i\varphi \rfloor$ is a singleton.

Suppose that $c_i = 1$ and $\lfloor h_i^{c_i}\varphi \rfloor$ is not a singleton. Then $\overline{\lfloor h_i^{c_i}\varphi \rfloor} = \cdots x_1x_2\cdots$ for some $x_1, x_2 \in \mathscr{A}$. Choose any $q \in \mathsf{con}(\overline{\lfloor x\varphi \rfloor})$ and $t \in \mathsf{con}(\overline{\lfloor y_i\varphi \rfloor})$. Then

$$\begin{aligned}\lfloor \mathtt{W}^{\uparrow}\varphi \rfloor &= \lfloor x\varphi \rfloor \cdots \lfloor y_i\varphi \rfloor \lfloor h_i^{c_i}\varphi \rfloor \lfloor (y_i\varphi)^{\star} \rfloor \cdots \lfloor (x\varphi)^{\star} \rfloor \\ &= \underbrace{(\cdots q^{\circledast_1}\cdots)}_{\lfloor x\varphi \rfloor}\cdots\underbrace{(\cdots t^{\circledast_2}\cdots)}_{\lfloor y_i\varphi \rfloor}\underbrace{(\cdots x_1^{\circledast_3}x_2^{\circledast_4}\cdots)}_{\lfloor h_i^{c_i}\varphi \rfloor}\underbrace{(\cdots t^{\circledast_5}\cdots)}_{\lfloor (y_i\varphi)^{\star} \rfloor}\cdots\underbrace{(\cdots q^{\circledast_6}\cdots)}_{\lfloor (x\varphi)^{\star} \rfloor}\end{aligned}$$

for some $\circledast_j \in \{1, \star\}$. Since $\mathtt{W}^{\uparrow}\varphi$ is a subterm of $\mathbf{t}_\ell$, by Remark 2.10(iii), either $\lfloor \mathtt{W}^{\uparrow}\varphi \rfloor$ or $\lfloor (\mathtt{W}^{\uparrow}\varphi)^{\star} \rfloor$ is a factor of the word $\lfloor \mathbf{t}_\ell \rfloor$, whence $\lfloor \mathbf{t}_\ell \rfloor$ contains a subterm of the form (7.6b), (7.6d), or (7.6e), depending on whether or not the variables x_1, x_2, q, and t are all distinct. But this is impossible by Lemma 7.22(ii). Therefore,

(C) if $c_i = 1$, then the word $\lfloor h_i^{c_i}\varphi \rfloor$ is a singleton.

Now since $\mathbf{u}_\ell\varphi = \mathtt{W}^{\uparrow}\varphi$ is a subterm of $\mathbf{t}_\ell$, by Remark 2.10(iii), either $\lfloor \mathtt{W}^{\uparrow}\varphi \rfloor$ or $\lfloor (\mathtt{W}^{\uparrow}\varphi)^{\star} \rfloor$ is a factor of the word $\lfloor \mathbf{t}_\ell \rfloor = \mathtt{V}^{\uparrow}_{\text{fix}}$, where

$$\lfloor \mathtt{W}^{\uparrow}\varphi \rfloor = \lfloor x\varphi \rfloor \bigg(\prod_{i=1}^{n} (\lfloor y_i\varphi \rfloor \lfloor h_i^{c_i}\varphi \rfloor \lfloor (y_i\varphi)^{\star} \rfloor) \bigg) \lfloor (x\varphi)^{\star} \rfloor$$

$$\text{and } \lfloor (\mathtt{W}^{\uparrow}\varphi)^{\star} \rfloor = \lfloor x\varphi \rfloor \bigg(\prod_{i=n}^{1} (\lfloor y_i\varphi \rfloor \lfloor (h_i^{c_i}\varphi)^{\star} \rfloor \lfloor (y_i\varphi)^{\star} \rfloor) \bigg) \lfloor (x\varphi)^{\star} \rfloor.$$

Therefore, (A)–(C) imply that $m = n$ and either $\mathtt{V}^{\uparrow}_{\text{fix}} = \lfloor \mathtt{W}^{\uparrow}\varphi \rfloor$ or $\mathtt{V}^{\uparrow}_{\text{fix}} = \lfloor (\mathtt{W}^{\uparrow}\varphi)^{\star} \rfloor$. It follows that $\mathtt{V}^{\uparrow}_{\text{fix}} \approx \mathtt{V}^{\downarrow}_{\text{fix}}$ coincides with the identity $\mathtt{W}^{\uparrow} \approx \mathtt{W}^{\downarrow}$ and so also the identity $\mathbf{u}_\ell \approx \mathbf{v}_\ell$. □

7.6 Smaller Examples

For any $n \geq 3$ and $\Re \in \mathsf{sq}(n)\backslash\{1\}$, although the involution semigroup $\langle \mathcal{L}_{3,n}, {}^{\star\Re} \rangle$ of order $6n$ with infinite irredundant identity basis can be considered small, smaller examples are available.

For each $n \geq 1$, the subsemigroup

$$\mathcal{L}_{3,n}^{\mathsf{sub}} = (\mathcal{L}_3 \times \{1\}) \cup (\{0\} \times \mathcal{Z}_n)$$

of $\mathcal{L}_{3,n}$ has $n + 5$ elements; this subsemigroup is proper if $n \geq 2$. It is easily checked that $\mathcal{L}_{3,n}^{\mathsf{sub}}$ is closed under the unary operation $(x, y) \mapsto (x^{\star}, y^{\Re})$ of $\mathcal{L}_{3,n}$. Therefore, $\langle \mathcal{L}_{3,n}^{\mathsf{sub}}, {}^{\star\Re} \rangle$

is an involution subsemigroup of $\langle \mathcal{L}_{3,n}, {}^{\star\mathfrak{R}} \rangle$, whence the inclusion $\mathscr{V}_{\mathsf{inv}}\{\langle \mathcal{L}_{3,n}^{\mathsf{sub}}, {}^{\star\mathfrak{R}} \rangle\} \subseteq \mathscr{V}_{\mathsf{inv}}\{\langle \mathcal{L}_{3,n}, {}^{\star\mathfrak{R}} \rangle\}$ holds; the reverse inclusion holds because

$$\begin{aligned}\langle \mathcal{L}_{3,n}, {}^{\star\mathfrak{R}} \rangle &= \langle \mathcal{L}_3, {}^{\star} \rangle \times \langle \mathcal{Z}_n, {}^{\mathfrak{R}} \rangle \\ &\cong \langle \mathcal{L}_3 \times \{1\}, {}^{\star\mathfrak{R}} \rangle \times \langle \{0\} \times \mathcal{Z}_n, {}^{\star\mathfrak{R}} \rangle \\ &\subseteq \langle \mathcal{L}_{3,n}^{\mathsf{sub}}, {}^{\star\mathfrak{R}} \rangle \times \langle \mathcal{L}_{3,n}^{\mathsf{sub}}, {}^{\star\mathfrak{R}} \rangle.\end{aligned}$$

Hence, the equality $\mathscr{V}_{\mathsf{inv}}\{\langle \mathcal{L}_{3,n}, {}^{\star\mathfrak{R}} \rangle\} = \mathscr{V}_{\mathsf{inv}}\{\langle \mathcal{L}_{3,n}^{\mathsf{sub}}, {}^{\star\mathfrak{R}} \rangle\}$ holds.

Consequently, the involution semigroup $\langle \mathcal{L}_{3,n}^{\mathsf{sub}}, {}^{\star\mathfrak{R}} \rangle$ of order $n + 5$ has an infinite irredundant identity basis if $n \geq 3$ and $\mathfrak{R} \in \mathsf{sq}(n) \backslash \{1\}$. In particular, $\langle \mathcal{L}_{3,3}^{\mathsf{sub}}, {}^{\star 2} \rangle$ is an involution semigroup of order eight with an infinite irredundant identity basis.

7.7 Summary

The main objective of the present chapter is to show that for each $\mathfrak{R} \in \mathsf{sq}(n)$ such that $\mathfrak{R} > 1$, the involution semigroup $\langle \mathcal{L}_{3,n}, {}^{\star\mathfrak{R}} \rangle$ possesses an infinite irredundant identity basis; see Theorem 7.21. An involution semigroup of order eight with an infinite irredundant identity basis is then exhibited. These results were published in Lee (2016a).

Finitely Based Involution Semigroups with Non-Finitely Based Reducts

8

This chapter is a detailed study of the identities satisfied by the involution semigroup

$$\langle \mathcal{L}_{3,n}, {}^{\star 1}\rangle = \langle \mathcal{L}_3, {}^\star\rangle \times \langle \mathcal{Z}_n, {}^1\rangle.$$

Refer to Sect. 1.5.1 for the roles $\langle \mathcal{L}_{3,n}, {}^{\star 1}\rangle$ plays in solving several open problems.

In Sects. 8.1 and 8.2, identities satisfied by the involution semigroup $\langle \mathcal{L}_{3,n}, {}^{\star 1}\rangle$ are examined. In Sect. 8.3, an explicit identity basis is exhibited for $\langle \mathcal{L}_{3,n}, {}^{\star 1}\rangle$. In Sect. 8.4, this identity basis is shown to be equivalent to some finite set of identities. Consequently, $\langle \mathcal{L}_{3,n}, {}^{\star 1}\rangle$ is a finitely based involution semigroup with a non-finitely based semigroup reduct.

8.1 Identities and $\star$-Sandwich Identities Satisfied by $\langle \mathcal{L}_{3,n}, {}^{\star 1}\rangle$

In this chapter, let $(7.1)_1$ denote the identities (7.1) with $\mathfrak{n} = 1$. In view of Lemma 7.1, the following result is routinely verified.

Lemma 8.1 *The involution semigroup $\langle \mathcal{L}_{3,n}, {}^{\star 1}\rangle$ satisfies the identities* $(7.1)_1$ *and*

$$xy^\star x \approx xyx, \quad xy^\star x^\star \approx xyx^\star, \quad x^\star y^\star x \approx x^\star yx. \tag{8.1}$$

Following Sect. 7.2.2, denote by

$$\mathsf{id}_{\mathrm{SAN}}\{\langle \mathcal{L}_{3,n}, {}^{\star 1}\rangle\}$$

the set of all $\star$-sandwich identities satisfied by the involution semigroup $\langle \mathcal{L}_{3,n}, {}^{\star 1}\rangle$.

E. W. H. Lee, *Advances in the Theory of Varieties of Semigroups*, Frontiers in Mathematics, https://doi.org/10.1007/978-3-031-16497-2_8

Lemma 8.2 *The identities* $\{(\text{inv}), (7.1)_1\} \cup \mathsf{id}_{\mathrm{SAN}}\{\langle \mathcal{L}_{3,n}, {}^{\star 1}\rangle\}$ *constitute an identity basis for the involution semigroup* $\langle \mathcal{L}_{3,n}, {}^{\star 1}\rangle$.

Proof This follows from Lemma 7.8 with $\Re = 1$. □

Lemma 8.3 *For any* $\star$*-sandwich* $\mathbf{s}$, *the deduction* $\{(\text{inv}), (7.1)_1, (8.1)\} \vdash \mathbf{s} \approx \mathbf{s}^\star$ *holds.*

Proof It suffices to convert the term $\mathbf{s}^\star$, using the identities $\{(\text{inv}), (7.1)_1, (8.1)\}$, into the word $\mathbf{s}$. If $\mathbf{s} = x\mathbf{u}x^\star$ is of type ($\star$S1), then

$$\begin{aligned} \mathbf{s}^\star &\overset{(\text{inv})}{\approx} x\mathbf{u}^\star x^\star \\ &\overset{(8.1)}{\approx} \mathbf{s}. \end{aligned}$$

The case when $\mathbf{s} = x^\star\mathbf{u}x$ is of type ($\star$S2) is similar. Therefore, it remains to assume that $\mathbf{s}$ is of type ($\star$S3), say

$$\mathbf{s} = \mathbf{x}\prod_{i=1}^{\ell}(\mathbf{u}_i\vec{\mathscr{X}})$$

with $\ell \geq 1$, $\mathscr{X} = \{x_1 \ll x_2 \ll \cdots \ll x_r\}$, $\mathbf{x} = x_1^{c_1}x_2^{c_2}\cdots x_r^{c_r} \in \mathscr{X}^{\boxplus}$, and $\mathbf{u}_i \in (\mathscr{A} \cup \mathscr{A}^\star)^+_\varnothing$. For convenience, write $\mathbf{w} = \prod_{i=1}^{\ell}(\mathbf{u}_i\vec{\mathscr{X}})$, so that $\mathbf{s} = \mathbf{x}\mathbf{w}$. If $r = 1$, then $\mathbf{s} = x_1^{c_1}\mathbf{w}$ and

$$\begin{aligned} \mathbf{s}^\star &\overset{(7.1\text{a})}{\approx} (x_1^{c_1}\mathbf{w}x_1^{2n})^\star \\ &\overset{(\text{inv})}{\approx} (x_1^\star)^{2n}\mathbf{w}^\star(x_1^\star)^{c_1} \\ &\overset{(7.1\text{r})_1}{\approx} x_1^{2n}\mathbf{w}^\star(x_1^\star)^{c_1} \\ &\overset{(8.1)}{\approx} x_1^{2n}\mathbf{w}(x_1^\star)^{c_1} \\ &\overset{(7.1\text{t})_1}{\approx} x_1^{2n}\mathbf{w}x_1^{c_1} \\ &\overset{(7.1\text{b})}{\approx} x_1^{c_1}\mathbf{w}x_1^{2n} \\ &\overset{(7.1\text{a})}{\approx} \mathbf{s}; \end{aligned}$$

if $r \geq 2$, then $\mathbf{s} = x_1^{c_1} x_2^{c_2} \cdots x_r^{c_r} \mathbf{w}$ and

$$\begin{aligned}
\mathbf{s}^\star &\overset{(7.1\mathrm{m})}{\approx} (x_1^{c_1} x_2^{c_2} x_3^{c_3} \cdots x_r^{c_r} \mathbf{w} x_1^{2n})^\star \\
&\overset{(\mathrm{inv})}{\approx} (x_1^\star)^{2n} (x_2^{c_2} x_3^{c_3} \cdots x_r^{c_r} \mathbf{w})^\star (x_1^\star)^{c_1} \\
&\overset{(7.1\mathrm{r})_1}{\approx} x_1^{2n} (x_2^{c_2} x_3^{c_3} \cdots x_r^{c_r} \mathbf{w})^\star (x_1^\star)^{c_1} \\
&\overset{(8.1)}{\approx} x_1^{2n} (x_2^{c_2} x_3^{c_3} \cdots x_r^{c_r} \mathbf{w}) (x_1^\star)^{c_1} \\
&\overset{(7.1\mathrm{t})_1}{\approx} x_1^{2n} (x_2^{c_2} x_3^{c_3} \cdots x_r^{c_r} \mathbf{w}) x_1^{c_1} \\
&\overset{(7.1\mathrm{b})}{\approx} x_1^{c_1} x_2^{c_2} x_3^{c_3} \cdots x_r^{c_r} \mathbf{w} x_1^{2n} \\
&\overset{(7.1\mathrm{m})}{\approx} \mathbf{s}.
\end{aligned}$$

Consequently, the deduction $\{(\mathrm{inv}), (7.1)_1, (8.1)\} \vdash \mathbf{s} \approx \mathbf{s}^\star$ holds in any case. □

8.2 Restrictions on ★-Sandwich Identities

Consider a word

$$\mathbf{r} = x \left(\prod_{i=1}^{k} \mathbf{p}_i \right) x^{\circledast}, \tag{8.2}$$

where $k \geq 1$, $\circledast \in \{1, \star\}$, $x \in \mathscr{A}$, and $\mathbf{p}_1, \mathbf{p}_2, \ldots, \mathbf{p}_k \in (\mathscr{A} \cup \mathscr{A}^\star)^+$ are such that $x, \mathbf{p}_1, \mathbf{p}_2, \ldots, \mathbf{p}_k$ are pairwise disjoint. Note that depending on $\circledast$, the word $\mathbf{r}$ is a level one ★-sandwich of type (★S1) or (★S3). This ★-sandwich is said to be *highly refined* if it satisfies all of the following conditions:

(★H1) each $\mathbf{p}_i$ is either a singleton or a ★-sandwich;
(★H2) if $k \geq 2$, then $\min(\mathbf{p}_1) \ll \min(\mathbf{p}_k)$;
(★H3) if $k = 1$ and $\mathbf{p}_1$ is a singleton, then $\mathbf{p}_1 \in \mathscr{A}$.

An identity $\mathbf{r} \approx \mathbf{r}'$ is a *highly refined ★-sandwich identity* if $\mathbf{r}$ and $\mathbf{r}'$ are highly refined ★-sandwiches. Denote by

$$\mathsf{id}_{\mathrm{HR}}\{\langle \mathcal{L}_{3,n}, {}^{\star 1} \rangle\}$$

the set of all highly refined ★-sandwich identities satisfied by the involution semigroup $\langle \mathcal{L}_{3,n}, {}^{\star 1} \rangle$.

Lemma 8.4 *Suppose that* $\mathbf{s} \approx \mathbf{s}' \in \mathsf{id}_{\mathsf{SAN}}\{\langle \mathcal{L}_{3,n}, {}^{\star 1}\rangle\}$. *Then there exists some finite subset* Σ *of* $\mathsf{id}_{\mathsf{HR}}\{\langle \mathcal{L}_{3,n}, {}^{\star 1}\rangle\}$ *such that*

$$\{(\mathrm{inv}), (7.1)_1, (8.1), \mathbf{s} \approx \mathbf{s}'\} \sim \{(\mathrm{inv}), (7.1)_1, (8.1)\} \cup \Sigma.$$

Further, each identity $\mathbf{r} \approx \mathbf{r}'$ *in* Σ *can be chosen so that* $|\mathsf{sim}(\overline{\mathbf{r}})| \leq |\mathsf{sim}(\overline{\mathbf{s}})|$ *and* $|\mathsf{non}(\overline{\mathbf{r}})| \leq |\mathsf{non}(\overline{\mathbf{s}})|$.

Proof There are three cases depending on the type of the $\star$-sandwich identity $\mathbf{s} \approx \mathbf{s}'$.

CASE 1: $\mathbf{s} \approx \mathbf{s}'$ is of type ($\star$S1). Then by Lemma 7.9,

$$\mathbf{s} = x\mathbf{u}x^\star \quad \text{and} \quad \mathbf{s}' = x\mathbf{u}'x^\star$$

for some $x \in \mathscr{A}$ and $\mathbf{u}, \mathbf{u}' \in (\mathscr{A} \cup \mathscr{A}^\star)^+_\varnothing$ such that $x \notin \mathsf{con}(\overline{\mathbf{u}}) = \mathsf{con}(\overline{\mathbf{u}}')$. If $\mathsf{con}(\overline{\mathbf{u}}) = \mathsf{con}(\overline{\mathbf{u}}') = \varnothing$, then the identity $\mathbf{s} \approx \mathbf{s}'$ is trivial, so that the result holds with $\Sigma = \varnothing$. Therefore, assume that $\mathsf{con}(\overline{\mathbf{u}}) = \mathsf{con}(\overline{\mathbf{u}'}) \neq \varnothing$. In what follows, it is shown that the identities $\{(\mathrm{inv}), (7.1)_1, (8.1)\}$ can be used to convert $\mathbf{s}$ into a highly refined $\star$-sandwich $\mathbf{r}$. Similarly, the identities $\{(\mathrm{inv}), (7.1)_1, (8.1)\}$ can be used to convert $\mathbf{s}'$ into a highly refined $\star$-sandwich $\mathbf{r}'$. Hence, the equivalence

$$\{(\mathrm{inv}), (7.1)_1, (8.1), \mathbf{s} \approx \mathbf{s}'\} \sim \{(\mathrm{inv}), (7.1)_1, (8.1), \mathbf{r} \approx \mathbf{r}'\}$$

holds. In particular, the deduction $\{(\mathrm{inv}), (7.1)_1, (8.1)\} \vdash \mathbf{s} \approx \mathbf{r}$ holds. Therefore, $\mathbf{s} \approx \mathbf{r} \in \mathsf{id}\{\langle \mathcal{L}_{3,n}, {}^{\star 1}\rangle\}$ by Lemma 8.1, so that $|\mathsf{sim}(\overline{\mathbf{r}})| = |\mathsf{sim}(\overline{\mathbf{s}})|$ and $|\mathsf{non}(\overline{\mathbf{r}})| = |\mathsf{non}(\overline{\mathbf{s}})|$ by Lemma 7.2.

Let $\mathbf{u} = \prod_{i=1}^k \mathbf{p}_i$ be the natural decomposition of $\mathbf{u}$, so that $x, \mathbf{p}_1, \mathbf{p}_2, \ldots, \mathbf{p}_k$ are pairwise disjoint and each $\mathbf{p}_i$ is either a singleton or a connected word. By Lemma 7.7, the identities $(7.1)_1$ can be used to convert any connected $\mathbf{p}_i$ into some $\star$-sandwich $\mathbf{s}_i$ with $\mathsf{sim}(\overline{\mathbf{p}}_i) = \mathsf{sim}(\overline{\mathbf{s}}_i)$ and $\mathsf{non}(\overline{\mathbf{p}}_i) = \mathsf{non}(\overline{\mathbf{s}}_i)$. Therefore, it can be assumed that each $\mathbf{p}_i$ is either a singleton or a $\star$-sandwich, so that $\mathbf{s}$ satisfies ($\star$H1).

Suppose that $\mathbf{s}$ does not satisfy ($\star$H2), that is, $\min(\mathbf{p}_1) \not\ll \min(\mathbf{p}_k)$ with $k \geq 2$. Then since $\mathbf{p}_1$ and $\mathbf{p}_k$ are disjoint, $\min(\mathbf{p}_k) \ll \min(\mathbf{p}_1)$. Note that

$$\begin{aligned} \mathbf{s} &\overset{(8.1)}{\approx} x\mathbf{u}^\star x^\star \\ &\overset{(\mathrm{inv})}{\approx} x\left(\prod_{i=k}^{1} \mathbf{p}_i^\star\right)x^\star. \end{aligned}$$

If $\mathbf{p}_i$ is a $\star$-sandwich, then the deduction $\{(\mathrm{inv}), (7.1)_1, (8.1)\} \vdash \mathbf{p}_i^\star \approx \mathbf{p}_i$ holds by Lemma 8.3. If $\mathbf{p}_i$ is a singleton in $\mathscr{A}$, then $\mathbf{p}_i^\star$ is clearly a singleton in $\mathscr{A}^\star$.

If $\mathbf{p}_i$ is a singleton in $\mathscr{A}^\star$, then the deduction (inv) $\vdash \mathbf{p}_i^\star \approx \overline{\mathbf{p}}_i$ holds, where $\overline{\mathbf{p}}_i$ is a singleton in $\mathscr{A}$. Therefore, the identities $\{(\text{inv}), (7.1)_1, (8.1)\}$ can be used to convert $\mathbf{s}$ into the $\star$-sandwich $\mathbf{r} = x(\prod_{i=k}^{1} \mathbf{q}_i)x^\star$, where

$$\mathbf{q}_i = \begin{cases} \mathbf{p}_i & \text{if } \mathbf{p}_i \text{ is a } \star\text{-sandwich,} \\ \mathbf{p}_i^\star & \text{if } \mathbf{p}_i \text{ is a singleton in } \mathscr{A}, \\ \overline{\mathbf{p}}_i & \text{if } \mathbf{p}_i \text{ is a singleton in } \mathscr{A}^\star. \end{cases}$$

Since $\min(\mathbf{q}_k) = \min(\mathbf{p}_k) \ll \min(\mathbf{p}_1) = \min(\mathbf{q}_1)$, the $\star$-sandwich $\mathbf{r}$ satisfies ($\star$H2) and so is highly refined.

Finally, if $\mathbf{s}$ does not satisfy ($\star$H3), then $\mathbf{s} = xy^\star x^\star$ for some $y \in \mathscr{A}$, whence the identities (8.1) can be used to convert it into the highly refined $\star$-sandwich $\mathbf{r} = xyx^\star$ that satisfies ($\star$H3).

Case 2: $\mathbf{s} \approx \mathbf{s}'$ is of type ($\star$S2). Then by Lemma 7.10,

$$\mathbf{s} = x^\star \mathbf{u} x \quad \text{and} \quad \mathbf{s}' = x^\star \mathbf{u}' x$$

for some $x \in \mathscr{A}$ and $\mathbf{u}, \mathbf{u}' \in (\mathscr{A} \cup \mathscr{A}^\star)^+_\varnothing$ such that $x \notin \mathsf{con}(\overline{\mathbf{u}}) = \mathsf{con}(\overline{\mathbf{u}'})$. It is clear that the equivalence $\{(\text{inv}), \mathbf{s} \approx \mathbf{s}'\} \sim \{(\text{inv}), x\mathbf{u}x^\star \approx x\mathbf{u}'x^\star\}$ holds. Since $x\mathbf{u}x^\star \approx x\mathbf{u}'x^\star$ is a $\star$-sandwich identity of type ($\star$S1), the result follows from Case 1.

Case 3: $\mathbf{s} \approx \mathbf{s}'$ is of type ($\star$S3). Then by Lemma 7.11,

$$\mathbf{s} = \mathbf{x} \prod_{i=1}^{\ell} (\mathbf{u}_i \vec{\mathscr{X}}) \quad \text{and} \quad \mathbf{s}' = \mathbf{x} \prod_{i=1}^{\ell} (\mathbf{u}_i' \vec{\mathscr{X}})$$

for some $\ell \geq 1$, nonempty $\mathscr{X} \subseteq \mathscr{A}$, $\mathbf{x} \in \mathscr{X}^\boxplus$, and $\mathbf{u}_i, \mathbf{u}_i' \in (\mathscr{A} \cup \mathscr{A}^\star)^+_\varnothing$ such that $\mathsf{con}(\overline{\mathbf{u}}_i) = \mathsf{con}(\overline{\mathbf{u}}_i')$ for each i and $\vec{\mathscr{X}}, \mathbf{u}_1, \mathbf{u}_2, \ldots, \mathbf{u}_\ell$ are pairwise disjoint. By Lemma 7.12, the equivalence

(A) $\{(7.1)_1, \mathbf{s} \approx \mathbf{s}'\} \sim \{(7.1)_1\} \cup \{x\mathbf{u}_i x \approx x\mathbf{u}_i' x \mid 1 \leq i \leq \ell\}$

holds. It is easily seen that

(B) $|\mathsf{sim}(\overline{x\mathbf{u}_i x})| \leq |\mathsf{sim}(\overline{\mathbf{s}})|$ and $|\mathsf{non}(\overline{x\mathbf{u}_i x})| \leq |\mathsf{non}(\overline{\mathbf{s}})|$ for all i.

For each i, the argument in Case 1 can be repeated to show that the equivalence

(C) $\{(\text{inv}), (7.1)_1, (8.1), x\mathbf{u}_i x \approx x\mathbf{u}_i' x\} \sim \{(\text{inv}), (7.1)_1, (8.1), \mathbf{r}_i \approx \mathbf{r}_i'\}$

holds for some $\mathbf{r}_i \approx \mathbf{r}_i' \in \mathsf{id}_{\mathrm{HR}}\{\langle \mathscr{L}_{3,n}, {}^{\star 1} \rangle\}$ such that

(D) $|\,\mathsf{sim}(\overline{\mathbf{r}}_i)| = |\,\mathsf{sim}(\overline{x\mathbf{u}_i x})|$ and $|\,\mathsf{non}(\overline{\mathbf{r}}_i)| = |\,\mathsf{non}(\overline{x\mathbf{u}_i x})|$.

Hence, the equivalence

$$\{(\text{inv}), (7.1)_1, (8.1), \mathbf{s} \approx \mathbf{s}'\} \sim \{(\text{inv}), (7.1)_1, (8.1)\} \cup \{\mathbf{r}_i \approx \mathbf{r}_i' \mid 1 \leq i \leq \ell\}$$

holds by (A) and (C), where $|\,\mathsf{sim}(\overline{\mathbf{r}}_i)| \leq |\,\mathsf{sim}(\overline{\mathbf{s}})|$ and $|\,\mathsf{non}(\overline{\mathbf{r}}_i)| \leq |\,\mathsf{non}(\overline{\mathbf{s}})|$ by (B) and (D). □

Lemma 8.5 *Suppose that* $\mathbf{r} \approx \mathbf{r}' \in \mathsf{id}_{\mathsf{HR}}\{\langle \mathcal{L}_{3,n}, {}^{\star 1}\rangle\}$, *where*

$$\mathbf{r} = x\left(\prod_{i=1}^{k} \mathbf{p}_i\right)x^{\circledast}$$

is the highly refined $\star$*-sandwich in* (8.2). *Then*

$$\mathbf{r}' = x\left(\prod_{i=1}^{k} \mathbf{p}_i'\right)x^{\circledast}$$

for some $\mathbf{p}_1', \mathbf{p}_2', \ldots, \mathbf{p}_k' \in (\mathscr{A} \cup \mathscr{A}^{\star})^+$ *such that each* $\mathbf{p}_i'$ *is either a singleton or a* $\star$*-sandwich,* $x, \mathbf{p}_1', \mathbf{p}_2', \ldots, \mathbf{p}_k'$ *are pairwise disjoint, and* $\mathsf{con}(\overline{\mathbf{p}}_i) = \mathsf{con}(\overline{\mathbf{p}}_i')$ *for all* i. *Further, for each* i,

(i) $\mathbf{p}_i$ *and* $\mathbf{p}_i'$ *are either both singletons or both* $\star$*-sandwiches*;
(ii) *if* $\mathbf{p}_i$ *and* $\mathbf{p}_i'$ *are both singletons, then* $\mathbf{p}_i = \mathbf{p}_i'$;
(iii) *if* $\mathbf{p}_i$ *and* $\mathbf{p}_i'$ *are both* $\star$*-sandwiches, then* $\mathbf{p}_i \approx \mathbf{p}_i' \in \mathsf{id}_{\mathsf{SAN}}\{\langle \mathcal{L}_{3,n}, {}^{\star 1}\rangle\}$.

Proof Repeating the proof of Lemma 7.15 yields

$$\mathbf{r}' = x\left(\prod_{i=1}^{k'} \mathbf{p}_i'\right)x^{\circledast},$$

where each $\mathbf{p}_i'$ is either a singleton or a $\star$-sandwich, $x, \mathbf{p}_1', \mathbf{p}_2', \ldots, \mathbf{p}_k'$ are pairwise disjoint, and

(H) $(\mathsf{con}(\overline{\mathbf{p}}_1'), \mathsf{con}(\overline{\mathbf{p}}_2'), \ldots, \mathsf{con}(\overline{\mathbf{p}}_k')) = (\mathsf{con}(\overline{\mathbf{p}}_1), \mathsf{con}(\overline{\mathbf{p}}_2), \ldots, \mathsf{con}(\overline{\mathbf{p}}_k))$ or
(H$'$) $(\mathsf{con}(\overline{\mathbf{p}}_1'), \mathsf{con}(\overline{\mathbf{p}}_2'), \ldots, \mathsf{con}(\overline{\mathbf{p}}_k')) = (\mathsf{con}(\overline{\mathbf{p}}_k), \mathsf{con}(\overline{\mathbf{p}}_{k-1}), \ldots, \mathsf{con}(\overline{\mathbf{p}}_1))$.

If $k = 1$, then (H) clearly holds. If $k \geq 2$, then since $\mathbf{r}$ and $\mathbf{r}'$ are highly refined $\star$-sandwiches, (H) holds due to ($\star$H2). Therefore, (H) holds in any case.

It remains to show that (i)–(iii) hold.

(i) This follows from (H) since $\mathsf{sim}(\overline{\mathbf{r}}) = \mathsf{sim}(\overline{\mathbf{r}}')$ and $\mathsf{non}(\overline{\mathbf{r}}) = \mathsf{non}(\overline{\mathbf{r}}')$ by Lemma 7.2.

(ii) Suppose that $\mathbf{p}_i$ and $\mathbf{p}_i'$ are both singletons. Since $\mathsf{con}(\overline{\mathbf{p}}_i) = \mathsf{con}(\overline{\mathbf{p}}_i')$ by (H), there exists some $y \in \mathscr{A}$ such that $\mathbf{p}_i = y^{\circledast_1}$ and $\mathbf{p}_i' = y^{\circledast_2}$ with $\circledast_i \in \{1, \star\}$. If $k = 1$, then $\mathbf{p}_i = y = \mathbf{p}_i'$ by ($\star$H3). Therefore, suppose that $k \geq 2$, so that either $1 < i$ or $i < k$. By symmetry, it suffices to assume that $1 < i$. Let $\chi : \mathscr{A} \to \mathcal{L}_3$ denote the substitution

$$z \mapsto \begin{cases} \mathsf{e} & \text{if } z \in \mathsf{con}(\overline{\mathbf{p}}_1\overline{\mathbf{p}}_2 \cdots \overline{\mathbf{p}}_{i-1}) = \mathsf{con}(\overline{\mathbf{p}}_1'\overline{\mathbf{p}}_2' \cdots \overline{\mathbf{p}}_{i-1}'), \\ \mathsf{ef} & \text{if } z = y, \\ \mathsf{f} & \text{otherwise.} \end{cases}$$

Then

$$\begin{aligned} \mathbf{r}\chi &= x\chi \cdot (\mathbf{p}_1\mathbf{p}_2 \cdots \mathbf{p}_{i-1})\chi \cdot y^{\circledast_1}\chi \cdot (\mathbf{p}_{i+1} \cdots \mathbf{p}_k x^{\circledast_1})\chi \\ &= \mathsf{fe}(\mathsf{ef})^{\circledast_1}\mathsf{f}. \end{aligned}$$

Similarly, $\mathbf{r}'\chi = \mathsf{fe}(\mathsf{ef})^{\circledast_2}\mathsf{f}$. Now $\mathbf{r}\chi = \mathbf{r}'\chi$, so that $\mathsf{fe}(\mathsf{ef})^{\circledast_1}\mathsf{f} = \mathsf{fe}(\mathsf{ef})^{\circledast_2}\mathsf{f}$. It is then easily shown that $\circledast_1 = \circledast_2$, whence $\mathbf{p}_i = \mathbf{p}_i'$.

(iii) Suppose that $\mathbf{p}_i$ and $\mathbf{p}_i'$ are $\star$-sandwiches. Let $\psi : \mathscr{A} \to \mathscr{A}^+$ denote the substitution

$$z \mapsto \begin{cases} z & \text{if } z \in \mathsf{con}(\overline{\mathbf{p}}_i) = \mathsf{con}(\overline{\mathbf{p}}_i'), \\ x^{2n} & \text{otherwise.} \end{cases}$$

Then the deduction $\{(\text{inv}), (7.1\text{a}), (7.1\text{p})\} \vdash \{x(\mathbf{r}\psi)x \approx x\mathbf{p}_i x, x(\mathbf{r}'\psi)x \approx x\mathbf{p}_i'x\}$ holds, so the deduction $\{(\text{inv}), (7.1\text{a}), (7.1\text{p}), \mathbf{r} \approx \mathbf{r}'\} \vdash x\mathbf{p}_i x \approx x\mathbf{p}_i'x$ follows. Therefore, $x\mathbf{p}_i x \approx x\mathbf{p}_i'x \in \mathsf{id}_{\mathsf{W}}\{\langle \mathcal{L}_{3,n}, {}^{\star 1}\rangle\}$ by Lemma 8.1, whence $\mathbf{p}_i \approx \mathbf{p}_i' \in \mathsf{id}_{\mathsf{W}}\{\langle \mathcal{L}_{3,n}, {}^{\star 1}\rangle\}$ by Lemma 7.14. □

8.3 An Explicit Identity Basis for $\langle \mathcal{L}_{3,n}, {}^{\star 1} \rangle$

Proposition 8.6 *The identities* $\{(\text{inv}), (7.1)_1, (8.1)\}$ *constitute an identity basis for the involution semigroup* $\langle \mathcal{L}_{3,n}, {}^{\star 1}\rangle$.

It follows from Lemma 8.2 that the identities $\{(\text{inv}), (7.1)_1, (8.1)\} \cup \mathsf{id}_{\mathrm{SAN}}\{\langle \mathcal{L}_{3,n}, {}^{\star 1}\rangle\}$ form an identity basis for the involution semigroup $\langle \mathcal{L}_{3,n}, {}^{\star 1}\rangle$. Therefore, to prove Proposition 8.6, it suffices to establish the deduction $\{(\text{inv}), (7.1)_1, (8.1)\} \vdash \mathsf{id}_{\text{SAN}}\{\langle \mathcal{L}_{3,n}, {}^{\star 1}\rangle\}$. This is achieved by verifying the following statement for each $m \geq 1$:

($\spadesuit_m$) If $\mathbf{s} \approx \mathbf{s}' \in \mathsf{id}_{\text{SAN}}\{\langle \mathcal{L}_{3,n}, {}^{\star 1}\rangle\}$ with $|\,\mathsf{non}(\overline{\mathbf{s}})| \leq m$, then $\{(\text{inv}), (7.1)\} \vdash \mathbf{s} \approx \mathbf{s}'$.

Lemma 8.7 *The statement* ($\spadesuit_1$) *holds.*

Proof Suppose that $\mathbf{s} \approx \mathbf{s}' \in \mathsf{id}_{\text{SAN}}\{\langle \mathcal{L}_{3,n}, {}^{\star 1}\rangle\}$ with $|\,\mathsf{non}(\overline{\mathbf{s}})| = 1$. Then by Lemma 8.4, there exists some finite $\Sigma \subseteq \mathsf{id}_{\text{HR}}\{\langle \mathcal{L}_{3,n}, {}^{\star 1}\rangle\}$ such that

(A) $\{(\text{inv}), (7.1)_1, (8.1), \mathbf{s} \approx \mathbf{s}'\} \sim \{(\text{inv}), (7.1)_1, (8.1)\} \cup \Sigma$;
(B) each $\mathbf{r} \approx \mathbf{r}' \in \Sigma$ satisfies $|\,\mathsf{sim}(\overline{\mathbf{r}})| \leq |\,\mathsf{sim}(\overline{\mathbf{s}})|$ and $|\,\mathsf{non}(\overline{\mathbf{r}})| \leq |\,\mathsf{non}(\overline{\mathbf{s}})| = 1$.

Consider any $\mathbf{r} \approx \mathbf{r}' \in \Sigma$. Generality is not lost by assuming that $\mathbf{r} = x(\prod_{i=1}^{k} \mathbf{p}_i)x^{\circledast}$ is the highly refined $\star$-sandwich in (8.2). Then it follows from Lemma 8.5 that $\mathbf{r}' = x(\prod_{i=1}^{k} \mathbf{p}_i')x^{\circledast}$ for some $\mathbf{p}_1', \mathbf{p}_2', \ldots, \mathbf{p}_k' \in (\mathscr{A} \cup \mathscr{A}^\star)^+$ such that $\mathsf{con}(\overline{\mathbf{p}}_i) = \mathsf{con}(\overline{\mathbf{p}}_i')$ for all i. Since $1 = |\,\mathsf{non}(\overline{\mathbf{r}})| = |\,\mathsf{non}(\overline{\mathbf{r}}')|$ by (B) and Lemma 7.2, it follows from (i) and (ii) of Lemma 8.5 that $\mathbf{p}_i$ and $\mathbf{p}_i'$ are the same singleton. Therefore, the identity $\mathbf{r} \approx \mathbf{r}'$ is trivial. Since the identity $\mathbf{r} \approx \mathbf{r}'$ is arbitrary in Σ, every identity in Σ is trivial. Hence, the deduction $\{(\text{inv}), (7.1)_1, (8.1)\} \vdash \mathbf{s} \approx \mathbf{s}'$ follows from (A). □

Lemma 8.8 *Suppose that the statement* ($\spadesuit_m$) *holds. Then the statement* ($\spadesuit_{m+1}$) *also holds.*

Proof Suppose that $\mathbf{s} \approx \mathbf{s}' \in \mathsf{id}_{\text{SAN}}\{\langle \mathcal{L}_{3,n}, {}^{\star 1}\rangle\}$ with $|\,\mathsf{non}(\overline{\mathbf{s}})| = m + 1$. Then by Lemma 8.4, there exists some finite $\Sigma \subseteq \mathsf{id}_{\text{HR}}\{\langle \mathcal{L}_{3,n}, {}^{\star 1}\rangle\}$ such that

(A) $\{(\text{inv}), (7.1)_1, (8.1), \mathbf{s} \approx \mathbf{s}'\} \sim \{(\text{inv}), (7.1)_1, (8.1)\} \cup \Sigma$;
(B) each $\mathbf{r} \approx \mathbf{r}' \in \Sigma$ satisfies $|\,\mathsf{sim}(\overline{\mathbf{r}})| \leq |\,\mathsf{sim}(\overline{\mathbf{s}})|$ and $|\,\mathsf{non}(\overline{\mathbf{r}})| \leq |\,\mathsf{non}(\overline{\mathbf{s}})| = m + 1$.

Consider any $\mathbf{r} \approx \mathbf{r}' \in \Sigma$. Generality is not lost by assuming that $\mathbf{r} = x(\prod_{i=1}^{k} \mathbf{p}_i)x^{\circledast}$ is the highly refined $\star$-sandwich in (8.2). Then it follows from Lemma 8.5 that $\mathbf{r}' = x(\prod_{i=1}^{k} \mathbf{p}_i')x^{\circledast}$ for some $\mathbf{p}_1', \mathbf{p}_2', \ldots, \mathbf{p}_k' \in (\mathscr{A} \cup \mathscr{A}^\star)^+$ such that $\mathsf{con}(\overline{\mathbf{p}}_i) = \mathsf{con}(\overline{\mathbf{p}}_i')$ for all i. By (i) of Lemma 8.5, the words $\mathbf{p}_i$ and $\mathbf{p}_i'$ are both singletons or both $\star$-sandwiches.

CASE 1: $\mathbf{p}_i$ and $\mathbf{p}_i'$ are singletons. Then $\mathbf{p}_i = \mathbf{p}_i'$ by (ii) of Lemma 8.5, so that the deduction $\{(\text{inv}), (7.1)_1, (8.1)\} \vdash \mathbf{p}_i \approx \mathbf{p}_i'$ holds vacuously.

CASE 2: $\mathbf{p}_i$ and $\mathbf{p}_i'$ are $\star$-sandwiches. Then $\mathbf{p}_i \approx \mathbf{p}_i' \in \mathsf{id}_{\text{SAN}}\{\langle \mathcal{L}_{3,n}, {}^{\star 1}\rangle\}$ by (iii) of Lemma 8.5. Since $|\,\mathsf{non}(\overline{\mathbf{p}}_i)| < |\,\mathsf{non}(\overline{\mathbf{r}})| \leq m + 1$ by (B), the deduction $\{(\text{inv}), (7.1)_1, (8.1)\} \vdash \mathbf{p}_i \approx \mathbf{p}_i'$ holds by ($\spadesuit_m$).

Therefore, the deduction $\{(\text{inv}), (7.1)_1, (8.1)\} \vdash \mathbf{p}_i \approx \mathbf{p}_i'$ holds in any case. Since

$$\mathbf{r} = x\left(\prod_{i=1}^{k} \mathbf{p}_i\right)x^{\circledast} \overset{(\text{inv}),(7.1)_1,(8.1)}{\approx} x\left(\prod_{i=1}^{k} \mathbf{p}_i'\right)x^{\circledast} = \mathbf{r}',$$

the deduction $\{(\text{inv}), (7.1)_1, (8.1)\} \vdash \mathbf{r} \approx \mathbf{r}'$ also holds. The identity $\mathbf{r} \approx \mathbf{r}'$ is arbitrary in Σ, so that $\{(\text{inv}), (7.1)_1, (8.1)\} \vdash \Sigma$. Consequently, $\{(\text{inv}), (7.1)_1, (8.1)\} \vdash \mathbf{s} \approx \mathbf{s}'$ by (A). □

8.4 A Finite Identity Basis for $\langle \mathcal{L}_{3,n}, {}^{\star 1} \rangle$

Theorem 8.9 *The identities* (inv) *and*

$$x^{n+2} \approx x^2, \tag{8.3a}$$

$$x^{n+1}yx^{n+1} \approx xyx, \tag{8.3b}$$

$$xhykxty \approx yhxkytx, \tag{8.3c}$$

$$xy^\star x^\star \approx xyx^\star, \tag{8.3d}$$

$$x^\star yx^\star \approx xyx, \tag{8.3e}$$

$$xhx^\star kx \approx xhxkx \tag{8.3f}$$

constitute an identity basis for the involution semigroup $\langle \mathcal{L}_{3,n}, {}^{\star 1} \rangle$.

Proof By Proposition 8.6, it suffices to show that $\{(\text{inv}), (7.1)_1, (8.1)\} \sim \{(\text{inv}), (8.3)\}$. It is easily checked that the involution semigroup $\langle \mathcal{L}_{3,n}, {}^{\star 1} \rangle$ satisfies the identities (8.3), so the deduction $\{(\text{inv}), (7.1)_1, (8.1)\} \vdash \{(\text{inv}), (8.3)\}$ follows. Verification of the deduction $\{(\text{inv}), (8.3)\} \vdash \{(\text{inv}), (7.1)_1, (8.1)\}$ is divided into eight cases.

Case 1: $\{(\text{inv}), (8.3)\} \vdash (8.1)$. The first identity in (8.1) is deducible from (8.3) because

$$\begin{aligned} xy^\star x &\overset{(8.3b)}{\approx} xx^n y^\star xx^n \\ &\overset{(8.3e)}{\approx} x^\star x^n y^\star x^\star x^n \\ &\overset{(8.3d)}{\approx} x^\star x^n yx^\star x^n \\ &\overset{(8.3e)}{\approx} xx^n yxx^n \\ &\overset{(8.3b)}{\approx} xyx. \end{aligned}$$

The second identity in (8.1) coincides with (8.3d), and the third identity in (8.1) is deducible from {(inv), (8.3)} because

$$\begin{aligned} x^\star y^\star x &\overset{\text{(inv)}}{\approx} x^\star y^\star (x^\star)^\star \\ &\overset{\text{(8.3d)}}{\approx} x^\star y (x^\star)^\star \\ &\overset{\text{(inv)}}{\approx} x^\star y x. \end{aligned}$$

CASE 2: {(inv), (8.3)} ⊢ {(7.1a), (7.4b), ..., (7.1o)}. Since

$$\begin{aligned} x\Bigg(\prod_{i=1}^{m}(y_i h_i y_i)\Bigg)x &\overset{\text{(8.1)}}{\approx} x\Bigg(\prod_{i=1}^{m}(y_i h_i y_i)\Bigg)^\star x \\ &\overset{\text{(inv)}}{\approx} x\Bigg(\prod_{i=m}^{1}(y_i^\star h_i^\star y_i^\star)\Bigg)x \\ &\overset{\text{(8.3e)}}{\approx} x\Bigg(\prod_{i=m}^{1}(y_i h_i^\star y_i)\Bigg)x \\ &\overset{\text{(8.1)}}{\approx} x\Bigg(\prod_{i=m}^{1}(y_i h_i y_i)\Bigg)x, \end{aligned}$$

the deduction {(inv), (8.1), (8.3)} ⊢ (6.6d) holds. Hence, the deduction {(inv), (8.3)} ⊢ (6.6d) also holds by Case 1. Since the identities (8.3a)–(8.3c) and (6.6a)–(6.6c) coincide, the deduction {(inv), (8.3)} ⊢ (6.6) holds, whence by Corollary 6.19, all plain identities satisfied by $\langle \mathcal{L}_{3,n}, {}^{\star 1}\rangle$ are deducible from {(inv), (8.3)}. In particular, the present case follows.

CASE 3: {(inv), (8.3)} ⊢ (7.1p). This deduction holds because

$$\begin{aligned} (x^{2n})^\star &\overset{\text{(inv)}}{\approx} (x^\star)^{2n} \\ &\overset{\text{(8.3a)}}{\approx} (x^\star)^{2n}(x^\star)^{n}(x^\star)^{2n} \\ &\overset{\text{(8.3e)}}{\approx} x^{2n}(x^\star)^{n}x^{2n} \\ &\overset{\text{(8.3f)}}{\approx} x^{2n}x^{n}x^{2n} \\ &\overset{\text{(8.3a)}}{\approx} x^{2n}. \end{aligned}$$

CASE 4: (8.3) ⊢ $(7.1\text{q})_1$. If $\mathbf{h}_1 = h_1$, then the deduction (8.3e) ⊢ $(7.1\text{q})_1$ clearly holds. If $\mathbf{h}_1 = \varnothing$, then the identity $(7.1\text{q})_1$ reduces to $(x^\star)^2 \approx x^2$, whence the deduction

(8.3) $\vdash$ (7.1q)$_1$ holds because

$$\begin{aligned} (x^\star)^2 &\overset{(8.3a)}{\approx} (x^\star)^{2n}(x^\star)^2(x^\star)^{2n} \\ &\overset{(8.3e)}{\approx} x^{2n}(x^\star)^2x^{2n} \\ &\overset{(8.3f)}{\approx} x^{2n}x^2x^{2n} \\ &\overset{(8.3a)}{\approx} x^2. \end{aligned}$$

CASE 5: (8.3) $\vdash$ (7.1s)$_1$. This deduction holds because

$$\begin{aligned} x\mathbf{h}_1x^\star\mathbf{h}_2x &\overset{(8.3b)}{\approx} xx^n\mathbf{h}_1x^\star\mathbf{h}_2x^nx \\ &\overset{(8.3f)}{\approx} xx^n\mathbf{h}_1x\mathbf{h}_2x^nx \\ &\overset{(8.3b)}{\approx} x\mathbf{h}_1x\mathbf{h}_2x. \end{aligned}$$

CASE 6: (8.3) $\vdash$ {(7.1r)$_1$, (7.1t)$_1$}. The deduction {(8.3), (7.1s)$_1$} $\vdash$ (7.1r)$_1$ holds because

$$\begin{aligned} x^\star\mathbf{h}_1x\mathbf{h}_2x &\overset{(8.3b)}{\approx} x^\star\mathbf{h}_1x^nx\mathbf{h}_2x^nx \\ &\overset{(8.3e)}{\approx} x^\star\mathbf{h}_1x^nx^\star\mathbf{h}_2x^nx^\star \\ &\overset{(8.3e)}{\approx} x\mathbf{h}_1x^nx^\star\mathbf{h}_2x^nx \\ &\overset{(7.1s)_1}{\approx} x\mathbf{h}_1x^nx\mathbf{h}_2x^nx \\ &\overset{(8.3b)}{\approx} x\mathbf{h}_1x\mathbf{h}_2x. \end{aligned}$$

In view of Case 5, the deduction (8.3) $\vdash$ (7.1r)$_1$ follows. By symmetry, the deduction (8.3) $\vdash$ (7.1t)$_1$ also holds.

CASE 7: {(inv), (8.3)} $\vdash$ {(7.1u)$_1$, (7.1v)$_1$, (7.1w)$_1$, (7.1x)$_1$}. It is easily seen that the identities {(7.1u)$_1$, (7.1v)$_1$, (7.1w)$_1$, (7.1x)$_1$} are deducible from

(A) $x^{\circledast_1}\mathbf{h}_1y^{\circledast_2}\mathbf{h}_2x^{\circledast_3}\mathbf{h}_3y^{\circledast_4} \approx x\mathbf{h}_1y\mathbf{h}_2x\mathbf{h}_3y$,

where $\mathbf{h}_i \in \{\varnothing, h_i\}$ and $\circledast_i \in \{1, \star\}$. Therefore, it suffices to establish the deduction {(inv), (8.3)} $\vdash$ (A). For this purpose, the following identities are required:

(B) $x^\star\mathbf{h}_1y^{\circledast_1}\mathbf{h}_2x^\star\mathbf{h}_3y^{\circledast_2} \approx x\mathbf{h}_1y^{\circledast_1}\mathbf{h}_2x\mathbf{h}_3y^{\circledast_2}$,

(C) $x\mathbf{h}_1y^{\circledast_1}\mathbf{h}_2x^\star\mathbf{h}_3y^{\circledast_2} \approx x\mathbf{h}_1y^{\circledast_1}\mathbf{h}_2x\mathbf{h}_3y^{\circledast_2}$,

(D) $x^\star\mathbf{h}_1y^{\circledast_1}\mathbf{h}_2x\mathbf{h}_3y^{\circledast_2} \approx x\mathbf{h}_1y^{\circledast_1}\mathbf{h}_2x\mathbf{h}_3y^{\circledast_2}$,

(E) $x^{\circledast_1}\mathbf{h}_1y^{\circledast_2}\mathbf{h}_2x^{\circledast_3}\mathbf{h}_3y^{\circledast_4} \approx x\mathbf{h}_1y^{\circledast_2}\mathbf{h}_2x\mathbf{h}_3y^{\circledast_4}$,

(F) $x^{\circledast_1}\mathbf{h}_1y^{\circledast_2}\mathbf{h}_2x^{\circledast_3}\mathbf{h}_3y^{\circledast_4} \approx x^{\circledast_1}\mathbf{h}_1y\mathbf{h}_2x^{\circledast_3}\mathbf{h}_3y$, where $\mathbf{h}_i \in \{\varnothing, h_i\}$ and $\circledast_i \in \{1, \star\}$. The deduction (8.3e) $\vdash$ `(B)` is obvious, and the deduction {(inv), (8.1), (8.3)} $\vdash$ `(C)` holds because

$$
\begin{aligned}
x\mathbf{h}_1 y^{\circledast_1}\mathbf{h}_2 x^\star \mathbf{h}_3 y^{\circledast_2} &\overset{(8.1)}{\approx} x\mathbf{h}_1 y^{\circledast_1}(\mathbf{h}_2 x^\star \mathbf{h}_3)^\star y^{\circledast_2} \\
&\overset{(\text{inv})}{\approx} x\mathbf{h}_1 y^{\circledast_1}\mathbf{h}_3^\star x \mathbf{h}_2^\star y^{\circledast_2} \\
&\overset{(8.3b)}{\approx} x^{n+1}\mathbf{h}_1 y^{\circledast_1}\mathbf{h}_3^\star x^{n+1} \mathbf{h}_2^\star y^{\circledast_2} \\
&\overset{(8.3a)}{\approx} x^{n+1}x^{2n-1}x\mathbf{h}_1 y^{\circledast_1}\mathbf{h}_3^\star x^{n+1} \mathbf{h}_2^\star y^{\circledast_2} \\
&\overset{(8.3b)}{\approx} xx^{2n-1}x\mathbf{h}_1 y^{\circledast_1}\mathbf{h}_3^\star x \mathbf{h}_2^\star y^{\circledast_2} \\
&\overset{(8.3f)}{\approx} xx^{2n-1}x^\star\mathbf{h}_1 y^{\circledast_1}\mathbf{h}_3^\star x \mathbf{h}_2^\star y^{\circledast_2} \\
&\overset{(8.3e)}{\approx} x^\star x^{2n-1}x^\star\mathbf{h}_1 y^{\circledast_1}\mathbf{h}_3^\star x^\star \mathbf{h}_2^\star y^{\circledast_2} \\
&\overset{(8.3e)}{\approx} xx^{2n-1}x\mathbf{h}_1 y^{\circledast_1}\mathbf{h}_3^\star x^\star \mathbf{h}_2^\star y^{\circledast_2} \\
&\overset{(\text{inv})}{\approx} x^{2n}x\mathbf{h}_1 y^{\circledast_1}(\mathbf{h}_2 x \mathbf{h}_3)^\star y^{\circledast_2} \\
&\overset{(8.1)}{\approx} x^{2n}x\mathbf{h}_1 y^{\circledast_1}\mathbf{h}_2 x \mathbf{h}_3 y^{\circledast_2} \\
&\overset{(8.3b)}{\approx} x^{2n}x^{n+1}\mathbf{h}_1 y^{\circledast_1}\mathbf{h}_2 x^{n+1} \mathbf{h}_3 y^{\circledast_2} \\
&\overset{(8.3a)}{\approx} x^{n+1}\mathbf{h}_1 y^{\circledast_1}\mathbf{h}_2 x^{n+1} \mathbf{h}_3 y^{\circledast_2} \\
&\overset{(8.3b)}{\approx} x\mathbf{h}_1 y^{\circledast_1}\mathbf{h}_2 x \mathbf{h}_3 y^{\circledast_2}.
\end{aligned}
$$

The deduction {(inv), (8.1), (8.3)} $\vdash$ `(D)` is similarly established. Therefore,

$$
\begin{aligned}
\{(\text{inv}), (8.1), (8.3)\} &\vdash \{\texttt{(B)}, \texttt{(C)}, \texttt{(D)}\} \\
&\vdash \texttt{(E)}.
\end{aligned}
$$

By symmetry, the deduction {(inv), (8.1), (8.3)} $\vdash$ `(F)` also holds. Hence,

$$
\begin{aligned}
\{(\text{inv}), (8.3)\} &\vdash \{(\text{inv}), (8.1), (8.3)\} \qquad \text{by Case 1} \\
&\vdash \{\texttt{(E)}, \texttt{(F)}\} \\
&\vdash \texttt{(A)}
\end{aligned}
$$

as required.

CASE 8: {(inv), (8.3d)} $\vdash$ (7.1y). Since

$$
\begin{aligned}
x\left(\prod_{i=1}^{m}(y_i h_i y_i^\star)\right)x^\star &\overset{(8.3d)}{\approx} x\left(\prod_{i=1}^{m}(y_i h_i y_i^\star)\right)^\star x^\star \\
&\overset{(\mathrm{inv})}{\approx} x\left(\prod_{i=m}^{1}(y_i h_i^\star y_i^\star)\right)x^\star \\
&\overset{(8.3d)}{\approx} x\left(\prod_{i=m}^{1}(y_i h_i y_i^\star)\right)x^\star,
\end{aligned}
$$

the present case follows. □

8.5 Summary

The main objective of the present chapter is to show that the involution semigroup $\langle \mathcal{L}_{3,n}, {}^{\star 1} \rangle$ is finitely based; see Theorem 8.9. This result was published in Lee (2016b).

Counterintuitive Examples of Involution Semigroups

9

The present chapter provides solutions to problems posed in Sects. 1.5.1 and 1.5.2. In Sect. 9.1, three finite involution semigroups sharing the same semigroup reduct are shown to possess different types of identity bases. In Sect. 9.2, a finite class of finite involution semigroups is used to demonstrate the drastic structural difference between the lattices $\mathfrak{L}_{\mathsf{inv}}$ and $\mathfrak{L}_{\mathsf{sem}}$.

9.1 Involution Semigroups with Different Types of Identity Bases

For any disjoint semigroups S_1 and S_2 that exclude the symbol 0, the *0-direct union* of S_1 and S_2 is the semigroup $S_1 \uplus S_2 = S_1 \cup S_2 \cup \{0\}$, where the product of any two elements from S_i is their usual product in S_i and $ab = ba = 0$ for all $a \in S_1 \cup \{0\}$ and $b \in S_2 \cup \{0\}$. More generally, it is possible to form the 0-direct union of any two semigroups S_1 and S_2 by first renaming their elements so that $S_1 \cap S_2 = \varnothing$ and $0 \notin S_1 \cup S_2$.

Theorem 9.1 *For each $n \geq 3$, there exist three finite involution semigroups, all sharing the semigroup reduct $\mathcal{L}_{3,n} \uplus \mathcal{L}_{3,n}$, such that one has a finite identity basis, one has an infinite irredundant identity basis, and one has no irredundant identity bases.*

In Sect. 9.1.1, involution semigroups having semigroup reduct $\mathcal{L}_{3,n} \uplus \mathcal{L}_{3,n}$ with an irredundant identity basis are exhibited; some of these involution semigroups have a finite identity basis, while others have an infinite irredundant identity basis. In Sect. 9.1.2, some involution semigroups having a semigroup reduct isomorphic to $\mathcal{L}_{3,n} \uplus \mathcal{L}_{3,n}$ are shown to have no irredundant identity bases. The proof of Theorem 9.1 is thus complete.

E. W. H. Lee, *Advances in the Theory of Varieties of Semigroups*, Frontiers in Mathematics, https://doi.org/10.1007/978-3-031-16497-2_9

9.1.1 Involution Semigroups with an Irredundant Identity Basis

Lemma 9.2

(i) $\mathscr{V}_{\mathsf{sem}}\{S^0\} = \mathscr{V}_{\mathsf{sem}}\{S, \mathcal{S}\ell_2\}$ *for any semigroup* S.
(ii) $\mathscr{V}_{\mathsf{sem}}\{S_1 \uplus S_2\} = \mathscr{V}_{\mathsf{sem}}\{S_1, S_2\}$ *for any semigroups* S_1 *and* S_2 *such that* $\mathcal{S}\ell_2 \in \mathscr{V}_{\mathsf{sem}}\{S_1, S_2\}$.

Proof

(i) Since the subset $I = \{(a, 0) \mid a \in S\}$ of $S \times \mathcal{S}\ell_2$ is an ideal such that $(S \times \mathcal{S}\ell_2)/I \cong S^0$, the inclusion $\mathscr{V}_{\mathsf{sem}}\{S^0\} \subseteq \mathscr{V}_{\mathsf{sem}}\{S, \mathcal{S}\ell_2\}$ holds. The inclusion $\mathscr{V}_{\mathsf{sem}}\{S^0\} \supseteq \mathscr{V}_{\mathsf{sem}}\{S, \mathcal{S}\ell_2\}$ is obvious.
(ii) The inclusion $\mathscr{V}_{\mathsf{sem}}\{S_1 \uplus S_2\} \subseteq \mathscr{V}_{\mathsf{sem}}\{S_1^0, S_2^0\}$ holds because $S_1 \uplus S_2$ is isomorphic to the subsemigroup $\{(a, 0) \mid a \in S_1\} \cup \{(0, a) \mid a \in S_2\}$ of $S_1^0 \times S_2^0$, and the reverse inclusion $\mathscr{V}_{\mathsf{sem}}\{S_1 \uplus S_2\} \supseteq \mathscr{V}_{\mathsf{sem}}\{S_1^0, S_2^0\}$ is obvious. Hence,

$$\begin{aligned}\mathscr{V}_{\mathsf{sem}}\{S_1 \uplus S_2\} &= \mathscr{V}_{\mathsf{sem}}\{S_1^0, S_2^0\}\\ &= \mathscr{V}_{\mathsf{sem}}\{S_1, S_2, \mathcal{S}\ell_2\}\\ &= \mathscr{V}_{\mathsf{sem}}\{S_1, S_2\},\end{aligned}$$

where the second equality holds by part (i) and the third equality holds because $\mathcal{S}\ell_2 \in \mathscr{V}_{\mathsf{sem}}\{S_1, S_2\}$. □

For any involution semigroups $\langle S_1, {}^{\star_1}\rangle$ and $\langle S_2, {}^{\star_2}\rangle$, their *0-direct union* is the involution semigroup $\langle S_1, {}^{\star_1}\rangle \uplus \langle S_2, {}^{\star_2}\rangle = \langle S_1 \uplus S_2, {}^\star\rangle$, where $0^\star = 0$ and $a^\star = a^{\star_i}$ for all $a \in S_i$. The proof of Lemma 9.2(ii) can be repeated to show that if $\langle \mathcal{S}\ell_2, {}^{\mathsf{tr}}\rangle \in \mathscr{V}_{\mathsf{inv}}\{\langle S_1, {}^{\star_1}\rangle, \langle S_2, {}^{\star_2}\rangle\}$, then the equality $\mathscr{V}_{\mathsf{inv}}\{\langle S_1, {}^{\star_1}\rangle \uplus \langle S_2, {}^{\star_2}\rangle\} = \mathscr{V}_{\mathsf{inv}}\{\langle S_1, {}^{\star_1}\rangle, \langle S_2, {}^{\star_2}\rangle\}$ holds. In particular,

$$\mathscr{V}_{\mathsf{inv}}\{\langle \mathcal{L}_{3,n}, {}^{\star\mathfrak{R}}\rangle \uplus \langle \mathcal{L}_{3,n}, {}^{\star\mathfrak{R}}\rangle\} = \mathscr{V}_{\mathsf{inv}}\{\langle \mathcal{L}_{3,n}, {}^{\star\mathfrak{R}}\rangle\}.$$

The following is thus a consequence of Theorems 7.21 and 8.9.

Proposition 9.3 *For any* $n \geq 3$ *and* $\mathfrak{R} \in \mathsf{sq}(n)$, *the 0-direct union* $\langle \mathcal{L}_{3,n}, {}^{\star\mathfrak{R}}\rangle \uplus \langle \mathcal{L}_{3,n}, {}^{\star\mathfrak{R}}\rangle$ *is an involution semigroup with reduct* $\mathcal{L}_{3,n} \uplus \mathcal{L}_{3,n}$ *that has a finite identity basis if* $\mathfrak{R} = 1$; *otherwise, it has an infinite irredundant identity basis.*

9.1.2 Involution Semigroups Without Irredundant Identity Bases

Recall that the *dual* of a semigroup S with binary operation $\cdot$ is the semigroup $S^{\triangleleft} = \{a^{\triangleleft} \mid a \in S\}$ with binary operation $\circ$ given by $a^{\triangleleft} \circ b^{\triangleleft} = (b \cdot a)^{\triangleleft}$ for all $a^{\triangleleft}, b^{\triangleleft} \in S^{\triangleleft}$. A semigroup that is isomorphic to its dual is said to be *self-dual*. Clearly, all commutative semigroups are self-dual. The semigroup $\mathcal{L}_3$ is an example of a noncommutative self-dual semigroup, so that $\mathcal{L}_{3,n}$ is also self-dual. Although not all semigroups are self-dual, the 0-direct union $S \uplus S^{\triangleleft}$ is self-dual for any semigroup S. Further, by defining $(a^{\triangleleft})^{\triangleleft} = a$ for all $a \in S$, the 0-direct union $S \uplus S^{\triangleleft}$ becomes an involution semigroup with unary operation $x \mapsto x^{\triangleleft}$ that fixes 0 and interchanges a in S with $a^{\triangleleft}$ in $S^{\triangleleft}$.

Lemma 9.4 (Crvenković et al. 2000, Corollary 8) *Let S be any self-dual semigroup such that $\mathcal{S}\ell_2 \in \mathscr{V}_{\mathsf{sem}}\{S\}$. Then $\langle S \uplus S^{\triangleleft}, {}^{\triangleleft}\rangle$ is finitely based if and only if S is finitely based.*

Corollary 9.5 *For each $n \geq 1$, the involution semigroup $\langle \mathcal{L}_{3,n} \uplus \mathcal{L}_{3,n}^{\triangleleft}, {}^{\triangleleft}\rangle$ is non-finitely based.*

Proof The subsemigroup $\{0, \mathsf{e}\}$ of $\mathcal{L}_3$ is isomorphic to $\mathcal{S}\ell_2$, so that $\mathcal{S}\ell_2 \in \mathscr{V}_{\mathsf{sem}}\{\mathcal{L}_{3,n}\}$. Therefore, since $\mathcal{L}_{3,n}$ is non-finitely based by Corollary 5.3, the involution semigroup $\langle \mathcal{L}_{3,n} \uplus \mathcal{L}_{3,n}^{\triangleleft}, {}^{\triangleleft}\rangle$ is also non-finitely based by Lemma 9.4. □

Recall the words

$$\mathsf{q}_m^{\uparrow} = x\left(\prod_{i=1}^{m}(y_i h_i y_i)\right)x \quad \text{and} \quad \mathsf{q}_m^{\downarrow} = x\left(\prod_{i=m}^{1}(y_i h_i y_i)\right)x$$

from the sets $\mathscr{Q}_m^{\uparrow}$ and $\mathscr{Q}_m^{\downarrow}$ introduced in Sect. 5.1.

Corollary 9.6 *The identities*

$$x^{n+2} \approx x^2, \quad x^{n+1}yx \approx xyx, \quad x^2yx \approx xyx^2, \quad xhykxty \approx yhxkytx \tag{9.1a}$$

$$\mathsf{q}_m^{\uparrow} \approx \mathsf{q}_m^{\downarrow}, \quad m \in \{2, 3, 4, \ldots\} \tag{9.1b}$$

constitute an identity basis for the semigroup $\mathcal{L}_{3,n} \uplus \mathcal{L}_{3,n}^{\triangleleft}$.

Proof It easily follows from Theorem 6.16 and Corollary 6.19 that the identities (9.1) constitute an identity basis for the semigroup $\mathcal{L}_{3,n}$. The equality $\mathscr{V}_{\mathsf{sem}}\{\mathcal{L}_{3,n}, \mathcal{L}_{3,n}^{\triangleleft}\} = \mathscr{V}_{\mathsf{sem}}\{\mathcal{L}_{3,n}\}$ holds since $\mathcal{L}_{3,n}$ is self-dual. Further, since $\mathcal{S}\ell_2 \in \mathscr{V}_{\mathsf{sem}}\{\mathcal{L}_{3,n}\}$, the equality $\mathscr{V}_{\mathsf{sem}}\{\mathcal{L}_{3,n} \uplus \mathcal{L}_{3,n}^{\triangleleft}\} = \mathscr{V}_{\mathsf{sem}}\{\mathcal{L}_{3,n}, \mathcal{L}_{3,n}^{\triangleleft}\}$ holds by Lemma 9.2(ii). Therefore, $\mathscr{V}_{\mathsf{sem}}\{\mathcal{L}_{3,n} \uplus \mathcal{L}_{3,n}^{\triangleleft}\} = \mathscr{V}_{\mathsf{sem}}\{\mathcal{L}_{3,n}\}$, so that the identities (9.1) also form an identity basis for the semigroup $\mathcal{L}_{3,n} \uplus \mathcal{L}_{3,n}^{\triangleleft}$. □

Lemma 9.7 (Crvenković et al. 2000, Theorem 6) *Let Σ be any identity basis for some 0-direct union $S \uplus S^{\triangleleft}$. Then the identities*

$$xx^{\triangleleft}y \approx xx^{\triangleleft}, \quad xyx^{\triangleleft} \approx xx^{\triangleleft} \tag{9.2}$$

and $\{(\text{inv})\} \cup \Sigma$ *constitute an identity basis for the involution semigroup* $\langle S \uplus S^{\triangleleft}, {}^{\triangleleft}\rangle$.

Lemma 9.8 *Let* $\mathbf{t} \approx \mathbf{t}' \in \mathsf{id}\{\langle \mathcal{L}_{3,n} \uplus \mathcal{L}_{3,n}^{\triangleleft}, {}^{\triangleleft}\rangle\}$. *Suppose that* $\lfloor \mathbf{t} \rfloor \in \mathcal{Q}_m^{\uparrow}$. *Then* $\lfloor \mathbf{t}' \rfloor \in \mathcal{Q}_m^{\uparrow} \cup \mathcal{Q}_m^{\downarrow}$.

Proof The assumptions imply that

(A) $\lfloor \mathbf{t} \rfloor \approx \lfloor \mathbf{t}' \rfloor \in \mathsf{id}\{\langle \mathcal{L}_{3,n} \uplus \mathcal{L}_{3,n}^{\triangleleft}, {}^{\triangleleft}\rangle\}$;
(B) $\lfloor \mathbf{t} \rfloor \in \mathscr{A}^{+}$.

Suppose that $x^{\triangleleft} \in \mathsf{con}(\lfloor \mathbf{t}' \rfloor)$ for some $x \in \mathscr{A}$. Let φ denote the substitution that maps every variable in $\mathscr{A}$ to $(\mathsf{e}, 1) \in \mathcal{L}_{3,n}$. Then $(x\lfloor \mathbf{t}' \rfloor)\varphi = (\mathsf{e}, 1) \cdots (\mathsf{e}, 1)^{\triangleleft} \cdots = 0$ and $(x\lfloor \mathbf{t} \rfloor)\varphi = (\mathsf{e}, 1)$ by (B). It follows that $x\lfloor \mathbf{t} \rfloor \approx x\lfloor \mathbf{t}' \rfloor \notin \mathsf{id}\{\langle \mathcal{L}_{3,n} \uplus \mathcal{L}_{3,n}^{\triangleleft}, {}^{\triangleleft}\rangle\}$, but this contradicts (A). Therefore, the variable x does not exist, so that $\lfloor \mathbf{t}' \rfloor \in \mathscr{A}^{+}$. Hence, by (A) and (B),

(C) $\lfloor \mathbf{t} \rfloor \approx \lfloor \mathbf{t}' \rfloor \in \mathsf{id}_{\mathsf{P}}\{\langle \mathcal{L}_{3,n} \uplus \mathcal{L}_{3,n}^{\triangleleft}, {}^{\triangleleft}\rangle\} = \mathsf{id}\{\mathcal{L}_{3,n} \uplus \mathcal{L}_{3,n}^{\triangleleft}\} \subseteq \mathsf{id}\{\mathcal{L}_3\}$.

Since $\lfloor \mathbf{t} \rfloor \in \mathcal{Q}_m^{\uparrow}$, it follows from (C) and Lemma 5.5(ii) that $\lfloor \mathbf{t}' \rfloor \in \mathcal{Q}_m^{\uparrow} \cup \mathcal{Q}_m^{\downarrow}$. □

Proposition 9.9 *For any $n \geq 1$, the involution semigroup $\langle \mathcal{L}_{3,n} \uplus \mathcal{L}_{3,n}^{\triangleleft}, {}^{\triangleleft}\rangle$, whose reduct is isomorphic to $\mathcal{L}_{3,n} \uplus \mathcal{L}_{3,n}$, has no irredundant identity bases.*

Proof The reduct $\mathcal{L}_{3,n} \uplus \mathcal{L}_{3,n}^{\triangleleft}$ is isomorphic to $\mathcal{L}_{3,n} \uplus \mathcal{L}_{3,n}$ because the semigroup $\mathcal{L}_{3,n}$ is self-dual. By Corollary 9.5, the involution semigroup $\langle \mathcal{L}_{3,n} \uplus \mathcal{L}_{3,n}^{\triangleleft}, {}^{\triangleleft}\rangle$ is non-finitely based. In what follows, Theorem 6.3 is used to establish the present proof.

Let $\Lambda_0 = \{(\text{inv}), (9.1\text{a}), (9.2)\}$, and let λ_k denote the identity $\mathsf{q}_{k+1}^{\uparrow} \approx \mathsf{q}_{k+1}^{\downarrow}$. Then by Corollary 9.6 and Lemma 9.7, the set $\Lambda = \Lambda_0 \cup \{\lambda_1, \lambda_2, \lambda_3, \ldots\}$ is an identity basis for $\langle \mathcal{L}_{3,n} \uplus \mathcal{L}_{3,n}^{\triangleleft}, {}^{\triangleleft}\rangle$. It is easily shown that the deduction $\Lambda_0 \cup \{\lambda_{k+1}\} \vdash \lambda_k$ holds for each $k \geq 1$. Hence, assumption (a) of Theorem 6.3 holds in Λ, and it remains to verify that assumption (b) of the same theorem also holds in Λ. As observed in the proof of Theorem 6.21, the following are easily established for any identity $\mathbf{w} \approx \mathbf{w}' \in \mathsf{id}\{\mathcal{Z}_n\}$:

(A) if $\mathbf{w}, \mathbf{w}' \in \mathcal{Q}_m^{\uparrow}$, then $(9.1\text{a}) \vdash \mathbf{w} \approx \mathbf{w}'$;
(B) if $\mathbf{w}, \mathbf{w}' \in \mathcal{Q}_m^{\downarrow}$, then $(9.1\text{a}) \vdash \mathbf{w} \approx \mathbf{w}'$.

Let Σ be any identity basis for $\langle \mathcal{L}_{3,n} \uplus \mathcal{L}_{3,n}^{\triangleleft}, {}^{\triangleleft} \rangle$. Since $\lambda_k \in \mathsf{id}\{\langle \mathcal{L}_{3,n} \uplus \mathcal{L}_{3,n}^{\triangleleft}, {}^{\triangleleft} \rangle\}$, some sequence

$$\mathsf{q}_{k+1}^{\uparrow} = \mathbf{t}_1, \mathbf{t}_2, \ldots, \mathbf{t}_r = \mathsf{q}_{k+1}^{\downarrow}$$

of terms exists, where each identity $\mathbf{t}_j \approx \mathbf{t}_{j+1}$ is directly deducible from some identity $\sigma_j \in \Sigma$. Further, since $\lfloor \mathsf{q}_{k+1}^{\uparrow} \rfloor = \mathsf{q}_{k+1}^{\uparrow} \in \mathscr{Q}_{k+1}^{\uparrow}$ and $\mathsf{q}_{k+1}^{\uparrow} \approx \mathbf{t}_j \in \mathsf{id}\{\langle \mathcal{L}_{3,n} \uplus \mathcal{L}_{3,n}^{\triangleleft}, {}^{\triangleleft} \rangle\}$ for all j, it follows from Lemma 9.8 that $\lfloor \mathbf{t}_j \rfloor \in \mathscr{Q}_{k+1}^{\uparrow} \cup \mathscr{Q}_{k+1}^{\downarrow}$ for all j. Now $\lfloor \mathbf{t}_r \rfloor = \lfloor \mathsf{q}_{k+1}^{\downarrow} \rfloor = \mathsf{q}_{k+1}^{\downarrow} \in \mathscr{Q}_{k+1}^{\downarrow}$ implies the existence of the least integer ℓ such that $\lfloor \mathbf{t}_\ell \rfloor \in \mathscr{Q}_{k+1}^{\downarrow}$. Therefore, $\lfloor \mathbf{t}_1 \rfloor, \lfloor \mathbf{t}_2 \rfloor, \ldots, \lfloor \mathbf{t}_{\ell-1} \rfloor \in \mathscr{Q}_{k+1}^{\uparrow}$. The deduction (9.1a) $\vdash \{\lfloor \mathbf{t}_1 \rfloor \approx \lfloor \mathbf{t}_{\ell-1} \rfloor, \lfloor \mathbf{t}_\ell \rfloor \approx \lfloor \mathbf{t}_r \rfloor\}$ thus holds by (A) and (B). Since

$$\begin{aligned} \mathsf{q}_{k+1}^{\uparrow} &= \lfloor \mathbf{t}_1 \rfloor \\ &\overset{(9.1a)}{\approx} \lfloor \mathbf{t}_{\ell-1} \rfloor \\ &\overset{(\mathrm{inv})}{\approx} \mathbf{t}_{\ell-1} \\ &\overset{\sigma_{\ell-1}}{\approx} \mathbf{t}_\ell \\ &\overset{(\mathrm{inv})}{\approx} \lfloor \mathbf{t}_\ell \rfloor \\ &\overset{(9.1a)}{\approx} \lfloor \mathbf{t}_r \rfloor \\ &= \mathsf{q}_{k+1}^{\downarrow}, \end{aligned}$$

the deduction $\Lambda_0 \cup \{\sigma_{\ell-1}\} \vdash \lambda_k$ holds. Consequently, assumption (b) of Theorem 6.3 holds with θ_k being $\sigma_{\ell-1}$. □

9.2 Two Incomparable Chains of Varieties of Involution Semigroups

Let * be any involution on the semigroup

$$\mathcal{L}_\ell = \langle \mathsf{e}, \mathsf{f} \mid \mathsf{e}^2 = \mathsf{e},\ \mathsf{f}^2 = \mathsf{f},\ \underbrace{\mathsf{efefe}\cdots}_{\text{length } \ell} = 0 \rangle.$$

It is routinely verified that any involution on a semigroup maps each idempotent to some idempotent and fixes any zero element. Therefore, given that e and f are the only nonzero idempotents of $\mathcal{L}_\ell$, the involution * either fixes or interchanges them, that is, * is a permutation of $\{\mathsf{e}, \mathsf{f}\}$. If $\ell = 2n$ is even, so that

$$\mathcal{L}_\ell = \langle \mathsf{e}, \mathsf{f} \mid \mathsf{e}^2 = \mathsf{e},\ \mathsf{f}^2 = \mathsf{f},\ (\mathsf{ef})^n = 0 \rangle,$$

then applying * to both sides of the relation $(\mathsf{ef})^n = 0$ results in $(\mathsf{f}^\star\mathsf{e}^\star)^n = 0$, and this forces * to interchange e and f because $(\mathsf{fe})^n \neq 0$ in $\mathcal{L}_\ell$. On the other hand, if $\ell = 2n + 1$ is odd, so that

$$\mathcal{L}_\ell = \langle \mathsf{e}, \mathsf{f} \mid \mathsf{e}^2 = \mathsf{e},\ \mathsf{f}^2 = \mathsf{f},\ (\mathsf{ef})^n\mathsf{e} = 0 \rangle,$$

then applying * to both sides of the relation $(\mathsf{ef})^n\mathsf{e} = 0$ results in $(\mathsf{e}^\star\mathsf{f}^\star)^n\mathsf{e}^\star = 0$, and this forces * to fix e and f because $(\mathsf{fe})^n\mathsf{f} \neq 0$ in $\mathcal{L}_\ell$. Therefore, the image of $\{\mathsf{e}, \mathsf{f}\}$ under * is given by

$$(\mathsf{e}^\star, \mathsf{f}^\star) = \begin{cases} (\mathsf{f}, \mathsf{e}) & \text{if } \ell \text{ is even,} \\ (\mathsf{e}, \mathsf{f}) & \text{if } \ell \text{ is odd.} \end{cases} \tag{9.3}$$

It is routinely checked that (9.3) uniquely determines the involution * on $\mathcal{L}_\ell$.

Theorem 9.10

(i) *The varieties* $\{\mathscr{V}_{\mathsf{sem}}\{\mathcal{L}_\ell\} \mid \ell = 2, 3, 4, \ldots\}$ *form a strictly increasing chain in the lattice* $\mathfrak{L}_{\mathsf{sem}}$.
(ii) *The varieties* $\{\mathscr{V}_{\mathsf{inv}}\{\langle \mathcal{L}_\ell, ^\star\rangle\} \mid \ell = 2, 4, 6, \ldots\}$ *and* $\{\mathscr{V}_{\mathsf{inv}}\{\langle \mathcal{L}_\ell, ^\star\rangle\} \mid \ell = 3, 5, 7, \ldots\}$ *form two incomparable strictly increasing chains in the lattice* $\mathfrak{L}_{\mathsf{inv}}$.

The involution semigroups $\{\langle \mathcal{L}_{2n}, ^\star\rangle \mid n \geq 1\}$ and $\{\langle \mathcal{L}_{2n+1}, ^\star\rangle \mid n \geq 1\}$ are examined in Sects. 9.2.1 and 9.2.2, respectively. The proof of Theorem 9.10 is then given in Sect. 9.2.3.

9.2.1 The Involution Semigroups $\langle \mathcal{L}_{2n}, ^\star\rangle$

For each $n \geq 1$, the semigroup $\mathcal{L}_{2n}$ coincides with the semigroup

$$\mathcal{H}_n = \langle \mathsf{e}, \mathsf{f} \mid \mathsf{e}^2 = \mathsf{e},\ \mathsf{f}^2 = \mathsf{f},\ (\mathsf{ef})^n = 0 \rangle$$

of order $4n$, whose elements are

$$\mathsf{e},\quad \mathsf{efe},\quad (\mathsf{ef})^2\mathsf{e},\quad \ldots,\quad (\mathsf{ef})^{n-2}\mathsf{e},\quad (\mathsf{ef})^{n-1}\mathsf{e}, \tag{9.4a}$$

$$\mathsf{f},\quad \mathsf{fef},\quad (\mathsf{fe})^2\mathsf{f},\quad \ldots,\quad (\mathsf{fe})^{n-2}\mathsf{f},\quad (\mathsf{fe})^{n-1}\mathsf{f}, \tag{9.4b}$$

$$\mathsf{ef},\quad (\mathsf{ef})^2,\quad (\mathsf{ef})^3,\quad \ldots,\quad (\mathsf{ef})^{n-1},\quad (\mathsf{ef})^n = 0,$$

$$\mathsf{fe},\quad (\mathsf{fe})^2,\quad (\mathsf{fe})^3,\quad \ldots,\quad (\mathsf{fe})^{n-1},\quad (\mathsf{fe})^n.$$

As observed at the beginning of this section, the nontrivial permutation of $\{\mathsf{e}, \mathsf{f}\}$ extends uniquely to the involution * on $\mathcal{H}_n$; this involution interchanges the i-th element in (9.4a) with the i-th element in (9.4b) and fixes every other element.

Lemma 9.11

(i) $\mathcal{H}_n \models (xy)^i x \approx (yx)^i y$ *if and only if* $i \geq n$.
(ii) $\mathcal{H}_n \not\models (xy)^n \approx (yx)^n$.

Proof

(i) If $i < n$, then $(\mathsf{ef})^i\mathsf{e} \neq (\mathsf{fe})^i\mathsf{f}$ in $\mathcal{H}_n$, so that $\mathcal{H}_n \not\models (xy)^i x \approx (yx)^i y$. Conversely, suppose that $i \geq n$. Let φ be any substitution into $\mathcal{H}_n$. It is clear that $((xy)^i x)\varphi = ((yx)^i y)\varphi$ if $x\varphi = y\varphi$. Therefore, assume that $x\varphi \neq y\varphi$. It is then easily seen that $((xy)^i x)\varphi = 0 = ((yx)^i y)\varphi$.
(ii) This holds because $(\mathsf{ef})^n = 0 \neq (\mathsf{fe})^n$ in $\mathcal{H}_n$. □

Lemma 9.12 *The inclusion $\mathscr{V}_{\mathsf{sem}}\{\mathcal{H}_n\} \subset \mathscr{V}_{\mathsf{sem}}\{\mathcal{H}_{n+1}\}$ holds and is proper. Consequently, the inclusion $\mathscr{V}_{\mathsf{inv}}\{\langle \mathcal{H}_n, ^\star\rangle\} \subset \mathscr{V}_{\mathsf{inv}}\{\langle \mathcal{H}_{n+1}, ^\star\rangle\}$ holds and is proper.*

Proof Identifying the element $(\mathsf{ef})^n$ in $\mathcal{H}_{n+1}$ with 0 results in the semigroup $\mathcal{H}_n$. Therefore, the inclusion $\mathscr{V}_{\mathsf{sem}}\{\mathcal{H}_n\} \subseteq \mathscr{V}_{\mathsf{sem}}\{\mathcal{H}_{n+1}\}$ holds; this inclusion is proper because

$$\mathcal{H}_n \models (xy)^n x \approx (yx)^n y \quad \text{and} \quad \mathcal{H}_{n+1} \not\models (xy)^n x \approx (yx)^n y$$

by Lemma 9.11(i). □

9.2.2 The Involution Semigroups $\langle \mathcal{L}_{2n+1}, ^\star\rangle$

For each $n \geq 1$, the semigroup $\mathcal{L}_{2n+1}$ coincides with the semigroup

$$\mathcal{K}_n = \langle \mathsf{e}, \mathsf{f} \,|\, \mathsf{e}^2 = \mathsf{e},\ \mathsf{f}^2 = \mathsf{f},\ (\mathsf{ef})^n\mathsf{e} = 0\rangle$$

of order $4n + 2$, whose elements are

e,	efe,	$(\mathsf{ef})^2\mathsf{e}$,	$\ldots$,	$(\mathsf{ef})^{n-2}\mathsf{e}$,	$(\mathsf{ef})^{n-1}\mathsf{e}$,	$(\mathsf{ef})^n\mathsf{e} = 0$,	
f,	fef,	$(\mathsf{fe})^2\mathsf{f}$,	$\ldots$,	$(\mathsf{fe})^{n-2}\mathsf{f}$,	$(\mathsf{fe})^{n-1}\mathsf{f}$,	$(\mathsf{fe})^n\mathsf{f}$,	
ef,	$(\mathsf{ef})^2$,	$(\mathsf{ef})^3$,	$\ldots$,	$(\mathsf{ef})^{n-1}$,	$(\mathsf{ef})^n$,		(9.5a)
fe,	$(\mathsf{fe})^2$,	$(\mathsf{fe})^3$,	$\ldots$,	$(\mathsf{fe})^{n-1}$,	$(\mathsf{fe})^n$.		(9.5b)

As observed at the beginning of this section, the trivial permutation of $\{\mathsf{e},\mathsf{f}\}$ extends uniquely to the involution * on $\mathcal{K}_n$; this involution interchanges the i-th element in (9.5a) with the i-th element in (9.5b) and fixes every other element.

Lemma 9.13

(i) $\mathcal{K}_n \models (xy)^i \approx (yx)^i$ *if and only if* $i \geq n+1$.
(ii) $\mathcal{K}_n \not\models (xy)^n x \approx (yx)^n y$.

Proof

(i) If $i \leq n$, then $(\mathsf{ef})^i \neq (\mathsf{fe})^i$ in $\mathcal{K}_n$, so that $\mathcal{K}_n \not\models (xy)^i \approx (yx)^i$. Conversely, suppose that $i \geq n+1$. Let φ be any substitution into $\mathcal{K}_n$. It is clear that $(xy)^i\varphi = (yx)^i\varphi$ if $x\varphi = y\varphi$. Therefore, assume that $x\varphi \neq y\varphi$. It is then easily seen that $(xy)^i\varphi = 0 = (yx)^i\varphi$.
(ii) This holds because $(\mathsf{ef})^n\mathsf{e} = 0 \neq (\mathsf{fe})^n\mathsf{f}$ in $\mathcal{K}_n$. □

Lemma 9.14 *The inclusion* $\mathscr{V}_{\mathsf{sem}}\{\mathcal{K}_n\} \subset \mathscr{V}_{\mathsf{sem}}\{\mathcal{K}_{n+1}\}$ *holds and is proper. Consequently, the inclusion* $\mathscr{V}_{\mathsf{inv}}\{\langle\mathcal{K}_n, {}^\star\rangle\} \subset \mathscr{V}_{\mathsf{inv}}\{\langle\mathcal{K}_{n+1}, {}^\star\rangle\}$ *holds and is proper.*

Proof Identifying the element $(\mathsf{ef})^n\mathsf{e}$ in $\mathcal{K}_{n+1}$ with 0 results in the semigroup $\mathcal{K}_n$. Therefore, the inclusion $\mathscr{V}_{\mathsf{sem}}\{\mathcal{K}_n\} \subseteq \mathscr{V}_{\mathsf{sem}}\{\mathcal{K}_{n+1}\}$ holds; this inclusion is proper because

$$\mathcal{K}_n \models (xy)^{n+1} \approx (yx)^{n+1} \text{ and } \mathcal{K}_{n+1} \not\models (xy)^{n+1} \approx (yx)^{n+1}$$

by Lemma 9.13(i). □

9.2.3 Proof of Theorem 9.10

(i) Identifying the element $(\mathsf{ef})^n$ in $\mathcal{K}_n$ with 0 results in the semigroup $\mathcal{H}_n$. Therefore, the inclusion $\mathscr{V}_{\mathsf{sem}}\{\mathcal{H}_n\} \subseteq \mathscr{V}_{\mathsf{sem}}\{\mathcal{K}_n\}$ holds; this inclusion is proper because

$$\mathcal{H}_n \models (xy)^n x \approx (yx)^n y \text{ and } \mathcal{K}_n \not\models (xy)^n x \approx (yx)^n y$$

by Lemmas 9.11(i) and 9.13(ii), respectively. Similarly, identifying the element $(\mathsf{ef})^n\mathsf{e}$ in $\mathcal{H}_{n+1}$ with 0 results in the semigroup $\mathcal{K}_n$. Therefore, the inclusion $\mathscr{V}_{\mathsf{sem}}\{\mathcal{K}_n\} \subseteq \mathscr{V}_{\mathsf{sem}}\{\mathcal{H}_{n+1}\}$ holds; this inclusion is proper because

$$\mathcal{K}_n \models (xy)^{n+1} \approx (yx)^{n+1} \text{ and } \mathcal{H}_{n+1} \not\models (xy)^{n+1} \approx (yx)^{n+1}$$

by Lemmas 9.13(i) and 9.11(ii), respectively. Consequently, the proper inclusions

$$\mathscr{V}_{\mathsf{sem}}\{\mathcal{H}_n\} \subset \mathscr{V}_{\mathsf{sem}}\{\mathcal{K}_n\} \subset \mathscr{V}_{\mathsf{sem}}\{\mathcal{H}_{n+1}\} \subset \mathscr{V}_{\mathsf{sem}}\{\mathcal{K}_{n+1}\}$$

hold for all $n \geq 1$.

(ii) In view of Lemmas 9.12 and 9.14, it suffices to show that the varieties $\mathscr{V}_{\mathsf{inv}}\{\langle \mathcal{H}_i, {}^\star \rangle\}$ and $\mathscr{V}_{\mathsf{inv}}\{\langle \mathcal{K}_j, {}^\star \rangle\}$ are incomparable for all $i, j \geq 1$. Since the finite semigroups $\mathcal{H}_i$ and $\mathcal{K}_j$ are aperiodic, the involution semigroups $\langle \mathcal{H}_i, {}^\star \rangle$ and $\langle \mathcal{K}_j, {}^\star \rangle$ satisfy the identity $x^{n+1} \approx x^n$ for all sufficiently large $n \geq 2$; in particular, $(\mathsf{ef})^n = \mathsf{0}$ in both $\mathcal{H}_i$ and $\mathcal{K}_j$. It is then routinely verified that

$$\langle \mathcal{K}_j, {}^\star \rangle \models (x^\star)^n \approx x^n \quad \text{and} \quad \langle \mathcal{H}_i, {}^\star \rangle \not\models (x^\star)^n \approx x^n,$$

whence $\mathscr{V}_{\mathsf{inv}}\{\langle \mathcal{H}_i, {}^\star \rangle\} \nsubseteq \mathscr{V}_{\mathsf{inv}}\{\langle \mathcal{K}_j, {}^\star \rangle\}$, and that

$$\langle \mathcal{H}_i, {}^\star \rangle \models (xx^\star)^n \approx (yy^\star)^n \quad \text{and} \quad \langle \mathcal{K}_j, {}^\star \rangle \not\models (xx^\star)^n \approx (yy^\star)^n,$$

whence $\mathscr{V}_{\mathsf{inv}}\{\langle \mathcal{K}_j, {}^\star \rangle\} \nsubseteq \mathscr{V}_{\mathsf{inv}}\{\langle \mathcal{H}_i, {}^\star \rangle\}$.

9.3 Summary

In Sect. 9.1, three finite involution semigroups sharing the semigroup reduct $\mathcal{L}_{3,n} \uplus \mathcal{L}_{3,n}$ are constructed so that one has a finite identity basis, one has an infinite irredundant identity basis, and one has no irredundant identity bases. In Sect. 9.2, it is shown that the varieties generated by the involution semigroups $\{\langle \mathcal{L}_\ell, {}^\star \rangle \mid \ell = 2, 3, 4, \ldots\}$ constitute two incomparable strictly increasing chains in $\mathfrak{L}_{\mathsf{inv}}$, while the varieties generated by their reducts $\{\mathcal{L}_\ell \mid \ell = 2, 3, 4, \ldots\}$ constitute a single strictly increasing chain in $\mathfrak{L}_{\mathsf{sem}}$. These results were published in Lee (2017b, 2018a).

Equational Theories of Twisted Involution Semigroups

10

The main result of the present chapter is the following restatement of Theorem 1.41.

Theorem 10.1 *Let* $\langle S, {}^\star\rangle$ *be any twisted involution semigroup. Suppose that the reduct* S *is non-finitely based. Then* $\langle S, {}^\star\rangle$ *is non-finitely based.*

In Sect. 10.1, each twisted involution semigroup is shown to possess some identity basis of a very specific form. This result is then used in Sect. 10.2 to establish the proof of Theorem 10.1.

10.1 Organized Identity Bases

Let $\mathbf{w} \in (\mathscr{A} \cup \mathscr{A}^\star)^+_\varnothing$ be any word. As in the case of plain words, the *content* of $\mathbf{w}$ is the set $\mathsf{con}(\mathbf{w})$ of variables from $\mathscr{A} \cup \mathscr{A}^\star$ occurring in $\mathbf{w}$. If $x, x^\star \in \mathsf{con}(\mathbf{w})$ for some $x \in \mathscr{A}$, then $\{x, x^\star\}$ is called a *mixed pair* of $\mathbf{w}$. A word is *mixed* if it has some mixed pair; otherwise, it is *unmixed*.

Lemma 10.2 *Suppose that* $\mathbf{w} \approx \mathbf{w}' \in \mathsf{id}_{\mathsf{W}}\{\langle \mathcal{S}\ell_3, {}^\star\rangle\}$. *Then*

(i) $\mathbf{w}$ *is mixed if and only if* $\mathbf{w}'$ *is mixed;*
(ii) $\mathsf{con}(\mathbf{w}) = \mathsf{con}(\mathbf{w}')$ *if* $\mathbf{w}$ *and* $\mathbf{w}'$ *are unmixed.*

Proof

(i) Seeking a contradiction, suppose that $\mathbf{w}$ and $\mathbf{w}'$ are not simultaneously mixed, say $\mathbf{w}$ is unmixed and $\mathbf{w}'$ is mixed. Then $\mathsf{con}(\mathbf{w}) = \mathscr{X} \cup \mathscr{Y}^\star$ for some disjoint finite sets

E. W. H. Lee, *Advances in the Theory of Varieties of Semigroups*, Frontiers in Mathematics, https://doi.org/10.1007/978-3-031-16497-2_10

$\mathscr{X}, \mathscr{Y} \subseteq \mathscr{A}$ such that $(\mathscr{X}, \mathscr{Y}) \neq (\varnothing, \varnothing)$. Let $\varphi_1 : \mathscr{A} \to \mathcal{S}\ell_3$ denote the substitution

$$x \mapsto \begin{cases} \mathsf{e} & \text{if } x \in \mathscr{X}, \\ \mathsf{f} & \text{otherwise.} \end{cases}$$

Then $\mathbf{w}\varphi_1 = \mathsf{e}$ and $\mathbf{w}'\varphi_1 = 0$.

(ii) Seeking a contradiction, suppose that $\mathbf{w}$ and $\mathbf{w}'$ are unmixed with $\mathsf{con}(\mathbf{w}) \neq \mathsf{con}(\mathbf{w}')$, say $\mathsf{con}(\mathbf{w}')\backslash \mathsf{con}(\mathbf{w}) \neq \varnothing$. Then $\mathsf{con}(\mathbf{w}) = \mathscr{X} \cup \mathscr{Y}^\star$ for some disjoint finite sets $\mathscr{X}, \mathscr{Y} \subseteq \mathscr{A}$ such that $(\mathscr{X}, \mathscr{Y}) \neq (\varnothing, \varnothing)$. Let $\varphi_2 : \mathscr{A} \to \mathcal{S}\ell_3$ denote the substitution

$$x \mapsto \begin{cases} \mathsf{e} & \text{if } x \in \mathscr{X}, \\ \mathsf{f} & \text{if } x \in \mathscr{Y}, \\ 0 & \text{otherwise.} \end{cases}$$

Then $\mathbf{w}\varphi_2 = \mathsf{e}$ and $\mathbf{w}'\varphi_2 = 0$. □

Lemma 10.3 *Let $\langle S, {}^\star \rangle$ be any involution semigroup. Suppose that $\mathbf{w}$ and $\mathbf{w}'$ are unmixed words such that* $\mathsf{con}(\mathbf{w}) = \mathsf{con}(\mathbf{w}')$. *Then $\mathbf{w} \approx \mathbf{w}' \in \mathsf{id}\{\langle S, {}^\star \rangle\}$ if and only if $\overline{\mathbf{w}} \approx \overline{\mathbf{w}'} \in \mathsf{id}\{\langle S, {}^\star \rangle\}$.*

Proof By assumption, $\mathsf{con}(\mathbf{w}) = \mathsf{con}(\mathbf{w}') = \mathscr{X} \cup \mathscr{Y}^\star$ for some disjoint finite sets $\mathscr{X}, \mathscr{Y} \subseteq \mathscr{A}$. Let φ denote the substitution given by $x \mapsto x^\star$ for all $x \in \mathscr{Y}$. If $\langle S, {}^\star \rangle$ satisfies the identity $\mathbf{w} \approx \mathbf{w}'$, then it satisfies $\overline{\mathbf{w}} = \lfloor \mathbf{w}\varphi \rfloor \approx \lfloor \mathbf{w}'\varphi \rfloor = \overline{\mathbf{w}}'$. Conversely, if $\langle S, {}^\star \rangle$ satisfies the identity $\overline{\mathbf{w}} \approx \overline{\mathbf{w}}'$, then it satisfies $\mathbf{w} = \overline{\mathbf{w}}\varphi \approx \overline{\mathbf{w}}'\varphi = \mathbf{w}'$. □

Lemma 10.4 *Let Σ be any identity basis for an involution semigroup $\langle S, {}^\star \rangle$. Then there exists some set $\Sigma_{\mathsf{W}}^{\mathsf{rev}} \subseteq \mathsf{id}_{\mathsf{W}}\{\langle S, {}^\star \rangle\}$ that is closed under reversal such that $\{(\text{inv})\} \cup \Sigma_{\mathsf{W}}^{\mathsf{rev}}$ is an identity basis for $\langle S, {}^\star \rangle$. Further, if Σ is finite, then $|\Sigma_{\mathsf{W}}^{\mathsf{rev}}| \leq 2|\Sigma|$.*

Proof In what follows, it is shown that $\Sigma_{\mathsf{W}}^{\mathsf{rev}} = \lfloor \Sigma \rfloor \cup \lfloor \Sigma \rfloor^{\triangleleft}$ is the required set, where

$$\lfloor \Sigma \rfloor = \{\lfloor \mathbf{t} \rfloor \approx \lfloor \mathbf{t}' \rfloor \mid \mathbf{t} \approx \mathbf{t}' \in \Sigma\} \text{ and } \lfloor \Sigma \rfloor^{\triangleleft} = \{\lfloor \mathbf{t} \rfloor^{\triangleleft} \approx \lfloor \mathbf{t}' \rfloor^{\triangleleft} \mid \mathbf{t} \approx \mathbf{t}' \in \Sigma\}.$$

First, it is easily seen that $\Sigma_{\mathsf{W}}^{\mathsf{rev}}$ is closed under reversal. Now since $\langle S, {}^\star \rangle$ satisfies an identity $\mathbf{t} \approx \mathbf{t}'$ if and only if it satisfies the word identity $\lfloor \mathbf{t} \rfloor \approx \lfloor \mathbf{t}' \rfloor$, the set $\{(\text{inv})\} \cup \lfloor \Sigma \rfloor$ is an identity basis for $\langle S, {}^\star \rangle$. Hence, by Lemma 2.12, the set $\{(\text{inv})\} \cup \Sigma_{\mathsf{W}}^{\mathsf{rev}}$ is also an identity basis for $\langle S, {}^\star \rangle$. □

An identity basis for an involution semigroup $\langle S, {}^\star\rangle$ is *organized* if it is of the form

$$\Sigma^{\mathsf{org}} = \{(\mathrm{inv})\} \cup \Sigma_{\mathsf{P}} \cup \Sigma_{\mathsf{W}}^{\mathsf{mix}},$$

where $\Sigma_{\mathsf{P}} \subseteq \mathsf{id}_{\mathsf{P}}\{\langle S, {}^\star\rangle\}$ is closed under reversal and $\Sigma_{\mathsf{W}}^{\mathsf{mix}} \subseteq \mathsf{id}_{\mathsf{W}}\{\langle S, {}^\star\rangle\}$ consists of identities formed by mixed words.

Lemma 10.5 *Given any identity basis Σ for a twisted involution semigroup $\langle S, {}^\star\rangle$, an organized identity basis $\Sigma^{\mathsf{org}} = \{(\mathrm{inv})\} \cup \Sigma_{\mathsf{P}} \cup \Sigma_{\mathsf{W}}^{\mathsf{mix}}$ for $\langle S, {}^\star\rangle$ can be constructed from Σ. Further, if Σ is finite, then Σ^{org} is also finite.*

Proof By Lemma 10.4, there exists some set $\Sigma_{\mathsf{W}}^{\mathsf{rev}} \subseteq \mathsf{id}_{\mathsf{W}}\{\langle S, {}^\star\rangle\}$ that is closed under reversal such that $\{(\mathrm{inv})\} \cup \Sigma_{\mathsf{W}}^{\mathsf{rev}}$ is an identity basis for $\langle S, {}^\star\rangle$. By Lemma 10.2, each identity $\mathbf{w} \approx \mathbf{w}'$ in $\Sigma_{\mathsf{W}}^{\mathsf{rev}}$ is of one of the following types:

(A) $\mathbf{w}$ and $\mathbf{w}'$ are mixed;
(B) $\mathbf{w}$ and $\mathbf{w}'$ are unmixed with $\mathsf{con}(\mathbf{w}) = \mathsf{con}(\mathbf{w}')$.

Suppose that $\mathbf{w} \approx \mathbf{w}' \in \Sigma_{\mathsf{W}}^{\mathsf{rev}}$ is of type (B). If $\mathbf{w} \approx \mathbf{w}'$ is not plain, then by Lemma 10.3, when $\mathbf{w} \approx \mathbf{w}'$ in $\{(\mathrm{inv})\} \cup \Sigma_{\mathsf{W}}^{\mathsf{rev}}$ is replaced by its plain projection $\overline{\mathbf{w}} \approx \overline{\mathbf{w}}'$, the resulting set

$$\{(\mathrm{inv})\} \cup (\Sigma_{\mathsf{W}}^{\mathsf{rev}} \backslash \{\mathbf{w} \approx \mathbf{w}'\}) \cup \{\overline{\mathbf{w}} \approx \overline{\mathbf{w}}'\}$$

remains an identity basis for $\langle S, {}^\star\rangle$. On the other hand, if the identity $\mathbf{w} \approx \mathbf{w}'$ is plain, then it is the same as $\overline{\mathbf{w}} \approx \overline{\mathbf{w}}'$. Therefore, when all word identities in $\{(\mathrm{inv})\} \cup \Sigma_{\mathsf{W}}^{\mathsf{rev}}$ of type (B) are replaced by their plain projections, then the resulting set remains an identity basis for $\langle S, {}^\star\rangle$. Hence, $\Sigma^{\mathsf{org}} = \{(\mathrm{inv})\} \cup \Sigma_{\mathsf{P}} \cup \Sigma_{\mathsf{W}}^{\mathsf{mix}}$ is an identity basis for $\langle S, {}^\star\rangle$, where $\Sigma_{\mathsf{W}}^{\mathsf{mix}}$ consists of identities in $\Sigma_{\mathsf{W}}^{\mathsf{rev}}$ of type (A) and

$$\Sigma_{\mathsf{P}} = \{\overline{\mathbf{w}} \approx \overline{\mathbf{w}}' \mid \mathbf{w} \approx \mathbf{w}' \in \Sigma_{\mathsf{W}}^{\mathsf{rev}} \text{ is of type (B)}\}.$$

Since $\Sigma_{\mathsf{W}}^{\mathsf{rev}}$ is closed under reversal, Σ_{P} is also closed under reversal. Consequently, Σ^{org} is organized.

If Σ is finite, then $|\Sigma^{\mathsf{org}}| = |(\mathrm{inv})| + |\Sigma_{\mathsf{W}}^{\mathsf{rev}}| \leq 2 + 2|\Sigma|$ by Lemma 10.4. □

10.2 Proof of Theorem 10.1

Lemma 10.6 *Let $\langle S, {}^\star\rangle$ be any twisted involution semigroup with some organized identity basis Σ^{org}. Suppose that $\mathbf{t} \approx \mathbf{t}'$ is any identity directly deducible from some identity in Σ^{org} with $\lfloor \mathbf{t} \rfloor, \lfloor \mathbf{t}' \rfloor \in \mathscr{A}^+$. Then the plain identity $\lfloor \mathbf{t} \rfloor \approx \lfloor \mathbf{t}' \rfloor$ is directly deducible from some identity in Σ_{P}.*

Proof Let $\mathbf{w} \approx \mathbf{w}'$ be an identity in $\Sigma^{\mathsf{org}} = \{(\mathrm{inv})\} \cup \Sigma_{\mathsf{P}} \cup \Sigma_{\mathsf{W}}^{\mathsf{mix}}$ from which the identity $\mathbf{t} \approx \mathbf{t}'$ is directly deducible. If $\mathbf{w} \approx \mathbf{w}'$ is from (inv), then by Lemma 2.11, the identity $\lfloor \mathbf{t} \rfloor \approx \lfloor \mathbf{t}' \rfloor$ is trivial and so is vacuously directly deducible from some identity in Σ_{P}. Therefore, assume that $\mathbf{w} \approx \mathbf{w}'$ is from $\Sigma_{\mathsf{P}} \cup \Sigma_{\mathsf{W}}^{\mathsf{mix}}$. Then there exists a substitution $\varphi : \mathscr{A} \to \mathsf{T}(\mathscr{A})$ such that $\mathbf{w}\varphi$ is a subterm of $\mathbf{t}$, and replacing this particular subterm $\mathbf{w}\varphi$ of $\mathbf{t}$ with $\mathbf{w}'\varphi$ results in the term $\mathbf{t}'$. As observed in Remark 2.10(iii), either $\lfloor \mathbf{w}\varphi \rfloor$ or $\lfloor (\mathbf{w}\varphi)^\star \rfloor$ is a factor of the plain word $\lfloor \mathbf{t} \rfloor$.

CASE 1: $\lfloor \mathbf{w}\varphi \rfloor$ is a factor of $\lfloor \mathbf{t} \rfloor$. Then $\lfloor \mathbf{t} \rfloor = \mathbf{a}\lfloor \mathbf{w}\varphi \rfloor\mathbf{b}$ for some $\mathbf{a}, \mathbf{b} \in \mathscr{A}_\varnothing^+$. Since $\mathbf{t}'$ is obtained by replacing $\mathbf{w}\varphi$ in $\mathbf{t}$ with $\mathbf{w}'\varphi$, it follows that $\lfloor \mathbf{t}' \rfloor = \mathbf{a}\lfloor \mathbf{w}'\varphi \rfloor\mathbf{b}$. If $\mathbf{w}$ has a mixed pair, then $\lfloor \mathbf{w}\varphi \rfloor$ has a mixed pair, whence the plainness of $\lfloor \mathbf{t} \rfloor$ is contradicted. Hence, $\mathbf{w}$ is unmixed, so that $\mathbf{w} \approx \mathbf{w}' \in \Sigma_{\mathsf{P}}$. Thus, $\mathsf{con}(\mathbf{w}) = \mathsf{con}(\mathbf{w}')$ by Lemma 10.2(ii), say $\mathbf{w} = \mathbf{w}_{\{x_1, x_2, \ldots, x_m\}}$ and $\mathbf{w}' = \mathbf{w}'_{\{x_1, x_2, \ldots, x_m\}}$ each involves all of the variables $x_1, x_2, \ldots, x_m$. Since the words $\lfloor \mathbf{t} \rfloor$ and $\mathbf{w}$ are plain with

$$\begin{aligned}\lfloor \mathbf{t} \rfloor &= \mathbf{a}\lfloor \mathbf{w}_{\{x_1\varphi, x_2\varphi, \ldots, x_m\varphi\}} \rfloor \mathbf{b} \\ &= \mathbf{a}\mathbf{w}_{\{\lfloor x_1\varphi \rfloor, \lfloor x_2\varphi \rfloor, \ldots, \lfloor x_m\varphi \rfloor\}}\mathbf{b},\end{aligned}$$

the words $\lfloor x_1\varphi \rfloor, \lfloor x_2\varphi \rfloor, \ldots, \lfloor x_m\varphi \rfloor$ are also plain. Let $\chi_1 : \mathsf{con}(\mathbf{w}) \to \mathscr{A}^+$ denote the substitution given by $x_i \mapsto \lfloor x_i\varphi \rfloor$ for all $i \in \{1, 2, \ldots, m\}$. Then $\lfloor \mathbf{t} \rfloor = \mathbf{a}(\mathbf{w}\chi_1)\mathbf{b}$ and $\lfloor \mathbf{t}' \rfloor = \mathbf{a}(\mathbf{w}'\chi_1)\mathbf{b}$. Hence, the identity $\lfloor \mathbf{t} \rfloor \approx \lfloor \mathbf{t}' \rfloor$ is directly deducible from $\mathbf{w} \approx \mathbf{w}' \in \Sigma_{\mathsf{P}}$.

CASE 2: $\lfloor (\mathbf{w}\varphi)^\star \rfloor$ is a factor of $\lfloor \mathbf{t} \rfloor$. Then following the argument in Case 1 yields $\lfloor \mathbf{t} \rfloor = \mathbf{a}\lfloor (\mathbf{w}\varphi)^\star \rfloor\mathbf{b}$ and $\lfloor \mathbf{t}' \rfloor = \mathbf{a}\lfloor (\mathbf{w}'\varphi)^\star \rfloor\mathbf{b}$ for some $\mathbf{a}, \mathbf{b} \in \mathscr{A}_\varnothing^+$, and $\mathbf{w} \approx \mathbf{w}' \in \Sigma_{\mathsf{P}}$ with $\mathsf{con}(\mathbf{w}) = \mathsf{con}(\mathbf{w}')$, say $\mathbf{w} = \mathbf{w}_{\{x_1, x_2, \ldots, x_m\}}$ and $\mathbf{w}' = \mathbf{w}'_{\{x_1, x_2, \ldots, x_m\}}$ each involves all of the variables $x_1, x_2, \ldots, x_m$. Since $\lfloor \mathbf{t} \rfloor$ and $\mathbf{w}$ are plain words with

$$\begin{aligned}\lfloor \mathbf{t} \rfloor &= \mathbf{a}\lfloor (\mathbf{w}_{\{x_1\varphi, x_2\varphi, \ldots, x_m\varphi\}})^\star \rfloor \mathbf{b} \\ &= \mathbf{a}\lfloor \mathbf{w}^{\triangleleft}_{\{(x_1\varphi)^\star, (x_2\varphi)^\star, \ldots, (x_m\varphi)^\star\}} \rfloor \mathbf{b} \\ &= \mathbf{a}(\mathbf{w}^{\triangleleft}_{\{\lfloor (x_1\varphi)^\star \rfloor, \lfloor (x_2\varphi)^\star \rfloor, \ldots, \lfloor (x_m\varphi)^\star \rfloor\}})\mathbf{b},\end{aligned}$$

the words $\lfloor (x_1\varphi)^\star \rfloor, \lfloor (x_2\varphi)^\star \rfloor, \ldots, \lfloor (x_m\varphi)^\star \rfloor$ are also plain. Let $\chi_2 : \mathsf{con}(\mathbf{w}) \to \mathscr{A}^+$ denote the substitution given by $x_i \mapsto \lfloor (x_i\varphi)^\star \rfloor$ for all $i \in \{1, 2, \ldots, m\}$. Then $\lfloor \mathbf{t} \rfloor = \mathbf{a}(\mathbf{w}^{\triangleleft}\chi_2)\mathbf{b}$ and $\lfloor \mathbf{t}' \rfloor = \mathbf{a}((\mathbf{w}')^{\triangleleft}\chi_2)\mathbf{b}$. Hence, the identity $\lfloor \mathbf{t} \rfloor \approx \lfloor \mathbf{t}' \rfloor$ is directly deducible from $\mathbf{w}^{\triangleleft} \approx (\mathbf{w}')^{\triangleleft}$. Since $\mathbf{w} \approx \mathbf{w}' \in \Sigma_{\mathsf{P}}$ and Σ_{P} is closed under reversal, $\mathbf{w}^{\triangleleft} \approx (\mathbf{w}')^{\triangleleft} \in \Sigma_{\mathsf{P}}$. □

Lemma 10.7 *Let $\langle S, {}^\star \rangle$ be any twisted involution semigroup. Suppose that Σ is any finite identity basis for $\langle S, {}^\star \rangle$. Then a finite organized identity basis* $\Sigma^{\mathsf{org}} = \{(\mathrm{inv})\} \cup \Sigma_{\mathsf{P}} \cup \Sigma_{\mathsf{W}}^{\mathsf{mix}}$

for $\langle S, {}^\star\rangle$ can be constructed from Σ with the property that Σ_{P} is an identity basis for the reduct S.

Proof By Lemma 10.5, a finite organized identity basis $\Sigma^{\mathsf{org}} = \{(\text{inv})\} \cup \Sigma_{\mathsf{P}} \cup \Sigma_{\mathsf{W}}^{\mathsf{mix}}$ for $\langle S, {}^\star\rangle$ can be constructed from Σ. Let $\mathbf{p} \approx \mathbf{p}' \in \mathsf{id}\{S\}$. Since $\mathbf{p} \approx \mathbf{p}'$ is deducible from Σ^{org}, there is a sequence

$$\mathbf{p} = \mathbf{t}_1, \mathbf{t}_2, \ldots, \mathbf{t}_r = \mathbf{p}'$$

of terms where each identity $\mathbf{t}_i \approx \mathbf{t}_{i+1}$ is directly deducible from some identity in Σ^{org}. Clearly, $\lfloor \mathbf{t}_1 \rfloor = \mathbf{p} \in \mathscr{A}^+$. Suppose that $\lfloor \mathbf{t}_i \rfloor \in \mathscr{A}^+$ for some $i \geq 1$. Then since $\lfloor \mathbf{t}_i \rfloor \approx \lfloor \mathbf{t}_{i+1} \rfloor \in \mathsf{id}_{\mathsf{W}}\{\langle \mathcal{S}\ell_3, {}^\star\rangle\}$, it follows from Lemma 10.2(i) that $\lfloor \mathbf{t}_{i+1} \rfloor \in \mathscr{A}^+$. By Lemma 10.6, the plain identity $\lfloor \mathbf{t}_i \rfloor \approx \lfloor \mathbf{t}_{i+1} \rfloor$ is directly deducible from some identity in Σ_{P}. Consequently, $\mathbf{p} \approx \mathbf{p}'$ is deducible from Σ_{P}. □

It follows from Lemma 10.7 that the semigroup reduct of any finitely based twisted involution semigroup is always finitely based. The proof of Theorem 10.1 is thus complete.

10.3 Summary

The main objective of the present chapter is to show that a twisted involution semigroup is non-finitely based if its semigroup reduct is non-finitely based; see Theorem 10.1. This result was published in Lee (2017a).

Part III

Monoids

Hereditarily Finitely Based Varieties of Monoids 11

Recall that $\mathbb{O}$ denotes the variety of monoids defined by the identities

$$xhxyty \approx xhyxty, \quad xhytxy \approx xhytyx. \tag{11.1}$$

The present chapter is concerned with the finite basis problem for subvarieties of $\mathbb{O}$.

Theorem 11.1 *Every subvariety of* $\mathbb{O}$ *is finitely based.*

In Sect. 11.1, identities that can be used to define noncommutative subvarieties of $\mathbb{O}$ are described. Based on this description, the proof of Theorem 11.1 is given in Sect. 11.2. In Sect. 11.3, hereditarily finitely based varieties of monoids from a large class are characterized. Theorem 1.58 is then deduced as a consequence.

Refer to Sects. 1.6.2 and 1.6.3 for the roles that the variety $\mathbb{O}$ plays in the description of other important subvarieties of $\mathbb{A}^{\mathsf{cen}}$.

11.1 Identities Satisfied by Noncommutative Subvarieties of $\mathbb{O}$

This section establishes restrictions on the type of identities that can be used to define noncommutative subvarieties of $\mathbb{O}$.

Proposition 11.2 *Each noncommutative subvariety of* $\mathbb{O}$ *can be defined by the identities* (11.1) *together with some of the following identities:*

$$x^{e_0} \prod_{i=1}^{r} (h_i x^{e_i}) \approx x^{f_0} \prod_{i=1}^{r} (h_i x^{f_i}), \tag{11.2a}$$

E. W. H. Lee, *Advances in the Theory of Varieties of Semigroups*, Frontiers in Mathematics, https://doi.org/10.1007/978-3-031-16497-2_11

where $e_0, f_0, e_1, f_1, \ldots, e_r, f_r \geq 0$ *and* $r \geq 0$;

$$x^{e_0} y^{f_0} \prod_{i=1}^{r} (h_i x^{e_i} y^{f_i}) \approx y^{f_0} x^{e_0} \prod_{i=1}^{r} (h_i x^{e_i} y^{f_i}), \tag{11.2b}$$

where $e_0, f_0 \geq 1, e_1, f_1, e_2, f_2, \ldots, e_r, f_r \geq 0$, $\sum_{i=0}^{r} e_i \geq 2$, $\sum_{i=0}^{r} f_i \geq 2$, *and* $r \geq 0$.

Some results are developed in Sects. 11.1.1 and 11.1.2 to put identities satisfied by subvarieties of $\mathbb{O}$ into specific forms. The proof of Proposition 11.2 is then given in Sect. 11.1.3.

11.1.1 Canonical Form

Let x be any variable in a word $\mathbf{w}$. Then $\mathbf{w}$ can be written as

$$\mathbf{w} = \mathbf{u}_0 \prod_{i=1}^{r} (x^{e_i} \mathbf{u}_i),$$

where $e_1, e_2, \ldots, e_r \geq 1$, $\mathbf{u}_0, \mathbf{u}_r \in \mathscr{A}_\varnothing^+$, and $\mathbf{u}_1, \mathbf{u}_2, \ldots, \mathbf{u}_{r-1} \in \mathscr{A}^+$ are such that $x \notin \mathsf{con}(\mathbf{u}_0\mathbf{u}_1 \cdots \mathbf{u}_r)$. The factors $x^{e_1}, x^{e_2}, \ldots, x^{e_r}$ of $\mathbf{w}$ are called *x-stacks*—or simply *stacks*—of $\mathbf{w}$. Specifically, x^{e_1} is the *primary x-stack* of $\mathbf{w}$ and $x^{e_2}, x^{e_3}, \ldots, x^{e_r}$ are the *secondary x-stacks* of $\mathbf{w}$.

For any set $\mathscr{X}$ of variables from $\mathscr{A}$, define

$$\mathscr{X}^\sharp = \left\{ x_1^{e_1} x_2^{e_2} \cdots x_r^{e_r} \;\middle|\; \begin{array}{l} x_1, x_2, \ldots, x_r \in \mathscr{X} \text{ are distinct,} \\ e_1, e_2, \ldots, e_r \geq 1, \text{ and } r \geq 0 \end{array} \right\}.$$

Note that every stack of every word in $\mathscr{X}^\sharp$ is primary.

Example 11.3

(i) $\{x\}^\sharp = \{\varnothing, x, x^2, \ldots\}$.
(ii) $\{x, y\}^\sharp = \{\varnothing, x^e, y^f, x^e y^f, y^e x^f \mid e, f \geq 1\}$.

In this chapter, a word $\mathbf{w}$ is said to be in *canonical form* if

$$\mathbf{w} = \mathbf{w}_0 \prod_{i=1}^{r} (h_i \mathbf{w}_i) \tag{11.3}$$

for some $r \geq 0$ such that $\mathsf{sim}(\mathbf{w}) = \{h_1, h_2, \ldots, h_r\}$ and $\mathbf{w}_0, \mathbf{w}_1, \ldots, \mathbf{w}_r \in \mathsf{non}(\mathbf{w})^\sharp$.

Lemma 11.4 *Any word can be converted by the identities* (11.1) *into a word in canonical form.*

Proof It is clear that any word $\mathbf{w}$ can be written in the form (11.3) for some $r \geq 0$ such that $\mathsf{sim}(\mathbf{w}) = \{h_1, h_2, \ldots, h_r\}$ and $\mathbf{w}_0, \mathbf{w}_1, \ldots, \mathbf{w}_r \in \mathsf{non}(\mathbf{w})^+ \cup \{\varnothing\}$. If a variable $x \in \mathsf{non}(\mathbf{w})$ occurs more than once in some $\mathbf{w}_i$, then the identities (11.1) can be used to gather any non-first occurrence of x in $\mathbf{w}_i$ with the first occurrence of x in $\mathbf{w}_i$. Repeat this gathering on every variable $x \in \mathsf{non}(\mathbf{w})$ that occurs more than once in $\mathbf{w}_i$ to result in $\mathbf{w}_i \in \mathsf{non}(\mathbf{w})^\sharp$. □

11.1.2 Fundamental Identities and Well-Balanced Identities

Recall the relation $\stackrel{\circ}{=}$ on $\mathscr{A}_\varnothing^+$ defined by $\mathbf{w} \stackrel{\circ}{=} \mathbf{w}'$ if $\mathsf{occ}(x, \mathbf{w}) = \mathsf{occ}(x, \mathbf{w}')$ for all $x \in \mathscr{A}$. Equivalently, two words $\mathbf{w}$ and $\mathbf{w}'$ are $\stackrel{\circ}{=}$-related if they can be obtained from one another by rearrangement of the variables.

An identity $\mathbf{u} \approx \mathbf{v}$ is said to be *fundamental* if the following conditions hold:

(F1) $\mathbf{u}$ and $\mathbf{v}$ are in canonical form;
(F2) $\mathsf{sim}(\mathbf{u}) = \mathsf{sim}(\mathbf{v})$ and $\mathsf{non}(\mathbf{u}) = \mathsf{non}(\mathbf{v})$;
(F3) $\mathbf{u}[\mathsf{sim}(\mathbf{u})] = \mathbf{v}[\mathsf{sim}(\mathbf{v})]$.

The words $\mathbf{u}$ and $\mathbf{v}$ that form a fundamental identity can thus be written as

$$\mathbf{u} = \mathbf{u}_0 \prod_{i=1}^{r} (h_i \mathbf{u}_i) \quad \text{and} \quad \mathbf{v} = \mathbf{v}_0 \prod_{i=1}^{r} (h_i \mathbf{v}_i), \tag{11.4}$$

where $\mathsf{sim}(\mathbf{u}) = \mathsf{sim}(\mathbf{v}) = \{h_1, h_2, \ldots, h_r\}$ and $\mathbf{u}_i, \mathbf{v}_i \in \mathsf{non}(\mathbf{u})^\sharp = \mathsf{non}(\mathbf{v})^\sharp$ for all i. Such a fundamental identity is said to be *well balanced* if $\mathbf{u}_i \stackrel{\circ}{=} \mathbf{v}_i$ for all i. The identity $\mathbf{u} \approx \mathbf{v}$ is *not well balanced at* x if $\mathsf{occ}(x, \mathbf{u}_i) \neq \mathsf{occ}(x, \mathbf{v}_i)$ for some i.

Lemma 11.5 *Let* $\sigma : \mathbf{u} \approx \mathbf{v}$ *be any fundamental identity that is not well balanced. Then*

$$\mathbb{O}\{\sigma\} = \mathbb{O}(\mathsf{A}^\sigma \cup \{\sigma'\})$$

for some finite set A^σ *of identities from* (11.2a) *and some well-balanced identity* σ'.

Proof By assumption, $\mathbf{u}$ and $\mathbf{v}$ can be assumed to be the words from (11.4). Suppose that σ is not well balanced at precisely the variables $x_1, x_2, \ldots, x_m$. For each $i \in \{0, 1, \ldots, r\}$, let $e_i = \mathsf{occ}(x_1, \mathbf{u}_i)$ and $f_i = \mathsf{occ}(x_1, \mathbf{v}_i)$, so that $\mathbf{u}_i = \mathbf{a}_i x_1^{e_i} \mathbf{b}_i$ and $\mathbf{v}_i = \mathbf{c}_i x_1^{f_i} \mathbf{d}_i$ for some $\mathbf{a}_i, \mathbf{b}_i, \mathbf{c}_i, \mathbf{d}_i \in \mathscr{A}_\varnothing^+$. Then

$$\mathbf{u} = \mathbf{a}_0 x_1^{e_0} \mathbf{b}_0 \prod_{i=1}^{r} (h_i \mathbf{a}_i x_1^{e_i} \mathbf{b}_i) \text{ and } \mathbf{v} = \mathbf{c}_0 x_1^{f_0} \mathbf{d}_0 \prod_{i=1}^{r} (h_i \mathbf{c}_i x_1^{f_i} \mathbf{d}_i).$$

The identity

$$\alpha_1 : x^{e_0} \prod_{i=1}^{r} (h_i x^{e_i}) \approx x^{f_0} \prod_{i=1}^{r} (h_i x^{f_i})$$

in (11.2a) is deducible from σ, whence $\mathbb{O}\{\sigma\} = \mathbb{O}\{\alpha_1, \sigma\}$. Let

$$\mathbf{v}^{(1)} = \mathbf{c}_0 x_1^{e_0} \mathbf{d}_0 \prod_{i=1}^{r} (h_i \mathbf{c}_i x_1^{e_i} \mathbf{d}_i).$$

Then the identity $\mathbf{v} \approx \mathbf{v}^{(1)}$ is clearly deducible from α_1, so that

$$\mathbb{O}\{\sigma\} = \mathbb{O}\{\alpha_1, \sigma^{(1)}\},$$

where $\sigma^{(1)} : \mathbf{u} \approx \mathbf{v}^{(1)}$ is a fundamental identity that is not well balanced at precisely $x_2, x_3, \ldots, x_m$.

Now for any j with $1 \le j < m$, suppose that the identity $\sigma^{(j)}$ is not well balanced at precisely the variables $x_{j+1}, x_{j+2}, \ldots, x_m$. Then the argument in the previous paragraph can be repeated on the variable x_{j+1} to obtain

$$\mathbb{O}\{\sigma^{(j)}\} = \mathbb{O}\{\alpha_{j+1}, \sigma^{(j+1)}\},$$

where α_{j+1} is some identity in (11.2a) and $\sigma^{(j+1)}$ is a fundamental identity that is not well balanced at precisely the variables $x_{j+2}, \ldots, x_m$. Hence,

$$\begin{aligned} \mathbb{O}\{\sigma\} &= \mathbb{O}\{\alpha_1, \sigma^{(1)}\} \\ &= \mathbb{O}\{\alpha_1, \alpha_2, \sigma^{(2)}\} \\ &\vdots \\ &= \mathbb{O}\{\alpha_1, \alpha_2, \ldots, \alpha_m, \sigma^{(m)}\}. \end{aligned}$$

Since $\sigma^{(m)}$ is a well-balanced identity, the lemma holds with $\mathsf{A}^\sigma = \{\alpha_1, \alpha_2, \ldots, \alpha_m\}$. □

Lemma 11.6 *Let σ be any nontrivial well-balanced identity. Then*

$$\mathbb{O}\{\sigma\} = \mathbb{O}\mathsf{B}^{\sigma}$$

for some finite set B^{σ} of identities from (11.2b).

For the remainder of this subsection, let $\sigma : \mathbf{u} \approx \mathbf{v}$ be the nontrivial well-balanced identity in Lemma 11.6. If either $\mathbf{u}$ or $\mathbf{v}$ is a simple word, then by (F2) and (F3), the identity σ is contradictorily trivial. Therefore, both words $\mathbf{u}$ and $\mathbf{v}$ are non-simple, and they can be assumed to be those from (11.4). The prefixes $\mathbf{u}_0$ and $\mathbf{v}_0$ are clearly products of primary stacks. For each $i \in \{1, 2, \ldots, r\}$, any secondary stack of $\mathbf{u}_i$ can be moved by the identities (11.1) to any position within $\mathbf{u}_i$. Specifically, all secondary stacks of $\mathbf{u}_i$ can be gathered to the left and arranged in alphabetical order. Hence, the word $\mathbf{u}$ can be rewritten as

$$\mathbf{u} \overset{(11.1)}{=} \mathbf{p}_0 \prod_{i=1}^{r} (h_i \mathbf{s}_i \mathbf{p}_i),$$

where the word $\mathbf{p}_i$ is a possibly empty product of primary stacks of $\mathbf{u}$ that occur in $\mathbf{u}_i$, the word $\mathbf{s}_i$ is a possibly empty product of secondary stacks of $\mathbf{u}$ that occur in $\mathbf{u}_i$, and the stacks in $\mathbf{s}_i$ are arranged in alphabetical order. Since $\mathbf{u}_i \overset{\circ}{=} \mathbf{v}_i$ for all i, gathering the stacks of $\mathbf{v}$ in a similar manner gives

$$\mathbf{v} \overset{(11.1)}{=} \mathbf{q}_0 \prod_{i=1}^{r} (h_i \mathbf{s}_i \mathbf{q}_i), \tag{11.5}$$

where the word $\mathbf{q}_i$ is a possibly empty product of some primary stacks of $\mathbf{v}$ that occur in $\mathbf{v}_i$.

Remark 11.7

(i) $\mathbf{p}_i \overset{\circ}{=} \mathbf{q}_i$ for all i since $\mathbf{u} \approx \mathbf{v}$ is well balanced.
(ii) The exponent of each stack in $\mathbf{p}_r$ and in $\mathbf{q}_r$ is at least two.

It is convenient to call $\mathbf{p}_i$ and $\mathbf{q}_i$ the i-th *primary stack products* of $\mathbf{u}$ and $\mathbf{v}$, respectively.

Lemma 11.8 *Let ℓ be the least integer such that the ℓ-th primary stack products of $\mathbf{u}$ and $\mathbf{v}$ are different, that is, $\mathbf{p}_0 = \mathbf{q}_0, \mathbf{p}_1 = \mathbf{q}_1, \ldots, \mathbf{p}_{\ell-1} = \mathbf{q}_{\ell-1}$, and $\mathbf{p}_\ell \neq \mathbf{q}_\ell$. Then*

$$\mathbb{O}\{\sigma\} = \mathbb{O}(\mathsf{B} \cup \{\mathbf{u} \approx \mathbf{v}^{\dagger}\})$$

for some set B *of identities from* (11.2b) *and some word* $\mathbf{v}^\dagger$ *of the form* (11.5) *such that for each* $i \in \{0, 1, \ldots, \ell\}$, *the* i*-th primary stack products of* $\mathbf{u}$ *and* $\mathbf{v}^\dagger$ *are identical.*

Proof Let $\mathbf{z} \in \mathscr{A}_\varnothing^+$ be the longest suffix that is common to both $\mathbf{p}_\ell$ and $\mathbf{q}_\ell$. Then

$$\mathbf{p}_\ell = \mathbf{a} y^f \mathbf{z} \text{ and } \mathbf{q}_\ell = \mathbf{b} y^f x_1^{e_1} x_2^{e_2} \cdots x_s^{e_s} \mathbf{z}$$

for some $\mathbf{a}, \mathbf{b} \in \mathscr{A}_\varnothing^+$ and distinct stacks $x_1^{e_1}, x_2^{e_2}, \ldots, x_s^{e_s}, y^f$ with $s \geq 1$. It suffices to show how $\mathbf{v}$ can be rewritten, by invoking some identities from $\operatorname{id}\mathbb{O}\{\sigma\}$, into a word $\mathbf{v}'$ such that the ℓ-th primary stack products of $\mathbf{u}$ and $\mathbf{v}'$ share the longer suffix $y^f\mathbf{z}$. This procedure can then be repeated to obtain the required word $\mathbf{v}^\dagger$.

Recall from Remark 11.7(i) that $\mathbf{p}_\ell \overset{\circ}{=} \mathbf{q}_\ell$. Hence, the stacks $x_1^{e_1}, x_2^{e_2}, \ldots, x_r^{e_r}$ of $\mathbf{q}_\ell$ must appear in the factor $\mathbf{a}$ of $\mathbf{p}_\ell$. Further, the secondary y-stacks and secondary x_1-stacks of both $\mathbf{u}$ and $\mathbf{v}$ must occur in some of $\mathbf{s}_{\ell+1}, \mathbf{s}_{\ell+2}, \ldots, \mathbf{s}_m$. Therefore, for any $i > \ell$, if either an x_1-stack (say x_1^p) or a y-stack (say y^q) or both occur in $\mathbf{s}_i$, then the identities (11.1) can be applied to gather these stacks to the left of $\mathbf{s}_i$ resulting in $\mathbf{w}_i\mathbf{s}_i'$, where $\mathbf{w}_i \in \{x_1^p, y^q, x_1^p y^q\}$ and $\mathbf{s}_i'$ is obtained from $\mathbf{s}_i$ by eliminating all occurrences of x_1 and y. Therefore,

$$\mathbf{u} \overset{(11.1)}{=} \mathbf{p}_0\Bigg(\prod_{i=1}^{\ell-1}(h_i\mathbf{s}_i\mathbf{p}_i)\Bigg)h_\ell\mathbf{s}_\ell\mathbf{a}y^f\mathbf{z}\prod_{i=\ell+1}^{r}(h_i\mathbf{w}_i\mathbf{s}_i'\mathbf{p}_i)$$

$$\text{and } \mathbf{v} \overset{(11.1)}{=} \mathbf{q}_0\Bigg(\prod_{i=1}^{\ell-1}(h_i\mathbf{s}_i\mathbf{q}_i)\Bigg)h_\ell\mathbf{s}_\ell\mathbf{b}y^f x_1^{e_1}x_2^{e_2}\cdots x_s^{e_s}\mathbf{z}\prod_{i=\ell+1}^{r}(h_i\mathbf{w}_i\mathbf{s}_i'\mathbf{q}_i). \tag{11.6}$$

Since the stack $x_1^{e_1}$ appears in $\mathbf{a}$, retaining only the variables $x_1, y, h_{\ell+1}, h_{\ell+2}, \ldots, h_r$ in the identity $\mathbf{u} \approx \mathbf{v}$ results in the identity

$$\beta_1 : x_1^{e_1}y^f \prod_{i=\ell+1}^{r}(h_i\mathbf{w}_i) \approx y^f x_1^{e_1} \prod_{i=\ell+1}^{r}(h_i\mathbf{w}_i)$$

from (11.2b). Hence, $\mathbb{O}\{\sigma\} = \mathbb{O}\{\beta_1, \sigma\}$. Now it is easily seen that the identity β_1 of $\mathbb{O}\{\sigma\}$ can be used to interchange the primary y-stack and primary x_1-stack of $\mathbf{v}$ in (11.6) to obtain

$$\mathbf{v}^{(1)} = \mathbf{q}_0\Bigg(\prod_{i=1}^{\ell-1}(h_i\mathbf{s}_i\mathbf{q}_i)\Bigg)h_\ell\mathbf{s}_\ell\mathbf{b}x_1^{e_1}y^f x_2^{e_2}x_3^{e_3}\cdots x_s^{e_s}\mathbf{z}\prod_{i=\ell+1}^{r}(h_i\mathbf{w}_i\mathbf{s}_i'\mathbf{q}_i).$$

Therefore, $\mathbb{O}\{\sigma\} = \mathbb{O}\{\beta_1, \mathbf{u} \approx \mathbf{v}^{(1)}\}$. If the factor $x_2^{e_2}x_3^{e_3}\cdots x_s^{e_s}$ of $\mathbf{v}^{(1)}$ is nonempty, then the procedure in this paragraph can be repeated to interchange the primary stacks y^f and

$x_2^{e_2}$ in $\mathbf{v}^{(1)}$. Specifically, $\mathbb{O}\{\mathbf{u} \approx \mathbf{v}^{(1)}\} = \mathbb{O}\{\beta_2, \mathbf{u} \approx \mathbf{v}^{(2)}\}$ for some identity β_2 from (11.2b) and

$$\mathbf{v}^{(2)} = \mathbf{q}_0\left(\prod_{i=1}^{\ell-1}(h_i\mathbf{s}_i\mathbf{q}_i)\right)h_\ell\mathbf{s}_\ell\mathbf{b}x_1^{e_1}x_2^{e_2}y^f x_3^{e_3}x_4^{e_4}\cdots x_s^{e_s}\mathbf{z}\prod_{i=\ell+1}^{r}(h_i\mathbf{w}_i\mathbf{s}_i'\mathbf{q}_i).$$

Continuing in this manner, the stack y^f can be moved to the right until it immediately precedes the factor $\mathbf{z}$, that is, $\mathbb{O}\{\mathbf{u} \approx \mathbf{v}^{(s-1)}\} = \mathbb{O}\{\beta_s, \mathbf{u} \approx \mathbf{v}^{(s)}\}$ for some identity β_s from (11.2b) and

$$\mathbf{v}^{(s)} = \mathbf{q}_0\left(\prod_{i=1}^{\ell-1}(h_i\mathbf{s}_i\mathbf{q}_i)\right)h_\ell\mathbf{s}_\ell\mathbf{b}x_1^{e_1}x_2^{e_2}\cdots x_s^{e_s}y^f\mathbf{z}\prod_{i=\ell+1}^{r}(h_i\mathbf{w}_i\mathbf{s}_i'\mathbf{q}_i).$$

Hence, $\mathbb{O}\{\sigma\} = \mathbb{O}(\mathsf{B} \cup \{\mathbf{u} \approx \mathbf{v}'\})$ with $\mathsf{B} = \{\beta_1, \beta_2, \ldots, \beta_s\}$ and $\mathbf{v}' = \mathbf{v}^{(s)}$. □

Proof of Lemma 11.6 Since the integer ℓ in Lemma 11.8 is arbitrary, the result can be repeated so that $\mathbb{O}\{\sigma\} = \mathbb{O}(\mathsf{B}^\sigma \cup \{\mathbf{u} \approx \mathbf{v}^\dagger\})$ for some set B^σ of identities from (11.2b) and some word $\mathbf{v}^\dagger$ of the form (11.5) such that for any ℓ, the ℓ-th primary stack products of $\mathbf{u}$ and $\mathbf{v}^\dagger$ are identical. The identity $\mathbf{u} \approx \mathbf{v}^\dagger$ is then trivial. Therefore, $\mathbb{O}\{\sigma\} = \mathbb{O}\mathsf{B}^\sigma$. □

11.1.3 Proof of Proposition 11.2

Let $\mathbb{V}$ be any noncommutative subvariety of $\mathbb{O}$. Then $\mathbb{V} = \mathbb{O}\Sigma$ for some set Σ of identities. To show that the identities in Σ can be chosen from (11.2), it suffices to show that if $\sigma : \mathbf{u} \approx \mathbf{v}$ is any identity in Σ, then $\mathbb{O}\{\sigma\} = \mathbb{O}(\mathsf{A}^\sigma \cup \mathsf{B}^\sigma)$ for some set A^σ of identities from (11.2a) and some set B^σ of identities from (11.2b). If $\mathsf{con}(\mathbf{u}) \neq \mathsf{con}(\mathbf{v})$, then the variety $\mathbb{V}$ satisfies the identity $\alpha : x^m \approx 1$ for some $m \geq 1$, whence $\mathbb{V}$ is contradictorily commutative because

$$xy \overset{\alpha}{\approx} x^m xyy^m \overset{(11.1)}{\approx} x^m yxy^m \overset{\alpha}{\approx} yx.$$

Therefore, assume that $\mathsf{con}(\mathbf{u}) = \mathsf{con}(\mathbf{v})$. There are two cases.

CASE 1: $\mathsf{sim}(\mathbf{u}) = \mathsf{sim}(\mathbf{v})$. Then the identity σ satisfies both (F2) and (F3). By Lemma 11.4, the words $\mathbf{u}$ and $\mathbf{v}$ can be chosen to be in canonical form, so that (F1) holds and the identity σ is fundamental. It then follows from Lemmas 11.5

and 11.6 that $\mathbb{O}\{\sigma\} = \mathbb{O}(\mathsf{A}^\sigma \cup \mathsf{B}^\sigma)$ for some finite set A^σ of identities from (11.2a) and some finite set B^σ of identities from (11.2b).

CASE 2: $\mathsf{sim}(\mathbf{u}) \neq \mathsf{sim}(\mathbf{v})$. Then there exists some $k \geq 2$ such that the identity $x^k \approx x$ is deducible from σ. Let $\mathbf{u}'$ be the word obtained by replacing every simple variable x in $\mathbf{u}$ with x^k, and let $\mathbf{v}'$ be the word similarly obtained from $\mathbf{v}$. Then $\mathbb{O}\{\sigma\} = \mathbb{O}\{x^k \approx x, \sigma'\}$, where the identity $x^k \approx x$ belongs to (11.2a) and the identity $\sigma' : \mathbf{u}' \approx \mathbf{v}'$ satisfies $\mathsf{sim}(\mathbf{u}') = \mathsf{sim}(\mathbf{v}')$. It follows from Case 1 that $\mathbb{O}\{\sigma'\} = \mathbb{O}(\mathsf{A}^{\sigma'} \cup \mathsf{B}^{\sigma'})$ for some finite set $\mathsf{A}^{\sigma'}$ of identities from (11.2a) and some finite set $\mathsf{B}^{\sigma'}$ of identities from (11.2b). Consequently, $\mathbb{O}\{\sigma\} = \mathbb{O}(\mathsf{A}^\sigma \cup \mathsf{B}^\sigma)$ with $\mathsf{A}^\sigma = \{x^k \approx x\} \cup \mathsf{A}^{\sigma'}$ and $\mathsf{B}^\sigma = \mathsf{B}^{\sigma'}$.

11.2 Finite Basis Property of Subvarieties of $\mathbb{O}$

Lemma 11.9 (Head 1968) *Every variety of commutative monoids is finitely based.*

Lemma 11.10 (Volkov 1990, Corollary 2) *Any set of identities from* (11.2a) *defines a finitely based variety.*

The main aim of the present section is to prove that every subvariety $\mathbb{V}$ of $\mathbb{O}$ is finitely based; by Lemma 11.9, it suffices to assume that $\mathbb{V}$ is noncommutative. By Proposition 11.2, there exists some set Σ of identities from (11.2) such that $\mathbb{V} = \mathbb{O}\Sigma$. By Lemma 11.10, it suffices to show that any set of identities from (11.2b) defines a finitely based subvariety of $\mathbb{O}$. This result is established in Proposition 11.15.

A *quasi-order* on a set X is a binary relation $\leq$ on X that is reflexive and transitive; in this case, the pair $(X, \leq)$ is called a *quasi-ordered set*. The *direct product* of two quasi-ordered sets $(\dot{X}, \dot{\leq})$ and $(\ddot{X}, \ddot{\leq})$ is $(\dot{X} \times \ddot{X}, \leq)$, where $\leq = \dot{\leq} \times \ddot{\leq}$ is given by $(\dot{a}, \ddot{a}) \leq (\dot{b}, \ddot{b})$ if $\dot{a} \mathrel{\dot{\leq}} \dot{b}$ and $\ddot{a} \mathrel{\ddot{\leq}} \ddot{b}$. A quasi-order on a set X is a *well-quasi-order* if any nonempty subset Y of X contains finitely positively many elements minimal in Y.

Lemma 11.11 (Higman 1952, Theorem 2.3) *The direct product of two well-quasi-ordered sets is well-quasi-ordered.*

In the present section, let $\leq$ denote the usual ordering on the set $\mathbb{N}_0 = \{0, 1, 2, \ldots\}$ of nonnegative integers. In what follows, since finite sequences of symbols from the direct product $\mathbb{N}_0^2 = \mathbb{N}_0 \times \mathbb{N}_0$ are often involved, it is less cumbersome to abbreviate each element (e, f) of $\mathbb{N}_0^2$ to ef. For instance, the order $\leq_2 = \leq \times \leq$ on $\mathbb{N}_0^2$ is given by $ef \leq_2 pq$ if $e \leq p$ and $f \leq q$. It follows from Lemma 11.11 that $(\mathbb{N}_0^2, \leq_2)$ is a well-quasi-ordered set.

Lemma 11.12 *Let*

$$\varepsilon : x^{e_0} y^{f_0} \prod_{i=1}^{r} (h_i x^{e_i} y^{f_i}) \approx y^{f_0} x^{e_0} \prod_{i=1}^{r} (h_i x^{e_i} y^{f_i})$$

$$\text{and } \pi : x^{p_0} y^{q_0} \prod_{i=1}^{s} (h_i x^{p_i} y^{q_i}) \approx y^{q_0} x^{p_0} \prod_{i=1}^{s} (h_i x^{p_i} y^{q_i})$$

be any identities from (11.2b). *Suppose that the following conditions hold*:

(a) $e_0 f_0 \leq_2 p_0 q_0$;
(b) $r \leq s$ *and there exists a subsequence* $j_1, j_2, \ldots, j_r$ *of* $1, 2, \ldots, s$ *such that*

$$e_1 f_1 \leq_2 p_{j_1} q_{j_1}, \quad e_2 f_2 \leq_2 p_{j_2} q_{j_2}, \quad \ldots, \quad e_r f_r \leq_2 p_{j_r} q_{j_r}.$$

Then the inclusion $\mathbb{O}\{\varepsilon\} \subseteq \mathbb{O}\{\pi\}$ *holds.*

Proof It follows from (a) that $p_0 - e_0, q_0 - f_0 \in \mathbb{N}_0$. First suppose that $r = 0$, so that the identity ε is $x^{e_0} y^{f_0} \approx y^{f_0} x^{e_0}$ with $e_0, f_0 \geq 2$. Since

$$\begin{aligned} x^{p_0} y^{q_0} &\overset{(11.1)}{\approx} x^{e_0} y^{f_0} x^{p_0 - e_0} y^{q_0 - f_0} \\ &\overset{\varepsilon}{\approx} y^{f_0} x^{e_0} x^{p_0 - e_0} y^{q_0 - f_0} \\ &\overset{(11.1)}{\approx} y^{q_0} x^{p_0}, \end{aligned}$$

the deductions $\{(11.1), \varepsilon\} \vdash x^{p_0} y^{q_0} \approx y^{q_0} x^{p_0} \vdash \pi$ hold, so that $\mathbb{O}\{\varepsilon\} \subseteq \mathbb{O}\{\pi\}$.

Now suppose that $r \geq 1$. Let φ denote the substitution $h_1 \mapsto x^{p_0 - e_0} y^{q_0 - f_0} h_1$. Then

$$\begin{aligned} x^{p_0} y^{q_0} \prod_{i=1}^{r} (h_i x^{e_i} y^{f_i}) &\overset{(11.1)}{\approx} x^{e_0} y^{f_0} x^{p_0 - e_0} y^{q_0 - f_0} \prod_{i=1}^{r} (h_i x^{e_i} y^{f_i}) \\ &= \left(x^{e_0} y^{f_0} \prod_{i=1}^{r} (h_i x^{e_i} y^{f_i}) \right) \varphi \\ &\overset{\varepsilon}{\approx} \left(y^{f_0} x^{e_0} \prod_{i=1}^{r} (h_i x^{e_i} y^{f_i}) \right) \varphi \\ &= y^{f_0} x^{e_0} x^{p_0 - e_0} y^{q_0 - f_0} \prod_{i=1}^{r} (h_i x^{e_i} y^{f_i}) \\ &\overset{(11.1)}{\approx} y^{q_0} x^{p_0} \prod_{i=1}^{r} (h_i x^{e_i} y^{f_i}). \end{aligned}$$

Hence, the identity

$$\varepsilon' : x^{p_0} y^{q_0} \prod_{i=1}^{r} (h_i x^{e_i} y^{f_i}) \approx y^{q_0} x^{p_0} \prod_{i=1}^{r} (h_i x^{e_i} y^{f_i})$$

is deducible from the identities $\{(11.1), \varepsilon\}$. By (b), it is easily seen that there is a substitution ψ that fixes both x and y such that $(\prod_{i=1}^{r} (h_i x^{e_i} y^{f_i}))\psi$ is a prefix of $\prod_{i=1}^{s} (h_i x^{p_i} y^{q_i})$, that is,

$$\left(\left(\prod_{i=1}^{r} (h_i x^{e_i} y^{f_i})\right)\psi\right)\mathbf{t} = \prod_{i=1}^{s} (h_i x^{p_i} y^{q_i})$$

for some $\mathbf{t} \in \mathscr{A}_{\varnothing}^{+}$. Then

$$\begin{aligned} x^{p_0} y^{q_0} \prod_{i=1}^{s} (h_i x^{p_i} y^{q_i}) &= \left(\left(x^{p_0} y^{q_0} \prod_{i=1}^{r} (h_i x^{e_i} y^{f_i})\right)\psi\right)\mathbf{t} \\ &\overset{\varepsilon'}{\approx} \left(\left(y^{q_0} x^{p_0} \prod_{i=1}^{r} (h_i x^{e_i} y^{f_i})\right)\psi\right)\mathbf{t} \\ &= y^{q_0} x^{p_0} \prod_{i=1}^{s} (h_i x^{p_i} y^{q_i}). \end{aligned}$$

Therefore, the identity π is deducible from ε', whence $\mathbb{O}\{\varepsilon\} \subseteq \mathbb{O}\{\pi\}$. □

Let $(\mathbb{N}_0^2)^*$ denote the set of all finite sequences of symbols from $\mathbb{N}_0^2$. For any two sequences $\mathbf{e} = (e_1 f_1, e_2 f_2, \ldots, e_r f_r)$ and $\mathbf{p} = (p_1 q_1, p_2 q_2, \ldots, p_s q_s)$ in $(\mathbb{N}_0^2)^*$, define $\mathbf{e} \leq_2^* \mathbf{p}$ if there exists a subsequence $j_1, j_2, \ldots, j_r$ of $1, 2, \ldots, s$ such that

$$e_1 f_1 \leq_2 p_{j_1} q_{j_1}, \quad e_2 f_2 \leq_2 p_{j_2} q_{j_2}, \quad \ldots, \quad e_r f_r \leq_2 p_{j_r} q_{j_r}.$$

Since $(\mathbb{N}_0^2, \leq_2)$ is a well-quasi-ordered set, it follows from Higman's lemma (1952) that $((\mathbb{N}_0^2)^*, \leq_2^*)$ is also a well-quasi-ordered set.

Lemma 11.13 *The pair* $(\mathbb{N}_0^2 \times (\mathbb{N}_0^2)^*, \preccurlyeq)$, *where* $\preccurlyeq \; = \; \leq_2 \times \leq_2^*$, *is a well-quasi-ordered set.*

Proof Since the pairs $(\mathbb{N}_0^2, \leq_2)$ and $((\mathbb{N}_0^2)^*, \leq_2^*)$ are well-quasi-ordered sets, by Lemma 11.11, their direct product $(\mathbb{N}_0^2, \leq_2) \times ((\mathbb{N}_0^2)^*, \leq_2^*) = (\mathbb{N}_0^2 \times (\mathbb{N}_0^2)^*, \preccurlyeq)$ is also a well-quasi-ordered set. □

Now associate each identity

$$\varepsilon : x^{e_0} y^{f_0} \prod_{i=1}^{r} (h_i x^{e_i} y^{f_i}) \approx y^{f_0} x^{e_0} \prod_{i=1}^{r} (h_i x^{e_i} y^{f_i})$$

from (11.2b) with the element $\overrightarrow{\varepsilon} = (e_0 f_0, (e_1 f_1, \ldots, e_r f_r))$ in $\mathbb{N}_0^2 \times (\mathbb{N}_0^2)^*$. Then the following result is a consequence of Lemma 11.12.

Lemma 11.14 *Let* ε *and* π *be any identities from* (11.2b). *Suppose that* $\overrightarrow{\varepsilon} \preccurlyeq \overrightarrow{\pi}$. *Then the inclusion* $\mathbb{O}\{\varepsilon\} \subseteq \mathbb{O}\{\pi\}$ *holds.*

Proposition 11.15 *Suppose that* B *is any set of identities from* (11.2b). *Then the variety* $\mathbb{O}\mathsf{B}$ *is finitely based.*

Proof Since $(\mathbb{N}_0^2 \times (\mathbb{N}_0^2)^*, \preccurlyeq)$ is well-quasi-ordered by Lemma 11.13, the subset $\overrightarrow{\mathsf{B}} = \{\overrightarrow{\varepsilon} \mid \varepsilon \in \mathsf{B}\}$ of $\mathbb{N}_0^2 \times (\mathbb{N}_0^2)^*$ contains finitely positively many elements that are minimal in $\overrightarrow{\mathsf{B}}$. Let $\varepsilon_1, \varepsilon_2, \ldots, \varepsilon_m \in \mathsf{B}$ be such that $\overrightarrow{\varepsilon_1}, \overrightarrow{\varepsilon_2}, \ldots, \overrightarrow{\varepsilon_m}$ are all the elements minimal in $\overrightarrow{\mathsf{B}}$. The inclusion $\mathbb{O}\mathsf{B} \subseteq \mathbb{O}\{\varepsilon_1, \varepsilon_2, \ldots, \varepsilon_m\}$ holds vacuously. If $\pi \in \mathsf{B}$, then $\overrightarrow{\varepsilon_i} \preccurlyeq \overrightarrow{\pi}$ for some i, so that by Lemma 11.14, the identity π is satisfied by the variety $\mathbb{O}\{\varepsilon_i\}$. The inclusion $\mathbb{O}\{\varepsilon_1, \varepsilon_2, \ldots, \varepsilon_m\} \subseteq \mathbb{O}\mathsf{B}$ thus follows. □

11.3 Distinguished Varieties

A variety $\mathbb{V}$ of monoids is *distinguished* if one of the following conditions holds: $\mathcal{R}_{\mathsf{Q}}\{xyx\} \in \mathbb{V}$, $\mathbb{V} \subseteq \mathbb{O}$, and $\mathbb{V} \subseteq \mathbb{O}^{\triangleleft}$. The present section is concerned with classifying distinguished varieties that are hereditarily finitely based.

Theorem 11.16 *The following statements on any distinguished variety* $\mathbb{V}$ *are equivalent*:

(a) $\mathbb{V}$ *is hereditarily finitely based*;
(b) $\mathbb{J}_1, \mathbb{J}_2 \nsubseteq \mathbb{V}$;
(c) $\mathbb{V} \subseteq \mathbb{O}$ *or* $\mathbb{V} \subseteq \mathbb{O}^{\triangleleft}$.

Proof The implication (c) ⇒ (a) follows from Theorem 11.1, and the implication (a) ⇒ (b) holds because the varieties $\mathbb{J}_1$ and $\mathbb{J}_2$ are non-finitely based (Jackson 2005b). Suppose that (b) holds. If $\mathcal{R}_{\mathsf{Q}}\{xyx\} \notin \mathbb{V}$, then (c) holds because $\mathbb{V}$ is distinguished. If $\mathcal{R}_{\mathsf{Q}}\{xyx\} \in \mathbb{V}$, then (c) holds by Lemma 2.7. □

Lemma 11.17 *A variety* $\mathbb{V}$ *of monoids is distinguished if one of the following conditions holds*:

(a) $\mathbb{V} \subseteq \mathbb{A}^{\mathsf{cen}}$;
(b) $\mathbb{V} \subseteq \mathbb{B}_2^1$;
(c) $\mathcal{R}_{\mathcal{Q}}\{xyx\} \in \mathbb{V}$;
(d) $\mathbb{C}\mathrm{OM} \subseteq \mathbb{V}$.

Proof Let $\mathbb{V}$ be any variety of monoids. If $\mathcal{R}_{\mathcal{Q}}\{xyx\} \in \mathbb{V}$, then $\mathbb{V}$ is distinguished by definition. Therefore, assume that $\mathcal{R}_{\mathcal{Q}}\{xyx\} \notin \mathbb{V}$, so that by Lemma 2.6, the variety $\mathbb{V}$ satisfies a nontrivial identity of the form $xyx \approx \mathbf{w}$.

CASE 1: $\mathbb{C}\mathrm{OM} \subseteq \mathbb{V}$. Then $\mathsf{occ}(x, \mathbf{w}) = 2$ and $\mathsf{occ}(y, \mathbf{w}) = 1$, so that $\mathbf{w} \in \{x^2y, yx^2\}$. It follows that either $\mathbb{V} \subseteq \mathbb{O}$ or $\mathbb{V} \subseteq \mathbb{O}^{\triangleleft}$, whence $\mathbb{V}$ is distinguished.
CASE 2: $\mathbb{V} \subseteq \mathbb{A}^{\mathsf{cen}}$. Then it is easily shown that $\mathbf{w} \in \{x^2y, yx^2\}$; see, for example, Jackson (2005b, proof of Lemma 4.1). Hence, $\mathbb{V}$ is distinguished as in Case 1.
CASE 3: $\mathbb{V} \subseteq \mathbb{B}_2^1$. Since the monoid $\mathcal{B}_2^1$ satisfies the identities

$$x^3 \approx x^2, \quad x^2yx \approx xyx^2, \quad x^2y^2 \approx y^2x^2, \tag{11.7}$$

the variety $\mathbb{V}$ also satisfies these identities. Since $\mathbb{V}$ satisfies the nontrivial identity $xyx \approx \mathbf{w}$, it is routinely shown to also satisfy the identities

$$x^2yx \approx xyx, \quad xyx^2 \approx xyx. \tag{11.8}$$

The identities (11.7) and (11.8) can be used to interchange any two adjacent non-simple variables of a word; specifically, if $\mathbf{u}, \mathbf{v} \in \mathscr{A}_\varnothing^+$ are such that $x, y \in \mathsf{con}(\mathbf{uv})$, then

$$\begin{aligned} \mathbf{u}xy\mathbf{v} &\overset{(11.8)}{\approx} \mathbf{u}x^2y^2\mathbf{v} \\ &\overset{(11.7)}{\approx} \mathbf{u}y^2x^2\mathbf{v} \\ &\overset{(11.8)}{\approx} \mathbf{u}yx\mathbf{v}. \end{aligned}$$

It follows that the variety $\mathbb{V}$ satisfies the identities (11.1), whence $\mathbb{V} \subseteq \mathbb{O}$. □

Theorem 1.58 now follows from Theorem 11.16 and Lemma 11.17.

Corollary 11.18 *The varieties* $\mathbb{J}_1$ *and* $\mathbb{J}_2$ *are the only limit varieties among all distinguished varieties of monoids. In particular,* $\mathbb{J}_1$ *and* $\mathbb{J}_2$ *are the only limit subvarieties of* $\mathbb{A}^{\mathsf{cen}}$, $\mathbb{B}_2^1$, *and* $\mathbb{P}_{25}$.

Proof The first part follows from Theorem 11.16. The second part holds because the inclusion $\mathbb{J}_1, \mathbb{J}_2 \subseteq \mathbb{A}^{\mathrm{cen}}$ is obvious, and the inclusions $\mathbb{J}_1, \mathbb{J}_2 \subseteq \mathbb{B}_2^1$ and $\mathbb{J}_1, \mathbb{J}_2 \subseteq \mathbb{P}_{25}$ hold by Lemma 2.9. □

11.4 Summary

The variety $\mathbb{O}$ is shown in Theorem 11.1 to be hereditarily finitely based. It follows that a distinguished variety is hereditarily finitely based if and only if it excludes the limit varieties $\mathbb{J}_1$ and $\mathbb{J}_2$; see Theorem 11.16. These results were published in Lee (2012b).

Varieties of Aperiodic Monoids with Central Idempotents

12

The objective of the present chapter is to establish the results stated in Sect. 1.6.3 regarding the class $\mathbb{A}^{\mathsf{cen}}$ of aperiodic monoids with central idempotents. For each $n \geq 1$, let $\mathbb{A}_n^{\mathsf{cen}}$ denote the variety of monoids defined by the identities

$$x^{n+1} \approx x^n, \quad x^n y \approx y x^n. \qquad (\blacktriangle_n)$$

Some preliminary results are first developed in Sect. 12.1. The variety $\mathbb{K} = \mathbb{A}_2^{\mathsf{cen}} \cap \mathbb{O} \cap \mathbb{O}^{\triangleleft}$ is then investigated in Sect. 12.2. It is shown that $\mathbb{K}$ is a non-finitely generated almost Cross variety and that the lattice $\mathfrak{L}(\mathbb{K})$ is a countably infinite chain. It follows that $\mathbb{K}$ is also a subvariety of $\mathbb{B}_2^{\frac{1}{2}}$.

In Sect. 12.3, the characterization of Cross subvarieties of $\mathbb{A}^{\mathsf{cen}}$, given in Theorem 1.64, is established. In Sect. 12.4, the characterization of varieties that are inherently non-finitely generated within $\mathbb{A}^{\mathsf{cen}}$, given in Theorem 1.65, is established. In Sect. 12.5, a non-finitely generated subvariety of $\mathbb{A}^{\mathsf{cen}}$ that is not inherently non-finitely generated within $\mathbb{A}^{\mathsf{cen}}$ is exhibited.

12.1 Rigid Identities

This section introduces some identities that are required to define subvarieties of $\mathbb{A}^{\mathsf{cen}}$. Some basic results regarding these identities are also established.

12.1.1 Definition and Basic Properties

Define a *rigid word* to be a word of the form

E. W. H. Lee, *Advances in the Theory of Varieties of Semigroups*, Frontiers in Mathematics, https://doi.org/10.1007/978-3-031-16497-2_12

$$\mathbf{w} = x^{e_0} \prod_{i=1}^{m} (h_i x^{e_i}),$$

where $e_0, e_1, \ldots, e_m \geq 0$ and $m \geq 0$; the number m is the *rank* of $\mathbf{w}$. Note that a rigid word of rank 0 is of the form x^e. The rigid word $\mathbf{w}$ above is *square-free* if $e_0, e_1, \ldots, e_m \leq 1$. A *rigid identity* is an identity that is formed by a pair of rigid words of the same rank.

Recall from Proposition 11.2 that rigid identities are one of two types of identities that can be used to define noncommutative subvarieties of $\mathbb{O}$; the following is a restatement of this result.

Lemma 12.1 *Each noncommutative subvariety of* $\mathbb{O}$ *is of the form* $\mathbb{O}\Sigma$, *where* Σ *is some set of the following identities*:

(a) *rigid identities*;
(b) *identities of the form*

$$x^{e_0} y^{f_0} \prod_{i=1}^{m} (h_i x^{e_i} y^{f_i}) \approx y^{f_0} x^{e_0} \prod_{i=1}^{m} (h_i x^{e_i} y^{f_i}),$$

where $e_0, f_0 \geq 1$, $e_1, f_1, e_2, f_2, \ldots, e_m, f_m \geq 0$, $\sum_{i=0}^{m} e_i \geq 2$, $\sum_{i=0}^{m} f_i \geq 2$, *and* $m \geq 0$.

Consider an identity of the form

$$\sigma : \mathbf{u}_0 \prod_{i=1}^{m} (h_i \mathbf{u}_i) \approx \mathbf{v}_0 \prod_{i=1}^{m} (h_i \mathbf{v}_i), \tag{12.1}$$

where $\mathbf{u}_0, \mathbf{v}_0, \mathbf{u}_1, \mathbf{v}_1, \ldots, \mathbf{u}_m, \mathbf{v}_m$ are words not containing the variables $h_1, h_2, \ldots, h_m$. Then the identity σ is said to be *efficient* if $(\mathbf{u}_0, \mathbf{v}_0), (\mathbf{u}_1, \mathbf{v}_1), \ldots, (\mathbf{u}_m, \mathbf{v}_m) \neq (\varnothing, \varnothing)$. Now if an identity σ is not already efficient, then there exists an efficient identity σ' such that σ and σ' define the same variety of monoids. This is easily demonstrated by a simple example.

Example 12.2 Suppose that $m = 5$ in the identity σ in (12.1) with $\mathbf{u}_0, \mathbf{u}_1, \mathbf{u}_3, \mathbf{v}_1, \mathbf{v}_3, \mathbf{v}_4 \neq \varnothing$ and $\mathbf{u}_2 = \mathbf{u}_4 = \mathbf{u}_5 = \mathbf{v}_0 = \mathbf{v}_2 = \mathbf{v}_5 = \varnothing$. Then the identity

$$\sigma : \mathbf{u}_0 h_1 \mathbf{u}_1 h_2 h_3 \mathbf{u}_3 h_4 h_5 \approx h_1 \mathbf{v}_1 h_2 h_3 \mathbf{v}_3 h_4 \mathbf{v}_4 h_5$$

is not efficient because $(\mathbf{u}_2, \mathbf{v}_2) = (\mathbf{u}_5, \mathbf{v}_5) = (\varnothing, \varnothing)$. The identity

$$\sigma' : \mathbf{u}_0 h_1 \mathbf{u}_1 h_3 \mathbf{u}_3 h_4 \approx h_1 \mathbf{v}_1 h_3 \mathbf{v}_3 h_4 \mathbf{v}_4,$$

obtained from σ by deleting the variables h_2 and h_5, is efficient. Conversely, by making the substitution $h_3 \mapsto h_2h_3$ in σ' and then post multiplying both sides of the resulting identity by h_5, the identity σ is obtained. Therefore, the identities σ and σ' define the same variety of monoids.

Remark 12.3 The identities in Lemma 12.1 can be assumed efficient.

Lemma 12.4 *The variety $\mathbb{A}_n^{\mathsf{cen}}$ satisfies the rigid identity*

$$x^{e_0} \prod_{i=1}^{m} (h_i x^{e_i}) \approx x^n \prod_{i=1}^{m} h_i \tag{12.2}$$

if $e_j \geq n$ for some $j \in \{0, 1, \ldots, m\}$.

Proof It is easily shown that if $e_j \geq n$ for some $j \in \{0, 1, \ldots, m\}$, then the identity (12.2) is deducible from the identity basis $\blacktriangle_n$ of $\mathbb{A}_n^{\mathsf{cen}}$. □

Lemma 12.5 *Let $\mathbb{V}$ be any subvariety of $\mathbb{A}^{\mathsf{cen}}$ that satisfies a nontrivial rigid identity*

$$x^{e_0} \prod_{i=1}^{m} (h_i x^{e_i}) \approx x^{f_0} \prod_{i=1}^{m} (h_i x^{f_i}),$$

where at least one side of the identity is a square-free word.

(i) *If $m = 0$, then $\mathbb{V}$ is commutative.*
(ii) *If $(e_0, e_1, \ldots, e_m) = (0, 0, \ldots, 0)$ or $(f_0, f_1, \ldots, f_m) = (0, 0, \ldots, 0)$, then $\mathbb{V}$ is trivial.*

Proof This is routinely verified based on the assumption that the variety $\mathbb{V}$ satisfies the identities $\blacktriangle_n$ for some $n \geq 1$. □

12.1.2 Straubing Identities

The *Straubing identities*

$$x \prod_{i=1}^{n-1} (h_i x) \approx x^n \prod_{i=1}^{n-1} h_i, \tag{$\mathbb{S}_n$}$$

where $n \geq 2$, play a significant role in the study of finitely generated subvarieties of $\mathbb{A}^{\mathsf{cen}}$. Note that $Ⓢ_n$ is a balanced rigid identity formed by rigid words of rank $n-1$. Let $\mathbb{S}_n$ denote the variety of monoids defined by $Ⓢ_n$.

Lemma 12.6 (Straubing 1982) *Each finite monoid in* $\mathbb{A}^{\mathsf{cen}}$ *belongs to* $\mathbb{A}_n^{\mathsf{cen}} \cap \mathbb{S}_n$ *for some* $n \geq 2$.

A word $\mathbf{w}$ is *n-limited* if $\mathsf{occ}(x, \mathbf{w}) \leq n$ for all $x \in \mathscr{A}$. Note that a word is simple if and only if it is 1-limited. An identity $\mathbf{w} \approx \mathbf{w}'$ is *n-limited* if the words $\mathbf{w}$ and $\mathbf{w}'$ are n-limited. For each $n \geq 1$, let $\mathscr{W}_{(n)}$ denote the set of all n-limited words in $\{x, y\}^+$.

Lemma 12.7 (Jackson and Sapir 2000, Corollary 3.1) *For each* $n \geq 2$, *the variety* $\mathbb{A}_n^{\mathsf{cen}} \cap \mathbb{S}_n$ *is generated by the finite monoid* $\mathcal{R}_{\mathcal{Q}}\mathscr{W}_{(n-1)}$.

Lemma 12.8 *Suppose that* $\mathbb{V}$ *is any subvariety of* $\mathbb{A}^{\mathsf{cen}}$ *that satisfies some nontrivial rigid identity* $\mathbf{u} \approx \mathbf{v}$, *where either* $\mathbf{u}$ *or* $\mathbf{v}$ *is square-free. Then* $\mathbb{V} \subseteq \mathbb{A}_n^{\mathsf{cen}} \cap \mathbb{S}_n$ *for some* $n \geq 2$.

Proof By assumption, $\mathbb{V} \subseteq \mathbb{A}_n^{\mathsf{cen}}$ for some $n \geq 2$, and

$$\mathbf{u} = x^{e_0} \prod_{i=1}^{m} (h_i x^{e_i}) \quad \text{and} \quad \mathbf{v} = x^{f_0} \prod_{i=1}^{m} (h_i x^{f_i})$$

for some $e_0, f_0, e_1, f_1, \ldots, e_m, f_m \geq 0$ with $(e_0, e_1, \ldots, e_m) \neq (f_0, f_1, \ldots, f_m)$. Further, it suffices to assume that $m \geq 1$ and $(e_0, e_1, \ldots, e_m), (f_0, f_1, \ldots, f_m) \neq (0, 0, \ldots, 0)$ since otherwise, $\mathbb{V} \subseteq \mathbb{S}_n$ by Lemma 12.5, whence the proof is complete.

Let $e = \sum_{i=0}^{m} e_i$ and $f = \sum_{i=0}^{m} f_i$. Generality is not lost by assuming one of the following cases:

(A) $\mathbf{u}$ is square-free and $\mathbf{v}$ is not square-free;
(B) $\mathbf{u}$ and $\mathbf{v}$ are both square-free with $0 < e \leq f$.

Then $e_0, e_1, \ldots, e_m \leq 1$ in both (A) and (B). Since $\mathbf{u}$ is a square-free rigid word and $\mathsf{occ}(x, \mathbf{u}) = e$, there is an appropriate deletion φ_1 of the variables h_i so that

$$\mathbf{u}\varphi_1 = x h_{j_1} x h_{j_2} x \cdots h_{j_{e-1}} x,$$

where $1 \leq j_1 < j_2 < \cdots < j_{e-1} \leq m$. Let φ_2 be the substitution

$$(h_{j_1}, h_{j_2}, \ldots, h_{j_{e-1}}) \mapsto (h_1, h_2, \ldots, h_{e-1}).$$

Then $\mathbf{u}\varphi_1\varphi_2 = x\prod_{i=1}^{e-1}(h_i x)$ is a square-free rigid word of rank $e-1$. (For instance, if $m = 7$ and $(e_0, e_1, e_2, e_3, e_4, e_5, e_6, e_7) = (0, 1, 0, 1, 0, 0, 1, 1)$, so that $\mathbf{u} = h_1 x h_2 h_3 x h_4 h_5 h_6 x h_7 x$ and $e = 4$, then $\mathbf{u}\varphi_1 = x h_2 x h_4 x h_7 x$ and $\mathbf{u}\varphi_1\varphi_2 = x h_1 x h_2 x h_3 x$.) Now perform the deletion φ_1 on $\mathbf{v}$, then followed by the substitution φ_2 on $\mathbf{v}\varphi_1$ to obtain $\mathbf{v}\varphi_1\varphi_2$. It is clear that in (A), the word $\mathbf{v}\varphi_1\varphi_2$ is a rigid word of rank $e-1$ that is not square-free. In (B), since the identity $\mathbf{u} \approx \mathbf{v}$ is nontrivial with $e \leq f$, the word $\mathbf{v}\varphi_1\varphi_2$ is also rigid and of rank $e-1$ that is not square-free. Hence, in both cases, $\mathbf{v}\varphi_1\varphi_2 = \mathbf{p}x^r\mathbf{q}$ for some $r \geq 2$ and $\mathbf{p}, \mathbf{q} \in \mathscr{A}_\varnothing^+$. It follows that the rigid identity

$$x\prod_{i=1}^{d}(h_i x) \approx \mathbf{p}x^r\mathbf{q}, \tag{12.3}$$

where $d = e-1$, is deducible from $\mathbf{u} \approx \mathbf{v}$. Let χ denote the substitution $x \mapsto x^r$. Then the identity

$$x^r\prod_{i=1}^{d}(h_i x^r) \approx (\mathbf{p}\chi)x^{r^2}(\mathbf{q}\chi) \tag{12.4}$$

is deducible from (12.3). Since

$$\begin{aligned} x\prod_{i=1}^{d^2+2d}(h_i x) &= \left(x\prod_{i=1}^{d}(h_i x)\right)h_{d+1}\left(x\prod_{i=d+2}^{2d+1}(h_i x)\right)h_{2d+2}\left(x\prod_{i=2d+3}^{3d+2}(h_i x)\right)\cdots \\ &\qquad \cdots h_{d^2+d}\left(x\prod_{i=d^2+d+1}^{d^2+2d}(h_i x)\right) \\ &\overset{(12.3)}{\approx} (\cdots x^r\cdots)h_{d+1}(\cdots x^r\cdots)h_{2d+2}(\cdots x^r\cdots)\cdots h_{d^2+d}(\cdots x^r\cdots) \\ &\overset{(12.4)}{\approx} \cdots x^{r^2}\cdots, \end{aligned}$$

the identity $\mathbf{u} \approx \mathbf{v}$ can be used to deduce a rigid identity of the form (12.3) with r replaced by r^2. The same procedure can be repeated sufficiently many times so that the identity $\mathbf{u} \approx \mathbf{v}$ can be used to deduce a rigid identity of the form (12.3) with r replaced by some number in $\{r, r^2, r^3, \ldots\}$ that is greater than n. Hence, generality is not lost by assuming $r \geq n$ in (12.3) to begin with. Since $\mathbf{p}x^r\mathbf{q}$ is a rigid word of rank d, it follows from Lemma 12.4 that

$$x^n\prod_{i=1}^{d}h_i \overset{(12.2)}{\approx} \mathbf{p}x^r\mathbf{q}$$

$$\stackrel{(12.3)}{\approx} x \prod_{i=1}^{d} (h_i x).$$

The variety $\mathbb{V}$ thus satisfies the identity

$$x \prod_{i=1}^{d} (h_i x) \approx x^n \prod_{i=1}^{d} h_i. \tag{12.5}$$

If $d \leq n-1$, then $\mathbb{V}$ satisfies the identity $\circledS_n$ because

$$\begin{aligned} x \prod_{i=1}^{n-1} (h_i x) &\stackrel{(12.5)}{\approx} \left(x^n \prod_{i=1}^{d} h_i\right)\left(\prod_{i=d+1}^{n-1} (h_i x)\right) \\ &\stackrel{(12.2)}{\approx} x^n \prod_{i=1}^{n-1} h_i, \end{aligned}$$

whence $\mathbb{V} \subseteq \mathbb{A}_n^{\mathsf{cen}} \cap \mathbb{S}_n$. If $d > n-1$, then $\mathbb{V}$ satisfies the identity $\circledS_d$ because

$$\begin{aligned} x \prod_{i=1}^{d} (h_i x) &\stackrel{(12.5)}{\approx} x^n \prod_{i=1}^{d} h_i \\ &\stackrel{\blacktriangle_n}{\approx} x^{d+1} \prod_{i=1}^{d} h_i, \end{aligned}$$

whence $\mathbb{V} \subseteq \mathbb{A}_n^{\mathsf{cen}} \cap \mathbb{S}_d \subseteq \mathbb{A}_d^{\mathsf{cen}} \cap \mathbb{S}_d$. □

12.1.3 Limiting Identities

Lemma 12.9 (Head 1968) *Every periodic variety of monoids contains finitely many commutative subvarieties.*

Lemma 12.10 *Any periodic subvariety of $\mathbb{O}$ is locally finite. Consequently, any small subvariety of $\mathbb{O}$ is Cross.*

Proof Since the variety $\mathbb{O}$ satisfies the identity $xyxhxyx \approx xyxhyx^2$ and is not periodic, it follows from M.V. Sapir (1987b, Proposition 3.1) that every periodic subvariety of $\mathbb{O}$ is locally finite.

Suppose that $\mathbb{V}$ is any small subvariety of $\mathbb{O}$. Then $\mathbb{V}$ is periodic and so also locally finite. It follows from Jackson (2005b, Lemma 6.1) that $\mathbb{V}$ is finitely generated. Since $\mathbb{V}$ is finitely based on Theorem 11.1, it is also Cross. □

A *limiting identity* is a rigid identity of the form

$$x\prod_{i=1}^{m}(h_i x) \approx x^{e_0}\prod_{i=1}^{m}(h_i x^{e_i}),$$

where $e_0, e_1, \ldots, e_m \geq 0$ and $m \geq 1$ are such that $e = e_0 + e_1 + \cdots + e_m \leq m$. Any variety of monoids that satisfies such a limiting identity also satisfies the periodicity identity $x^{m+1} \approx x^e$.

Proposition 12.11 *For any limiting identity σ, the variety $\mathbb{O}\{\sigma\}$ is Cross.*

Proof Let $\mathbb{V}$ be any noncommutative subvariety of $\mathbb{O}\{\sigma\}$. Then by Lemma 12.1 and Remark 12.3, the variety $\mathbb{V}$ is of the form $\mathbb{O}(\{\sigma\} \cup \Sigma)$, where Σ is some set of efficient identities from Lemma 12.1. In view of the limiting identity σ, the identities in Σ can be chosen to be n-limited for some fixed $n \geq 1$. Up to renaming of variables, there can only be finitely many n-limited efficient identities from Lemma 12.1. Hence, there are only finitely many choices for Σ. It follows that $\mathbb{O}\{\sigma\}$ contains finitely many noncommutative subvarieties. But the variety $\mathbb{O}\{\sigma\}$ is periodic and, so by Lemma 12.9, contains only finitely many commutative subvarieties. Consequently, the variety $\mathbb{O}\{\sigma\}$ is small and so is Cross by Lemma 12.10. □

Corollary 12.12 *For any $m \geq 1$ and $n \geq 2$, the variety $\mathbb{A}_m^{\mathsf{cen}} \cap \mathbb{O} \cap \mathbb{S}_n$ is Cross.*

Proof Since

$$x\prod_{i=1}^{mn-1}(h_i x) \overset{\mathbb{S}_{mn}}{\approx} x^{mn}\prod_{i=1}^{mn-1} h_i$$

$$\overset{\blacktriangle_m}{\approx} x^{m}\prod_{i=1}^{mn-1} h_i,$$

the variety $\mathbb{A}_m^{\mathsf{cen}} \cap \mathbb{O} \cap \mathbb{S}_{mn}$ satisfies the limiting identity $x\prod_{i=1}^{mn-1}(h_i x) \approx x^m \prod_{i=1}^{mn-1} h_i$ and so is Cross by Proposition 12.11. Therefore, the subvariety $\mathbb{A}_m^{\mathsf{cen}} \cap \mathbb{O} \cap \mathbb{S}_n$ of $\mathbb{A}_m^{\mathsf{cen}} \cap \mathbb{O} \cap \mathbb{S}_{mn}$ is also Cross. □

12.2 The Variety $\mathbb{K}$

The present section is an in-depth study of the variety $\mathbb{K} = \mathbb{A}_2^{\mathsf{cen}} \cap \mathbb{O} \cap \mathbb{O}^{\triangleleft}$ and its subvarieties. Note that $\mathbb{K}$ is the subvariety of $\mathbb{A}_2^{\mathsf{cen}}$ defined by the identities

$$xyhxty \approx yxhxty, \quad xhxyty \approx xhyxty, \quad xhytxy \approx xhytyx. \tag{12.6}$$

In Sect. 12.2.1, it is shown that $\mathbb{K}$ is a non-finitely generated almost Cross variety. In Sect. 12.2.2, the subvarieties of $\mathbb{K}$ are shown to constitute a countably infinite chain.

12.2.1 Almost Cross Property

Lemma 12.13 *The variety $\mathbb{K}$ satisfies a nontrivial rigid identity $\mathbf{u} \approx \mathbf{v}$ if and only if both $\mathbf{u}$ and $\mathbf{v}$ are not square-free.*

Proof Suppose that $\mathbf{u}$ and $\mathbf{v}$ are non-square-free rigid words, say of rank m. Then by Lemma 12.4, the variety $\mathbb{A}_2^{\mathsf{cen}}$ satisfies the identities $\mathbf{u} \approx x^2 \prod_{i=1}^m h_i$ and $\mathbf{v} \approx x^2 \prod_{i=1}^m h_i$. Therefore, both the variety $\mathbb{A}_2^{\mathsf{cen}}$ and its subvariety $\mathbb{K}$ satisfy the identity $\mathbf{u} \approx \mathbf{v}$. Conversely, if either $\mathbf{u}$ or $\mathbf{v}$ is a square-free rigid word, then it is easily shown that the identity $\mathbf{u} \approx \mathbf{v}$ is not deducible from the identity basis $\{\blacktriangle_2, (12.6)\}$ for $\mathbb{K}$, whence $\mathbb{K}$ violates $\mathbf{u} \approx \mathbf{v}$. □

Lemma 12.14 *The following statements on any subvariety $\mathbb{V}$ of $\mathbb{A}^{\mathsf{cen}} \cap \mathbb{O}$ are equivalent:*

(a) *$\mathbb{V}$ is Cross;*
(b) *$\mathbb{V}$ is finitely generated;*
(c) *$\mathbb{V} \subseteq \mathbb{S}_n$ for some $n \geq 2$;*
(d) *$\mathbb{K} \nsubseteq \mathbb{V}$.*

Proof The implication (a) ⇒ (b) is obvious, and the implication (b) ⇒ (c) follows from Lemma 12.6. The implication (c) ⇒ (d) holds because by Lemma 12.13, the variety $\mathbb{K}$ violates the identity Ⓢ$_n$ that defines $\mathbb{S}_n$. Therefore, it remains to establish the implication (d) ⇒ (a).

There exists some $m \geq 1$ such that $\mathbb{V} \subseteq \mathbb{A}_m^{\mathsf{cen}} \cap \mathbb{O}$. Suppose that $\mathbb{K} \nsubseteq \mathbb{V}$. Since by Lemma 12.9, the variety $\mathbb{V}$ is already Cross if it is commutative, it suffices to further assume that $\mathbb{V}$ is noncommutative, so that $m \geq 2$. It then follows from Lemma 12.1 that $\mathbb{V} = (\mathbb{A}_m^{\mathsf{cen}} \cap \mathbb{O})\Sigma$ for some set Σ of identities from Lemma 12.1. Since $\mathbb{K} \subseteq \mathbb{A}_m^{\mathsf{cen}} \cap \mathbb{O}$ and $\mathbb{K} \nsubseteq (\mathbb{A}_m^{\mathsf{cen}} \cap \mathbb{O})\Sigma$, the variety $\mathbb{K}$ must violate some identity in Σ. However, it is easily seen that $\mathbb{K}$ satisfies all identities from Lemma 12.1(b). Therefore, $\mathbb{K}$ violates some nontrivial rigid identity $\mathbf{u} \approx \mathbf{v}$ in Σ. By Lemma 12.13, either $\mathbf{u}$ or $\mathbf{v}$ is square-free. Since

$\mathbb{V}$ satisfies the identity $\mathbf{u} \approx \mathbf{v}$, it follows from Lemma 12.8 that $\mathbb{V} \subseteq \mathbb{A}_n^{\mathsf{cen}} \cap \mathbb{S}_n$ for some $n \geq 2$. Consequently, $\mathbb{V} \subseteq \mathbb{A}_n^{\mathsf{cen}} \cap \mathbb{O} \cap \mathbb{S}_n$ is Cross by Corollary 12.12. □

Proposition 12.15 *The variety $\mathbb{K}$ is non-finitely generated and almost Cross.*

Proof Since $\mathbb{K} \subseteq \mathbb{A}_2^{\mathsf{cen}} \cap \mathbb{O}$, it follows from Lemma 12.14 that $\mathbb{K}$ is non-finitely generated and every proper subvariety of $\mathbb{K}$ is Cross. □

12.2.2 Subvarieties of $\mathbb{K}$

In the present subsection, the subvarieties of $\mathbb{K}$ are shown to constitute a countably infinite chain. For this purpose, the *reduced Straubing identities*

$$x \prod_{i=1}^{n-1} (h_i x) \approx x^2 \prod_{i=1}^{n-1} h_i, \tag{Ⓡ$_n$}$$

where $n \geq 2$, play a crucial role. Let $\mathscr{K} = \{\mathbf{k}_2, \mathbf{k}_3, \mathbf{k}_4, \ldots\}$, where

$$\mathbf{k}_n = x \prod_{i=1}^{n-1} (h_i x)$$

is the word on the left side of the identity Ⓡ$_n$.

Lemma 12.16 *Let $e_0, e_1, \ldots, e_m \geq 0$ and $\ell \geq 2$ be such that $\ell \leq \sum_{i=0}^m e_i$. Then the identity*

$$x^{e_0} \prod_{i=1}^{m} (h_i x^{e_i}) \approx x^2 \prod_{i=1}^{m} h_i$$

is deducible from the identities $\{\blacktriangle_2, Ⓡ_\ell\}$.

Proof Let $e = \sum_{i=0}^m e_i$. Then $e_0, e_1, \ldots, e_m \geq 0$ and $\ell \leq e$ imply that

$$x^{e_0} \prod_{i=1}^{m} (h_i x^{e_i}) = \mathbf{q}_0 \bigg(x \prod_{i=1}^{\ell-1} (\mathbf{q}_i x) \bigg) \bigg(\prod_{i=\ell}^{e-1} (\mathbf{q}_i x) \bigg) \mathbf{q}_e$$

for some $\mathbf{q}_0, \mathbf{q}_1, \ldots, \mathbf{q}_e \in \mathscr{A}_\varnothing^+$ such that $\mathbf{q}_0 \mathbf{q}_1 \cdots \mathbf{q}_e = h_1 h_2 \cdots h_m$. Hence,

$$
\begin{aligned}
x^{e_0}\prod_{i=1}^{m}(h_i x^{e_i}) &\overset{Ⓡ_\ell}{\approx} \mathbf{q}_0\Big(x^2\prod_{i=1}^{\ell-1}\mathbf{q}_i\Big)\Big(\prod_{i=\ell}^{e-1}(\mathbf{q}_i x)\Big)\mathbf{q}_e \\
&\overset{(12.2)}{\approx} x^2\prod_{i=0}^{e}\mathbf{q}_i \qquad \text{by Lemma 12.4 with } n=2 \\
&= x^2\prod_{i=1}^{m}h_i.
\end{aligned}
$$

The result holds since by Lemma 12.4, the identity (12.2) is deducible from $\blacktriangle_2$. □

Lemma 12.17 *The noncommutative subvarieties of* $\mathbb{K}$ *form the chain*

$$\mathbb{K}\{Ⓡ_2\} \subseteq \mathbb{K}\{Ⓡ_3\} \subseteq \cdots \subseteq \mathbb{K}.$$

Proof The inclusions in this chain hold by Lemma 12.16. Let $\mathbb{V}$ be any noncommutative proper subvariety of $\mathbb{K}$. Then the variety $\mathbb{V}$ is Cross because $\mathbb{K}$ is almost Cross by Proposition 12.15. Since $\mathbb{V} \subsetneq \mathbb{K} = \mathbb{A}_2^{\mathsf{cen}} \cap \mathbb{O} \cap \mathbb{O}^{\triangleleft}$, it follows from Lemma 12.1 that $\mathbb{V} = \mathbb{K}\Sigma$ for some set Σ of rigid identities that are violated by $\mathbb{K}$. Since $\mathbb{V}$ is finitely generated, it follows from Lemma 12.6 that $\mathbb{V} \subseteq \mathbb{S}_k$ for some $k \geq 2$. Therefore, $\mathbb{V} \subseteq \mathbb{K} \cap \mathbb{S}_k \subseteq \mathbb{K}\{Ⓡ_k\}$. Let ℓ be the least integer such that $\mathbb{V} \subseteq \mathbb{K}\{Ⓡ_\ell\}$.

Consider any identity $\mathbf{u} \approx \mathbf{v}$ from Σ. Then

$$\mathbf{u} = x^{e_0}\prod_{i=1}^{m}(h_i x^{e_i}) \text{ and } \mathbf{v} = x^{f_0}\prod_{i=1}^{m}(h_i x^{f_i})$$

for some $e_0, f_0, e_1, f_1, \ldots, e_m, f_m \geq 0$ with $(e_0, e_1, \ldots, e_m) \neq (f_0, f_1, \ldots, f_m)$. Further, it suffices to assume that $m \geq 1$ and $(e_0, e_1, \ldots, e_m), (f_0, f_1, \ldots, f_m) \neq (0, 0, \ldots, 0)$ since otherwise, the variety $\mathbb{V}$ is contradictorily commutative by Lemma 12.5. Now the identity $\mathbf{u} \approx \mathbf{v}$ is violated by $\mathbb{K}$ so that by Lemma 12.13, either $\mathbf{u}$ or $\mathbf{v}$ is square-free. Let $e = \sum_{i=0}^{m} e_i$ and $f = \sum_{i=0}^{m} f_i$. Generality is not lost by assuming one of the following cases:

(A) $\mathbf{u}$ is square-free and $\mathbf{v}$ is not square-free;
(B) $\mathbf{u}$ and $\mathbf{v}$ are both square-free with $0 < e \leq f$.

Following the proof of Lemma 12.8, the identity $\mathbf{u} \approx \mathbf{v}$ can be used to deduce the rigid identity

$$x\prod_{i=1}^{e-1}(h_i x) \approx \mathbf{p}x^r\mathbf{q} \tag{12.7}$$

for some $\mathbf{p}, \mathbf{q} \in \mathscr{A}_\varnothing^+$ and $r \geq 2$. Since $\mathbb{V}$ is a subvariety of $\mathbb{A}_2^{\mathsf{cen}}$, it follows that

$$\begin{aligned} x \prod_{i=1}^{e-1} (h_i x) &\stackrel{(12.7)}{\approx} \mathbf{p} x^r \mathbf{q} \\ &\stackrel{(12.2)}{\approx} x^2 \prod_{i=0}^{e-1} \mathbf{q}_i \end{aligned}$$

by Lemma 12.4 with $n = 2$, whence $\mathbb{V}$ satisfies the identity $\circledR_e$. The minimality of ℓ implies that $\ell \leq e$. In (A), since $f_j \geq 2$ for some j,

$$\begin{aligned} \mathbf{u} &= x^{e_0} \prod_{i=1}^{m} (h_i x^{e_i}) \\ &\stackrel{\blacktriangle_2, \circledR_\ell}{\approx} x^2 \prod_{i=1}^{m} h_i && \text{by Lemma 12.16} \\ &\stackrel{(12.2)}{\approx} x^{f_0} \prod_{i=1}^{m} (h_i x^{f_i}) && \text{by Lemma 12.4 with } n = 2 \\ &= \mathbf{v}. \end{aligned}$$

In (B),

$$\begin{aligned} \mathbf{u} &= x^{e_0} \prod_{i=1}^{m} (h_i x^{e_i}) \\ &\stackrel{\blacktriangle_2, \circledR_\ell}{\approx} x^2 \prod_{i=1}^{m} h_i && \text{by Lemma 12.16} \\ &\stackrel{\blacktriangle_2, \circledR_\ell}{\approx} x^{f_0} \prod_{i=1}^{m} (h_i x^{f_i}) && \text{by Lemma 12.16} \\ &= \mathbf{v}. \end{aligned}$$

Therefore, in both cases, the identity $\mathbf{u} \approx \mathbf{v}$ is deducible from $\{\blacktriangle_2, \circledR_\ell\}$. Since the identity $\mathbf{u} \approx \mathbf{v}$ from Σ is arbitrarily chosen, $\mathbb{K}\{\circledR_\ell\} = \mathbb{K}\Sigma = \mathbb{V}$. □

Proposition 12.18 *The subvarieties of* $\mathbb{K}$ *constitute the chain*

$$\mathbf{0} \subset \mathbb{R}_Q \varnothing \subset \mathbb{R}_Q\{x\} \subset \mathbb{R}_Q\{xy\} \subset \mathbb{R}_Q\{\mathbf{k}_2\} \subset \mathbb{R}_Q\{\mathbf{k}_3\} \subset \cdots \subset \mathbb{R}_Q \mathscr{K} = \mathbb{K}. \tag{12.8}$$

Proof An identity basis for $\mathcal{R}_Q\{x\}$ is easily shown to be $\{x^3 \approx x^2, xy \approx yx\}$; it is then routinely verified that $\mathbb{K}\{xy \approx yx\} = \mathbb{R}_Q\{x\}$. As shown in Theorem 1.56, the variety $\mathbb{R}_Q\varnothing$ is the only nontrivial proper subvariety of $\mathbb{R}_Q\{x\}$. Therefore, $\mathbb{R}_Q\varnothing$ and $\mathbb{R}_Q\{x\}$ are the only nontrivial commutative subvarieties of $\mathbb{K}$. It then follows from Lemma 12.17 that the subvarieties of $\mathbb{K}$ constitute the chain

$$\mathbf{0} \subset \mathbb{R}_Q\varnothing \subset \mathbb{R}_Q\{x\} \subset \mathbb{K}\{Ⓡ_2\} \subseteq \mathbb{K}\{Ⓡ_3\} \subseteq \cdots \subseteq \mathbb{K}.$$

It is known that $\mathbb{R}_Q\{xy\} = \mathbb{K}\{Ⓡ_2\}$ (Jackson 2005b, Lemma 4.5(ii)). For each $n \geq 2$, it is routinely checked that the monoid $\mathcal{R}_Q\{\mathbf{k}_n\}$ satisfies the identities $\{\blacktriangle_2, (12.6), Ⓡ_{n+1}\}$ but violates the identity $Ⓡ_n$, whence $\mathbb{R}_Q\{\mathbf{k}_n\} = \mathbb{K}\{Ⓡ_{n+1}\} \neq \mathbb{K}\{Ⓡ_n\}$ and $\mathbb{R}_Q\mathscr{K} = \mathbb{K} \neq \mathbb{K}\{Ⓡ_n\}$. Consequently, the subvarieties of $\mathbb{K}$ constitute the chain (12.8). □

Corollary 12.19 *The variety $\mathbb{K}$ is a subvariety of $\mathbb{B}_2^1$.*

Proof It is routinely checked that for each $n \geq 2$, the word $\mathbf{k}_n = x\prod_{i=1}^{n-1}(h_i x)$ is an isoterm for $\mathcal{B}_2^1$, so that $\mathcal{R}_Q\{\mathbf{k}_n\} \in \mathbb{B}_2^1$ by Lemma 2.6. Hence, the inclusion $\mathbb{K} \subseteq \mathbb{B}_2^1$ holds by Proposition 12.18. □

12.3 Cross Subvarieties of $\mathbb{A}^{\mathsf{cen}}$

In the present section, the following statements on any subvariety $\mathbb{V}$ of $\mathbb{A}^{\mathsf{cen}}$ are shown to be equivalent:

(a) $\mathbb{V}$ is Cross;
(b) $\mathbb{J}_1, \mathbb{J}_2, \mathbb{K} \nsubseteq \mathbb{V}$;
(c) $\mathbb{V} \subseteq \mathbb{A}_n^{\mathsf{cen}} \cap \mathbb{O} \cap \mathbb{S}_n$ or $\mathbb{V} \subseteq \mathbb{A}_n^{\mathsf{cen}} \cap \mathbb{O}^{\lhd} \cap \mathbb{S}_n$ for some $n \geq 2$.

The proof of Theorem 1.64 is thus established.

The implication (c) $\Rightarrow$ (a) holds by Corollary 12.12, and the implication (a) $\Rightarrow$ (b) holds because $\mathbb{J}_1$, $\mathbb{J}_2$, and $\mathbb{K}$ are non-Cross varieties. Hence, it remains to establish the implication (b) $\Rightarrow$ (c). Let $m \geq 1$ be such that $\mathbb{V} \subseteq \mathbb{A}_m^{\mathsf{cen}}$. Suppose that $\mathbb{J}_1, \mathbb{J}_2, \mathbb{K} \nsubseteq \mathbb{V}$. Then by Theorem 11.16, either $\mathbb{V} \subseteq \mathbb{O}$ or $\mathbb{V} \subseteq \mathbb{O}^{\lhd}$. It thus follows from Lemma 12.14 that $\mathbb{V} \subseteq \mathbb{S}_r$ for some $r \geq 2$. Consequently, either $\mathbb{V} \subseteq \mathbb{A}_n^{\mathsf{cen}} \cap \mathbb{O} \cap \mathbb{S}_n$ or $\mathbb{V} \subseteq \mathbb{A}_n^{\mathsf{cen}} \cap \mathbb{O}^{\lhd} \cap \mathbb{S}_n$ with $n = \max\{m, r\}$.

12.4 Varieties Inherently Non-Finitely Generated Within $\mathbb{A}^{\mathsf{cen}}$

In the present section, the following statements on any subvariety $\mathbb{V}$ of $\mathbb{A}^{\mathsf{cen}}$ are shown to be equivalent:

(a) $\mathbb{V}$ is inherently non-finitely generated within $\mathbb{A}^{\mathsf{cen}}$;
(b) $\mathbb{K} \subseteq \mathbb{V}$;
(c) $\mathbb{V} \nsubseteq \mathbb{S}_n$ for all $n \geq 2$.

The proof of Theorem 1.65 is thus established.

(a) $\Rightarrow$ (c). Suppose that $\mathbb{V} \subseteq \mathbb{S}_n$ for some $n \geq 2$. Then $\mathbb{V} \subseteq \mathbb{A}_m^{\mathsf{cen}} \cap \mathbb{S}_m$ for some $m \geq n$. Since the variety $\mathbb{A}_m^{\mathsf{cen}} \cap \mathbb{S}_m$ is finitely generated by Lemma 12.7, its subvariety $\mathbb{V}$ is not inherently non-finitely generated within $\mathbb{A}^{\mathsf{cen}}$.

(c) $\Rightarrow$ (a). By Lemma 12.6, every finitely generated subvariety of $\mathbb{A}^{\mathsf{cen}}$ is contained in some $\mathbb{S}_n$. Therefore, if $\mathbb{V} \nsubseteq \mathbb{S}_n$ for all $n \geq 2$, then the variety $\mathbb{V}$ is not contained in any finitely generated subvariety of $\mathbb{A}^{\mathsf{cen}}$, whence $\mathbb{V}$ is inherently non-finitely generated within $\mathbb{A}^{\mathsf{cen}}$.

(b) $\Rightarrow$ (c). Suppose that $\mathbb{K} \subseteq \mathbb{V}$. Then $\mathcal{R}_{\mathbb{Q}}\mathscr{K} \in \mathbb{V}$ by Proposition 12.18, so that $\mathscr{K} \subseteq \mathsf{iso}\,\mathbb{V}$ by Lemma 2.6. It follows that $\mathbb{V}$ violates the identity Ⓢ$_n$ for all $n \geq 2$, whence $\mathbb{V} \nsubseteq \mathbb{S}_n$ for all $n \geq 2$.

(c) $\Rightarrow$ (b). Suppose that $\mathbb{K} \nsubseteq \mathbb{V}$. Then it follows from Proposition 12.18 that $\mathcal{R}_{\mathbb{Q}}\{\mathbf{k}_n\} \notin \mathbb{V}$ for some $n \geq 2$. Therefore, $\mathbf{k}_n = x\prod_{i=1}^{n-1}(h_i x) \notin \mathsf{iso}\,\mathbb{V}$ by Lemma 2.6, so that the variety $\mathbb{V}$ satisfies some nontrivial identity $x\prod_{i=1}^{n-1}(h_i x) \approx \mathbf{w}$. Consider the following conditions on $\mathbf{w}$:

(A) $\mathsf{con}(\mathbf{w}) = \{x, h_1, h_2, \ldots, h_{n-1}\}$;
(B) $\mathsf{sim}(\mathbf{w}) = \{h_1, h_2, \ldots, h_{n-1}\}$;
(C) $\mathbf{w}[h_1, h_2, \ldots, h_{n-1}] = h_1 h_2 \cdots h_{n-1}$.

If (A)–(C) hold, then $\mathbf{w}$ is a rigid word of rank $n-1$ different from $\mathbf{k}_n$, so that by Lemma 12.8, the inclusion $\mathbb{V} \subseteq \mathbb{S}_m$ holds for some $m \geq 2$. If any of (A)–(C) fails, then it is straightforwardly shown that the variety $\mathbb{V}$ is either commutative or idempotent, whence $\mathbb{V} \subseteq \mathbb{S}_2$.

12.5 A Non-Finitely Generated Subvariety of $\mathbb{R}_{\mathbb{Q}}\{x^2y^2\}$

Let $\mathbb{W}$ denote the variety of monoids defined by the identities

$$x^2 y \approx y x^2, \quad xhxtx \approx x^2 ht, \tag{12.9a}$$

$$xyhxty \approx yxhxty, \tag{12.9b}$$

$$xhytxy \approx xhytyx. \tag{12.9c}$$

In the present section, the variety $\mathbb{W}$ is shown to be non-finitely generated. But $\mathbb{W}$ is not inherently non-finitely generated within $\mathbb{A}^{\mathsf{cen}}$ because

$$\begin{aligned}\mathbb{W} &\subseteq (\mathbb{A}_3^{\mathsf{cen}} \cap \mathbb{S}_3)\{(12.9\text{b}), (12.9\text{c})\}\\ &= \mathbb{R}_{\mathcal{Q}}\{x^2y^2\},\end{aligned}$$

where the inclusion follows from the easily established deduction (12.9a) $\vdash \{\blacktriangle_3, \circledS_3\}$ and the equality holds by O.B. Sapir (2019, Lemma 3.2(i)).

Lemma 12.20 *The variety* $\mathbb{W}$ *satisfies the identities*

$$x^2hxtx \approx xhxtx, \quad xhx^2tx \approx xhxtx, \quad xhxtx^2 \approx xhxtx, \tag{12.10a}$$

$$xh^2yxty \approx x^2h^2yx^2ty, \tag{12.10b}$$

$$xhyxt^2y \approx xhy^2xt^2y^2. \tag{12.10c}$$

Proof It is easily shown that the variety $\mathbb{W}$ satisfies the identities (12.10a). Since

$$\begin{aligned}xh^2yxty &\overset{(12.9\text{a})}{\approx} h^2xyxty\\ &\overset{(12.9\text{b})}{\approx} h^2yx^2ty\\ &\overset{(12.9\text{a})}{\approx} h^2yx^4ty\\ &\overset{(12.9\text{a})}{\approx} x^2h^2yx^2ty,\end{aligned}$$

the variety $\mathbb{W}$ satisfies the identity (12.10b). By symmetry, the variety $\mathbb{W}$ also satisfies the identity (12.10c). □

Lemma 12.21 *For each* $n \geq 2$, *the variety* $\mathbb{W}$ *violates the identity* $\mathbf{g}_n \approx \mathbf{g}_n'$, *where*

$$\mathbf{g}_n = x_0h\Bigg(\prod_{i=0}^{n}(x_{i+1}x_i)\Bigg)tx_{n+1} = x_0h \cdot x_1x_0 \cdot x_2x_1 \cdots x_{n+1}x_n \cdot tx_{n+1}$$

$$\textit{and } \mathbf{g}_n' = x_0hx_0\Bigg(\prod_{i=0}^{n}x_i^2\Bigg)x_{n+1}tx_{n+1} = x_0hx_0 \cdot x_1^2x_2^2 \cdots x_n^2 \cdot x_{n+1}tx_{n+1}.$$

Proof First observe that the 2-limited word $\mathbf{g}_n$ excludes factors of the form $\mathbf{x}^2$, $\mathbf{xhytxy}$, and $\mathbf{xyhxty}$, where $\mathbf{x}, \mathbf{y} \in \mathscr{A}^+$ and $\mathbf{h}, \mathbf{t} \in \mathscr{A}_\varnothing^+$. It is then easily seen that it is impossible to convert $\mathbf{g}_n$ into a different word by applying only the identities (12.9). It follows that the variety $\mathbb{W}$ violates the identity $\mathbf{g}_n \approx \mathbf{g}_n'$. □

Lemma 12.22 *Any finite monoid in the variety* $\mathbb{W}$ *satisfies the identity* $\mathbf{g}_n \approx \mathbf{g}_n'$ *for all sufficiently large* $n \geq 2$.

Proof Let M be any finite monoid in the variety $\mathbb{W}$ and fix any $n > |M|$. Suppose that φ is any substitution from $\mathscr{A}$ into M. Then it is shown in the following that $\mathbf{g}_n\varphi = \mathbf{g}_n'\varphi$ in M. Consequently, the monoid M satisfies the identity $\mathbf{g}_n \approx \mathbf{g}_n'$.

For notational brevity, write $x\varphi = \hat{x}$ for any $x \in \mathscr{A}$. Since $n > |M|$, the list $\hat{x}_1, \hat{x}_2, \ldots, \hat{x}_n$ of elements from M must contain some repetition, say $\hat{x}_i = \hat{x}_j$ with $1 \leq i < j \leq n$.

CASE 1: $1 < i < j \leq n$. Note that the variable x_i occurs twice in the word $\mathbf{g}_n$. Since $\hat{x}_i = \hat{x}_j$, the element $\hat{x}_j$ appears at least thrice in the product $\mathbf{g}_n\varphi$, whence the identities (12.10a) can be applied to replace any $\hat{x}_i$ in $\mathbf{g}_n\varphi$ by $\hat{x}_i^2$:

$$\begin{aligned}\mathbf{g}_n\varphi &= \cdots \hat{x}_{i-1}\hat{x}_{i-2} \cdot \hat{x}_i\hat{x}_{i-1} \cdot \hat{x}_{i+1}\hat{x}_i \cdot \hat{x}_{i+2}\hat{x}_{i+1} \cdots \\ &\overset{(12.10a)}{=} \cdots \hat{x}_{i-1}\hat{x}_{i-2} \cdot \hat{x}_i^2\hat{x}_{i-1} \cdot \hat{x}_{i+1}\hat{x}_i^2 \cdot \hat{x}_{i+2}\hat{x}_{i+1} \cdots .\end{aligned}$$

Then the identity (12.10b) can be applied to replace $\hat{x}_{i+1}$ by $\hat{x}_{i+1}^2$, and the identity (12.10c) can be applied to replace $\hat{x}_{i-1}$ by $\hat{x}_{i-1}^2$:

$$\begin{aligned}&\cdots \hat{x}_{i-1}\hat{x}_{i-2} \cdot \hat{x}_i^2\hat{x}_{i-1} \cdot \hat{x}_{i+1}\hat{x}_i^2 \cdot \hat{x}_{i+2}\hat{x}_{i+1} \cdots \\ \overset{(12.10b)}{=}\; &\cdots \hat{x}_{i-1}\hat{x}_{i-2} \cdot \hat{x}_i^2\hat{x}_{i-1} \cdot \hat{x}_{i+1}^2\hat{x}_i^2 \cdot \hat{x}_{i+2}\hat{x}_{i+1}^2 \cdots \\ \overset{(12.10c)}{=}\; &\cdots \hat{x}_{i-1}^2\hat{x}_{i-2} \cdot \hat{x}_i^2\hat{x}_{i-1}^2 \cdot \hat{x}_{i+1}^2\hat{x}_i^2 \cdot \hat{x}_{i+2}\hat{x}_{i+1}^2 \cdots .\end{aligned}$$

This procedure can be repeated until $\hat{x}_1, \hat{x}_2, \ldots, \hat{x}_n$ are replaced by $\hat{x}_1^2, \hat{x}_2^2, \ldots, \hat{x}_n^2$. Hence,

$$\begin{aligned}\mathbf{g}_n\varphi &= \hat{x}_0\hat{h} \cdot \hat{x}_1^2\hat{x}_0 \cdot \hat{x}_2^2\hat{x}_1^2 \cdot \hat{x}_3^2\hat{x}_2^2 \cdots \hat{x}_n^2\hat{x}_{n-1}^2 \cdot \hat{x}_{n+1}\hat{x}_n^2 \cdot \hat{t}\hat{x}_{n+1} \\ &\overset{(12.9a)}{=} \hat{x}_0\hat{h}\hat{x}_0 \cdot \hat{x}_1^2\hat{x}_2^2 \cdots \hat{x}_n^2 \cdot \hat{x}_{n+1}\hat{t}\hat{x}_{n+1} \\ &= \mathbf{g}_n'\varphi.\end{aligned}$$

CASE 2: $1 = i < j \leq n$. Note that the variable x_1 occurs twice in the word $\mathbf{g}_n$. Since $\hat{x}_1 = \hat{x}_j$, the element $\hat{x}_1$ appears at least thrice in the product $\mathbf{g}_n\varphi$, whence the identities (12.10a) can be applied to replace any $\hat{x}_1$ in $\mathbf{g}_n\varphi$ by $\hat{x}_1^2$:

$$\begin{aligned}\mathbf{g}_n\varphi &= \hat{x}_0\hat{h}\cdot\hat{x}_1\hat{x}_0\cdot\hat{x}_2\hat{x}_1\cdot\hat{x}_3\hat{x}_2\cdots \\ &\overset{(12.10a)}{=} \hat{x}_0\hat{h}\cdot\hat{x}_1^2\hat{x}_0\cdot\hat{x}_2\hat{x}_1^2\cdot\hat{x}_3\hat{x}_2\cdots.\end{aligned}$$

The identity (12.10b) can then be applied to replace $\hat{x}_2$ by $\hat{x}_2^2$:

$$\hat{x}_0\hat{h}\cdot\hat{x}_1^2\hat{x}_0\cdot\hat{x}_2\hat{x}_1^2\cdot\hat{x}_3\hat{x}_2\cdots \overset{(12.10b)}{=} \hat{x}_0\hat{h}\cdot\hat{x}_1^2\hat{x}_0\cdot\hat{x}_2^2\hat{x}_1^2\cdot\hat{x}_3\hat{x}_2^2\cdots.$$

This procedure can be repeated until $\hat{x}_1, \hat{x}_2, \ldots, \hat{x}_n$ are replaced by $\hat{x}_1^2, \hat{x}_2^2, \ldots, \hat{x}_n^2$. The equality $\mathbf{g}_n\varphi = \mathbf{g}_n'\varphi$ is then obtained as in Case 1. □

Proposition 12.23 *The variety* $\mathbb{W}$ *is non-finitely generated.*

Proof Suppose that the variety $\mathbb{W}$ is finitely generated. Then by Lemma 12.22, it satisfies the identity $\mathbf{g}_n \approx \mathbf{g}_n'$ for some $n \geq 2$. But this is impossible by Lemma 12.21. □

12.6 Summary

The varieties $\mathbb{J}_1$ and $\mathbb{J}_2$ were first shown in Lee (2009a) to be the only finitely generated limit subvarieties of $\mathbb{A}^{\mathsf{cen}}$, and they were later shown in Lee (2011a) to be the only finitely generated Cross subvarieties of $\mathbb{A}^{\mathsf{cen}}$. The discovery of the non-finitely generated almost Cross variety $\mathbb{K}$ and the characterization in Sect. 12.3 of all Cross subvarieties of $\mathbb{A}^{\mathsf{cen}}$ were published in Lee (2013a). The description in Proposition 12.18 of the lattice of subvarieties of $\mathbb{K}$, the characterization in Sect. 12.4 of all varieties that are inherently non-finitely generated within $\mathbb{A}^{\mathsf{cen}}$, and the non-finitely generated variety $\mathbb{W}$ in Sect. 12.5 were published in Lee (2014a).

13 Certain Cross Varieties of Aperiodic Monoids with Commuting Idempotents

The present chapter is concerned with the class $\mathbb{A}^{\mathsf{com}}$ of aperiodic monoids with commuting idempotents. Recall from Theorem 1.64 that the exclusion of the almost Cross varieties $\mathbb{J}_1$, $\mathbb{J}_2$, and $\mathbb{K}$ is both necessary and sufficient for any subvariety of $\mathbb{A}^{\mathsf{cen}}$ to be Cross. The main goal of the chapter is to generalize this theorem to a result on subvarieties of $\mathbb{A}^{\mathsf{com}}$ that satisfy the identity

$$x^2yx \approx xyx^2. \qquad (\rightleftarrows)$$

Theorem 13.1 *A subvariety of* $\mathbb{A}^{\mathsf{com}}\{\rightleftarrows\}$ *is Cross if and only if it excludes* $\mathbb{J}_1$, $\mathbb{J}_2$, *and* $\mathbb{K}$.

The proof of Theorem 13.1 relies crucially upon the variety $\mathbb{Q}^1$ generated by the monoid $\mathcal{Q}^1$ obtained from the $\mathscr{J}$-trivial semigroup

$$\mathcal{Q} = \langle \mathsf{a}, \mathsf{b}, \mathsf{e} \mid \mathsf{ae} = \mathsf{ba} = \mathsf{eb} = 0,\ \mathsf{be} = \mathsf{b},\ \mathsf{ea} = \mathsf{a},\ \mathsf{e}^2 = \mathsf{e} \rangle;$$

see Table 13.1. In Sect. 13.1, the exclusion of the variety $\mathbb{Q}^1$ from any subvariety of $\mathbb{A}^{\mathsf{com}}$ is investigated, and $\mathbb{Q}^1$ is shown to be Cross. Subvarieties of $\mathbb{A}^{\mathsf{com}}\{\rightleftarrows\}$ that contain the variety $\mathbb{Q}^1$ are examined in Sect. 13.2, and subvarieties of $\mathbb{A}^{\mathsf{com}}\{\rightleftarrows\}$ that exclude the almost Cross variety $\mathbb{K}$ are examined in Sect. 13.3. Based on the results in Sects. 13.1–13.3, the proof of Theorem 13.1 is established in Sect. 13.4.

E. W. H. Lee, *Advances in the Theory of Varieties of Semigroups*, Frontiers in Mathematics, https://doi.org/10.1007/978-3-031-16497-2_13

Table 13.1 Multiplication table of $\mathcal{Q}$

$\mathcal{Q}$	0	ab	b	a	e
0	0	0	0	0	0
ab	0	0	0	0	ab
b	0	0	0	0	b
a	0	0	ab	0	0
e	0	ab	0	a	e

13.1 The Variety $\mathbb{Q}^1$

For each $n \geq 1$, let $\mathbb{A}_n^{\mathsf{com}}$ denote the variety of monoids defined by the identities

$$x^{n+1} \approx x^n, \quad x^n y^n \approx y^n x^n. \tag{$\blacklozenge_n$}$$

The variety $\mathbb{Q}^1$ is defined by the identities

$$x^3 \approx x^2, \quad x^2yx \approx xyx, \quad xyx^2 \approx xyx, \quad x^2y^2 \approx y^2x^2 \tag{13.1}$$

and is hereditarily finitely based (Lee and Li 2011, Chap. 4). It is routinely checked that $\mathbb{Q}^1$ is a subvariety of $\mathbb{A}_2^{\mathsf{com}}\{\rightleftarrows\} \cap \mathbb{O} \cap \mathbb{O}^{\triangleleft}$. In this section, some results regarding subvarieties of $\mathbb{A}^{\mathsf{com}}$ that exclude $\mathbb{Q}^1$ are established. The variety $\mathbb{Q}^1$ is also shown to be Cross.

A word $\mathbf{w}$ is said to be in *regular form* if

$$\mathbf{w} = \mathbf{w}_0 \prod_{i=1}^{m} (h_i \mathbf{w}_i),$$

where $\mathsf{sim}(\mathbf{w}) = \{h_1, h_2, \ldots, h_m\}$, $\mathbf{w}_0, \mathbf{w}_1, \ldots, \mathbf{w}_m \in \mathsf{non}(\mathbf{w})^+ \cup \{\varnothing\}$, and $m \geq 0$.

Lemma 13.2 *Suppose that the words*

$$\mathbf{u} = \mathbf{u}_0 \prod_{i=1}^{m} (h_i \mathbf{u}_i) \quad \textit{and} \quad \mathbf{v} = \mathbf{v}_0 \prod_{i=1}^{k} (t_i \mathbf{v}_i)$$

are in regular form. Then the monoid $\mathcal{Q}^1$ *satisfies the identity* $\mathbf{u} \approx \mathbf{v}$ *if and only if the following conditions hold*:

(a) $m = k$ *and* $h_i = t_i$ *for all* $i \geq 1$;
(b) $\mathsf{con}(\mathbf{u}_i) = \mathsf{con}(\mathbf{v}_i)$ *for all* $i \geq 0$.

Proof This is routine; see Lee and Li (2011, proof of Propsition 4.3). □

Corollary 13.3 *The monoid $\mathcal{Q}^1$ satisfies an efficient rigid identity*

$$x^{e_0}\prod_{i=1}^{m}(h_i x^{e_i}) \approx x^{f_0}\prod_{i=1}^{m}(h_i x^{f_i})$$

of rank $m \geq 1$ if and only if $e_0, f_0, e_1, f_1, \ldots, e_m, f_m \geq 1$; it satisfies a rigid identity $x^{e_0} \approx x^{f_0}$ of rank 0 if and only if either $e_0 = f_0$ or $e_0, f_0 \geq 2$.

Lemma 13.4 *A subvariety of $\mathbb{A}_n^{\mathsf{com}}$ excludes the variety $\mathbb{Q}^1$ if and only if it satisfies the identity*

$$x^n yxzx^n \approx x^n yzx^n. \tag{13.2}$$

Proof The monoid $\mathcal{Q}^1$ violates the identity (13.2) under the substitution $(x, y, z) \mapsto (\mathsf{e}, \mathsf{a}, \mathsf{b})$. Therefore, any variety that satisfies the identity (13.2) must exclude $\mathcal{Q}^1$.

Conversely, let $\mathbb{V}$ be any subvariety of $\mathbb{A}_n^{\mathsf{com}}$ such that $\mathcal{Q}^1 \notin \mathbb{V}$. If the variety $\mathbb{V}$ is either commutative or idempotent, then it satisfies the identity (13.2) because

$$\{\blacklozenge_n, xy \approx yx\} \vdash x^n yxzx^n \approx x^{n+1}yzx^n \approx x^n yzx^n \vdash (13.2);$$
$$\{\blacklozenge_n, x^2 \approx x\} \vdash x^n yxzx^n \approx x^n y^n x^n zx^n \approx x^n x^n y^n zx^n \approx x^n yzx^n \vdash (13.2).$$

Therefore, it suffices to assume that $\mathbb{V}$ is neither commutative nor idempotent, whence $n \geq 2$. By assumption, there exists an identity $\mathbf{u} \approx \mathbf{v}$ that is satisfied by $\mathbb{V}$ but is violated by $\mathcal{Q}^1$. Suppose that when written in regular form, $\mathbf{u}$ and $\mathbf{v}$ are the words in Lemma 13.2, so that either (a) or (b) of this lemma does not hold. If (a) does not hold, then $\mathsf{sim}(\mathbf{u}) \neq \mathsf{sim}(\mathbf{v})$ or $\mathbf{u}[\mathsf{sim}(\mathbf{u})] \neq \mathbf{v}[\mathsf{sim}(\mathbf{v})]$, whence $\mathbb{V}$ is contradictorily idempotent or commutative. Thus, assume that (a) holds but (b) does not hold. Then $\mathsf{con}(\mathbf{u}_i) \neq \mathsf{con}(\mathbf{v}_i)$ for some i, say $x \in \mathsf{con}(\mathbf{u}_i)\backslash \mathsf{con}(\mathbf{v}_i)$. There are three cases.

CASE 1: $0 < i < m$. Then $\mathbf{u}[x, h_i, h_{i+1}] = x^p h_i x^q h_{i+1} x^r$ and $\mathbf{v}[x, h_i, h_{i+1}] = x^s h_i h_{i+1} x^t$ for some $p, q, r, s, t \geq 1$. Since $\mathbb{V}$ satisfies the identity $\mathbf{u}[x^n, h_i, h_{i+1}] \approx \mathbf{v}[x^n, h_i, h_{i+1}]$, where

$$\mathbf{u}[x^n, h_i, h_{i+1}] \overset{\blacklozenge_n}{\approx} x^n h_i x^n h_{i+1} x^n \text{ and } \mathbf{v}[x^n, h_i, h_{i+1}] \overset{\blacklozenge_n}{\approx} x^n h_i h_{i+1} x^n,$$

it also satisfies the identity $\sigma : x^n h_i x^n h_{i+1} x^n \approx x^n h_i h_{i+1} x^n$. Then $\mathbb{V}$ satisfies (13.2) because

$$x^n yxzx^n \overset{\sigma}{\approx} x^n yx^{n+1}zx^n$$

$$\overset{\blacklozenge_n}{\approx} x^n y x^n z x^n$$

$$\overset{\sigma}{\approx} x^n y z x^n.$$

CASE 2: $i = 0$. Then $\mathbf{u}[x, h_1] = x^p h_1 x^q$ and $\mathbf{v}[x, h_1] = h_1 x^r$ for some $p, q, r \geq 1$. Since $\mathbb{V}$ satisfies the identity $x^n h_0(\mathbf{u}[x^n, h_1]) \approx x^n h_0(\mathbf{v}[x^n, h_1])$, where

$$x^n h_0(\mathbf{u}[x^n, h_1]) \overset{\blacklozenge_n}{\approx} x^n h_0 x^n h_1 x^n \quad \text{and} \quad x^n h_0(\mathbf{v}[x^n, h_1]) \overset{\blacklozenge_n}{\approx} x^n h_0 h_1 x^n,$$

it also satisfies the identity σ in Case 1. Therefore, $\mathbb{V}$ satisfies the identity (13.2).

CASE 3: $i = m$. This is symmetrical to Case 2. □

Proposition 13.5 *The variety* $\mathbb{Q}^1$ *is Cross.*

Proof By Lemma 13.4, any proper subvariety of $\mathbb{Q}^1$ satisfies the identities $\{(13.1), (13.2)\}$. Since

$$xyxzx \overset{(13.1)}{\approx} x^n yxzx^n \overset{(13.2)}{\approx} x^n yzx^n \overset{(13.1)}{\approx} xyzx,$$

any proper subvariety of $\mathbb{Q}^1$ satisfies the limiting identity $\sigma : xyxzx \approx xyzx$. Therefore, $\mathbb{Q}^1\{\sigma\}$ is the unique maximal proper subvariety of $\mathbb{Q}^1$. As observed at the beginning of this section, the inclusion $\mathbb{Q}^1 \subseteq \mathbb{O}$ holds, so that $\mathbb{Q}^1\{\sigma\} \subseteq \mathbb{O}\{\sigma\}$. Since the variety $\mathbb{O}\{\sigma\}$ is Cross by Proposition 12.11, its subvariety $\mathbb{Q}^1\{\sigma\}$ and the variety $\mathbb{Q}^1$ are also Cross. □

By referring to the identity basis for the monoid $\mathcal{B}_0^1$ (Edmunds 1977, Proposition 3.1(i)), it is straightforwardly deduced that the maximal subvariety $\mathbb{Q}^1\{xyxzx \approx xyzx\}$ of $\mathbb{Q}^1$ coincides with the variety $\mathbb{B}_0^1 = \mathscr{V}_{\mathsf{mon}}\{\mathcal{B}_0^1\}$. A description of the lattice $\mathfrak{L}(\mathbb{B}_0^1)$ can be found in Lee (2008b, Fig. 4). It follows that the subvarieties of $\mathbb{Q}^1$ constitute the lattice in Fig. 13.1. The varieties $\mathbb{U}_1$ and $\mathbb{U}_2$ are generated by the submonoids $\mathcal{B}_0^1\backslash\{\mathsf{e}\}$ and $\mathcal{B}_0^1\backslash\{\mathsf{f}\}$ of $\mathcal{B}_0^1$, respectively.

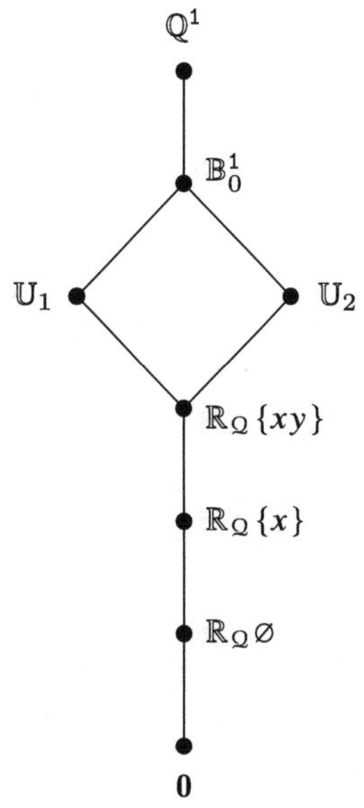

Fig. 13.1 The lattice $\mathfrak{L}(\mathbb{Q}^1)$

13.2 Varieties that Contain $\mathbb{Q}^1$

The present section is concerned with subvarieties of $\mathbb{A}^{\mathsf{com}}\{\rightleftarrows\}$ that satisfy the rigid identity

$$x^{e+1}\prod_{i=1}^{r}(h_i x) \approx x^{f+1}\prod_{i=1}^{r}(h_i x), \qquad (\rho_r^{e,f})$$

where $e, f, r \geq 0$ with $e \geq f$. Note that each rigid identity $\rho_r^{e,f}$ is efficient.

Remark 13.6

(i) By Corollary 13.3, the identity $\rho_r^{e,f}$ is violated by the variety $\mathbb{Q}^1$ if and only if $e > f = r = 0$.
(ii) By referring to the identity basis (13.1) for the variety $\mathbb{Q}^1$, it is easily shown that $\mathbb{Q}^1$ satisfies every identity from Lemma 12.1(b). It follows that all results in the present section concern varieties that contain $\mathbb{Q}^1$.

Lemma 13.7 *Let θ be any efficient rigid identity that is satisfied by the variety $\mathbb{Q}^1$. Then*

$$\mathbb{A}_n^{\mathsf{com}}\{\rightleftarrows, \theta\} = \mathbb{A}_n^{\mathsf{com}}\{\rightleftarrows, \rho_r^{e,f}\}$$

for some $e, f, r \geq 0$ such that $r \geq 1$ or $e = f$ or $f \geq 1$.

Proof By assumption, θ is an efficient rigid identity of rank $r \geq 0$:

$$x^{e_0} \prod_{i=1}^{r} (h_i x^{e_i}) \approx x^{f_0} \prod_{i=1}^{r} (h_i x^{f_i}),$$

where $e_0, f_0, e_1, f_1, \ldots, e_r, f_r \geq 0$ with $(e_0, f_0), (e_1, f_1), \ldots, (e_r, f_r) \neq (0, 0)$.

CASE 1: $r \geq 1$. Then $e_0, f_0, e_1, f_1, \ldots, e_r, f_r \geq 1$ by Corollary 13.3. For any $j \geq 1$ such that $e_j \geq 2$, the identity $\rightleftarrows$ can be used to group the prefix $x^{e_j - 1}$ of x^{e_j} with x^{e_0}:

$$\begin{aligned} & x^{e_0} \left(\prod_{i=1}^{j-1} (h_i x^{e_i}) \right) h_j x^{e_j} \left(\prod_{i=j+1}^{r} (h_i x^{e_i}) \right) \\ \overset{\rightleftarrows}{\approx} \; & x^{e_0 + e_j - 1} \left(\prod_{i=1}^{j-1} (h_i x^{e_i}) \right) h_j x \left(\prod_{i=j+1}^{r} (h_i x^{e_i}) \right). \end{aligned}$$

This can be repeated on any of the exponents $e_1, f_1, e_2, f_2, \ldots, e_m, f_m$ that is greater than 1, so that the identity θ is converted by $\rightleftarrows$ into the identity $\rho_r^{e,f}$ with $e = (\sum_{i=0}^{r} e_i) - r - 1$ and $f = (\sum_{i=0}^{r} f_i) - r - 1$, whence $\mathbb{A}_n^{\mathsf{com}}\{\rightleftarrows, \theta\} = \mathbb{A}_n^{\mathsf{com}}\{\rightleftarrows, \rho_r^{e,f}\}$.

CASE 2: $r = 0$. Then the identity θ is $x^{e_0} \approx x^{f_0}$ with $(e_0, f_0) \neq (0, 0)$. Therefore, by Corollary 13.3, either $e_0 = f_0$ or $e_0, f_0 \geq 2$. It follows that $\mathbb{A}_n^{\mathsf{com}}\{\rightleftarrows, \theta\} = \mathbb{A}_n^{\mathsf{com}}\{\rightleftarrows, \rho_0^{e,f}\}$ for some $e, f \geq 0$ such that $e = f$ or $f \geq 1$. □

Lemma 13.8 *For any fixed $n \geq 2$ and $r \geq 0$, there exist only finitely many varieties of the form $\mathbb{A}_n^{\mathsf{com}}(\{\rightleftarrows, \rho_r^{1,0}\} \cup \Theta)$, where Θ is any set of efficient rigid identities satisfied by $\mathbb{Q}^1$.*

Proof Let θ be any efficient rigid identity satisfied by the variety $\mathbb{Q}^1$. Then Lemma 13.7 implies that $\mathbb{A}_n^{\mathsf{com}}\{\rightleftarrows, \theta\} = \mathbb{A}_n^{\mathsf{com}}\{\rightleftarrows, \rho_s^{e,f}\}$ for some $e, f, s \geq 0$. Therefore, generality is not lost by assuming that $\Theta \subseteq \{\rho_s^{e,f} \mid e, f, s \geq 0\}$ to begin with. Let $\mathbf{u} = x^{e+1} \prod_{i=1}^{s}(h_i x)$ and $\mathbf{v} = x^{f+1} \prod_{i=1}^{s}(h_i x)$, so that $\mathbf{u} \approx \mathbf{v}$ is the identity $\rho_s^{e,f}$. If $\mathsf{occ}(x, \mathbf{u}) = e + 1 + s > r + 1$, then the identity $\rho_r^{1,0}$ can be used to reduce the exponent e in $\mathbf{u}$ to some e' with $e' + 1 + s = r + 1$:

$$\mathbf{u} = x^{e+1} \prod_{i=1}^{s}(h_i x) \overset{\rho_r^{1,0}}{\approx} x^{e'+1} \prod_{i=1}^{s}(h_i x).$$

Similarly, if $\mathsf{occ}(x, \mathbf{v}) = f + 1 + s > r + 1$, then the identity $\rho_r^{1,0}$ can be used to reduce the exponent f in $\mathbf{v}$ to some f' with $f' + 1 + s = r + 1$. Hence,

$$\mathbb{A}_n^{\mathsf{com}}(\{\rightleftarrows, \rho_r^{1,0}\} \cup \Theta) = \mathbb{A}_n^{\mathsf{com}}(\{\rightleftarrows, \rho_r^{1,0}\} \cup \Theta')$$

for some $\Theta' \subseteq \{\rho_s^{e,f} \mid e, f, s \geq 0 \text{ and } e+s,\ f+s \leq r\}$. Since there are only finitely many choices for Θ', there are only finitely many varieties of the form $\mathbb{A}_n^{\mathsf{com}}(\{\rightleftarrows, \rho_r^{1,0}\} \cup \Theta)$. □

Recall from Lemma 12.1 and Remark 12.3 that each noncommutative subvariety of $\mathbb{O}$ is of the form $\mathbb{O}\Sigma$, where Σ is some set of efficient rigid identities and efficient identities of the form

$$x^{e_0} y^{f_0} \prod_{i=1}^{m}(h_i x^{e_i} y^{f_i}) \approx y^{f_0} x^{e_0} \prod_{i=1}^{m}(h_i x^{e_i} y^{f_i}), \tag{13.3}$$

where $e_0, f_0 \geq 1$, $e_1, f_1, e_2, f_2, \ldots, e_m, f_m \geq 0$, $\sum_{i=0}^{m} e_i \geq 2$, $\sum_{i=0}^{m} f_i \geq 2$, and $m \geq 0$.

Let λ be any identity of the form (13.3). Then $(e_1, f_1), (e_2, f_2), \ldots, (e_m, f_m) \neq (0, 0)$ by the efficiency of λ. Define the *weight* of λ to be (e, f), where $e = \sum_{i=0}^{m} e_i$ and $f = \sum_{i=0}^{m} f_i$, and define

$$\begin{aligned} \exp_x^0(\lambda) &= \{(i, i+1) \mid e_i = e_{i+1} = 0\} \\ \text{and } \exp_y^0(\lambda) &= \{(i, i+1) \mid f_i = f_{i+1} = 0\}. \end{aligned}$$

Lemma 13.9 *Let λ be any efficient identity of the form* (13.3) *with weight (e, f). Suppose that $e \geq r + 1$ and $\exp_y^0(\lambda) \neq \varnothing$. Then there exists some efficient identity λ' of the form* (13.3) *with weight (e, f) and $\exp_y^0(\lambda') = \varnothing$ such that*

$$\mathbb{A}_n^{\mathsf{com}}\{\rightleftarrows, \rho_r^{1,0}, \lambda\} = \mathbb{A}_n^{\mathsf{com}}\{\rightleftarrows, \rho_r^{1,0}, \lambda'\}.$$

Proof Generality is not lost by assuming that λ is the efficient identity (13.3). Since $\exp_y^0(\lambda) \neq \varnothing$, there exists a largest possible ℓ such that $(\ell, \ell+1) \in \exp_y^0(\lambda)$, that is, $f_\ell = f_{\ell+1} = 0$. Further, the efficiency of λ implies that $e_\ell, e_{\ell+1} \geq 1$. Hence, the identity λ is

$$\begin{aligned} & x^{e_0} y^{f_0} \bigg(\prod_{i=1}^{\ell-1} (h_i x^{e_i} y^{f_i}) \bigg) h_\ell x^{e_\ell} h_{\ell+1} x^{e_{\ell+1}} \bigg(\prod_{i=\ell+2}^{m} (h_i x^{e_i} y^{f_i}) \bigg) \\ \approx\; & y^{f_0} x^{e_0} \bigg(\prod_{i=1}^{\ell-1} (h_i x^{e_i} y^{f_i}) \bigg) h_\ell x^{e_\ell} h_{\ell+1} x^{e_{\ell+1}} \bigg(\prod_{i=\ell+2}^{m} (h_i x^{e_i} y^{f_i}) \bigg) \end{aligned}$$

with weight (e, f), where $e = \sum_{i=0}^{m} e_i$ and $f = \sum_{i=0}^{\ell-1} f_i + \sum_{i=\ell+2}^{m} f_i$. It is easily seen that if the suffix $\prod_{i=\ell+2}^{m} (h_i x^{e_i} y^{f_i})$ is nonempty, then $f_{\ell+2} \geq 1$ by the maximality of ℓ.

Let λ_1 denote the identity obtained from λ by replacing e_0 with $e_0 + e_{\ell+1}$ and eliminating the factor $h_{\ell+1} x^{e_{\ell+1}}$:

$$\begin{aligned} & x^{e_0+e_{\ell+1}} y^{f_0} \bigg(\prod_{i=1}^{\ell-1} (h_i x^{e_i} y^{f_i}) \bigg) h_\ell x^{e_\ell} \bigg(\prod_{i=\ell+2}^{m} (h_i x^{e_i} y^{f_i}) \bigg) \\ \approx\; & y^{f_0} x^{e_0+e_{\ell+1}} \bigg(\prod_{i=1}^{\ell-1} (h_i x^{e_i} y^{f_i}) \bigg) h_\ell x^{e_\ell} \bigg(\prod_{i=\ell+2}^{m} (h_i x^{e_i} y^{f_i}) \bigg). \end{aligned}$$

Then λ_1 is, after making the substitution $(h_{\ell+2}, h_{\ell+3}, \ldots, h_m) \mapsto (h_{\ell+1}, h_{\ell+2}, \ldots, h_{m-1})$, an efficient identity of the form (13.3) with weight (e, f) and $|\exp_y^0(\lambda)| > |\exp_y^0(\lambda_1)|$. Since

$$\begin{aligned} & x^{e_0+e_{\ell+1}} y^{f_0} \bigg(\prod_{i=1}^{\ell-1} (h_i x^{e_i} y^{f_i}) \bigg) h_\ell x^{e_\ell} \bigg(\prod_{i=\ell+2}^{m} (h_i x^{e_i} y^{f_i}) \bigg) \\ \overset{\rightleftarrows}{\approx}\; & x^{e_0} y^{f_0} \bigg(\prod_{i=1}^{\ell-1} (h_i x^{e_i} y^{f_i}) \bigg) h_\ell x^{e_\ell} x^{e_{\ell+1}} \bigg(\prod_{i=\ell+2}^{m} (h_i x^{e_i} y^{f_i}) \bigg) \\ \overset{\lambda}{\approx}\; & y^{f_0} x^{e_0} \bigg(\prod_{i=1}^{\ell-1} (h_i x^{e_i} y^{f_i}) \bigg) h_\ell x^{e_\ell} x^{e_{\ell+1}} \bigg(\prod_{i=\ell+2}^{m} (h_i x^{e_i} y^{f_i}) \bigg) \\ \overset{\rightleftarrows}{\approx}\; & y^{f_0} x^{e_0+e_{\ell+1}} \bigg(\prod_{i=1}^{\ell-1} (h_i x^{e_i} y^{f_i}) \bigg) h_\ell x^{e_\ell} \bigg(\prod_{i=\ell+2}^{m} (h_i x^{e_i} y^{f_i}) \bigg), \end{aligned}$$

the identity λ_1 is deducible from $\{\rightleftarrows, \lambda\}$. On the other hand, since

$$
\begin{aligned}
& x^{e_0} y^{f_0} \Big(\prod_{i=1}^{\ell-1} (h_i x^{e_i} y^{f_i}) \Big) h_\ell x^{e_\ell} h_{\ell+1} x^{e_{\ell+1}} \Big(\prod_{i=\ell+2}^{m} (h_i x^{e_i} y^{f_i}) \Big) \\
\overset{\rho_r^{1,0}}{\approx}\; & x^{e_0+e_{\ell+1}} y^{f_0} \Big(\prod_{i=1}^{\ell-1} (h_i x^{e_i} y^{f_i}) \Big) h_\ell x^{e_\ell} h_{\ell+1} x^{e_{\ell+1}} \Big(\prod_{i=\ell+2}^{m} (h_i x^{e_i} y^{f_i}) \Big) \\
& \qquad\qquad\qquad\qquad \text{because } \sum_{i=0}^{m} e_i = e \geq r+1 \\
\overset{\lambda_1}{\approx}\; & y^{f_0} x^{e_0+e_{\ell+1}} \Big(\prod_{i=1}^{\ell-1} (h_i x^{e_i} y^{f_i}) \Big) h_\ell x^{e_\ell} h_{\ell+1} x^{e_{\ell+1}} \Big(\prod_{i=\ell+2}^{m} (h_i x^{e_i} y^{f_i}) \Big) \\
\overset{\rho_r^{1,0}}{\approx}\; & y^{f_0} x^{e_0} \Big(\prod_{i=1}^{\ell-1} (h_i x^{e_i} y^{f_i}) \Big) h_\ell x^{e_\ell} h_{\ell+1} x^{e_{\ell+1}} \Big(\prod_{i=\ell+2}^{m} (h_i x^{e_i} y^{f_i}) \Big),
\end{aligned}
$$

the identity λ is deducible from $\{\rho_r^{1,0}, \lambda_1\}$. Thus, $\mathbb{A}_n^{\mathsf{com}}\{\rightleftarrows, \rho_r^{1,0}, \lambda\} = \mathbb{A}_n^{\mathsf{com}}\{\rightleftarrows, \rho_r^{1,0}, \lambda_1\}$.

If $|\exp_y^0(\lambda_i)| \neq 0$, then the argument from the previous paragraph can be repeated to obtain an efficient identity λ_{i+1} of the form (13.3) with weight (e, f) and $|\exp_y^0(\lambda_i)| > |\exp_y^0(\lambda_{i+1})|$ such that $\mathbb{A}_n^{\mathsf{com}}\{\rightleftarrows, \rho_r^{1,0}, \lambda_i\} = \mathbb{A}_n^{\mathsf{com}}\{\rightleftarrows, \rho_r^{1,0}, \lambda_{i+1}\}$. The sequence $\lambda_1, \lambda_2, \lambda_3, \ldots$ is finite since $|\exp_y^0(\lambda_k)| = 0$ for some k. The identity λ_k is then the required identity λ'. □

Lemma 13.10 *For any fixed $n \geq 2$ and $r \geq 0$, there exist only finitely many varieties of the form $\mathbb{A}_n^{\mathsf{com}}(\{\rightleftarrows, \rho_r^{1,0}\} \cup \Lambda)$, where Λ is any set of efficient identities of the form* (13.3).

Proof Let λ be the identity (13.3). There are four cases.

CASE 1: $e, f \leq r$. Then clearly there are only finitely many varieties of the form $\mathbb{A}_n^{\mathsf{com}}\{\rightleftarrows, \rho_r^{1,0}, \lambda\}$.

CASE 2: $e, f \geq r+1$. Let $k \geq 1$ be the least integer such that $e_0, f_0 \leq kn$. Since

$$
\begin{aligned}
x^{e_0} y^{f_0} \prod_{i=1}^{m} (h_i x^{e_i} y^{f_i}) &\overset{\rho_r^{1,0}}{\approx} x^{kn} y^{kn} \prod_{i=1}^{m} (h_i x^{e_i} y^{f_i}) \\
&\overset{\blacklozenge_n}{\approx} y^{kn} x^{kn} \prod_{i=1}^{m} (h_i x^{e_i} y^{f_i}) \\
&\overset{\rho_r^{1,0}}{\approx} y^{f_0} x^{e_0} \prod_{i=1}^{m} (h_i x^{e_i} y^{f_i}),
\end{aligned}
$$

the identity λ is deducible from $\{\blacklozenge_n, \rho_r^{1,0}\}$. Therefore, $\mathbb{A}_n^{\mathsf{com}}\{\rightleftarrows, \rho_r^{1,0}, \lambda\} = \mathbb{A}_n^{\mathsf{com}}\{\rightleftarrows, \rho_r^{1,0}\}$.

CASE 3: $e \geq r+1$ and $f \leq r$. By Lemma 13.9, there exists some efficient identity λ' of the form (13.3) with weight (e, f) and $\exp_y^0(\lambda') = \varnothing$ such that $\mathbb{A}_n^{\mathsf{com}}\{\rightleftarrows, \rho_r^{1,0}, \lambda\} = \mathbb{A}_n^{\mathsf{com}}\{\rightleftarrows, \rho_r^{1,0}, \lambda'\}$. Hence, generality is not lost by assuming that $\exp_y^0(\lambda) = \varnothing$ to begin with. In the presence of the identities $\blacklozenge_n$, it can further be assumed that $e_0, f_0, e_1, f_1, \ldots, e_m, f_m \leq n$. It is easily seen that the restrictions $f \leq r$ and $\exp_y^0(\lambda) = \varnothing$ place a bound on the value of m. It then follows from the assumption $e_0, f_0, e_1, f_1, \ldots, e_m, f_m \leq n$ that there exist only finitely many varieties of the form $\mathbb{A}_n^{\mathsf{com}}\{\rightleftarrows, \rho_r^{1,0}, \lambda\}$.

CASE 4: $e \leq r$ and $f \geq r+1$. This is symmetrical to Case 3.

In any case, there exist only finitely many varieties of the form $\mathbb{A}_n^{\mathsf{com}}\{\rightleftarrows, \rho_r^{1,0}, \lambda\}$, say

$$\left\{ \mathbb{A}_n^{\mathsf{com}}\{\rightleftarrows, \rho_r^{1,0}, \lambda\} \;\middle|\; \begin{array}{l} \lambda \text{ is an efficient identity} \\ \text{of the form (13.3)} \end{array} \right\} = \{\mathbb{V}_1, \mathbb{V}_2, \ldots, \mathbb{V}_k\}.$$

Therefore, $\mathbb{A}_n^{\mathsf{com}}(\{\rightleftarrows, \rho_r^{1,0}\} \cup \Lambda)$ is equal to the intersection of some of the varieties $\mathbb{V}_1, \mathbb{V}_2, \ldots, \mathbb{V}_k$, whence there exist only finitely many varieties of the form $\mathbb{A}_n^{\mathsf{com}}(\{\rightleftarrows, \rho_r^{1,0}\} \cup \Lambda)$. □

13.3 Varieties that Exclude $\mathbb{K}$

Lemma 13.11 *Let $\mathbb{V}$ be any variety such that $\mathbb{K} \nsubseteq \mathbb{V}$. Then for any $n \geq 2$, the variety $\mathbb{V}$ satisfies a rigid identity of the form*

$$x \prod_{i=1}^{m} (h_i x) \approx x^{e_0} \prod_{i=1}^{m} (h_i x^{e_i}), \tag{13.4}$$

where $e_0, e_1, \ldots, e_m \geq 0$ with $e_j \geq n$ for some $j \in \{0, 1, \ldots, m\}$.

Proof Any commutative variety satisfies the rigid identity $x \prod_{i=1}^{n-1} (h_i x) \approx x^n \prod_{i=1}^{n-1} h_i$, and any idempotent variety satisfies the rigid identity $x h_1 x \approx x^n h_1 x$. Therefore, it suffices to assume that the variety $\mathbb{V}$ is neither commutative nor idempotent.

Now since $\bigvee_{m\geq 2} \mathbb{R}_{\mathcal{Q}}\{\mathbf{k}_m\} = \mathbb{K}$ by Proposition 12.18, it follows from the assumption $\mathbb{K} \nsubseteq \mathbb{V}$ that $\mathcal{R}_{\mathcal{Q}}\{\mathbf{k}_{m+1}\} \notin \mathbb{V}$ for some $m \geq 1$. Therefore, $x \prod_{i=1}^{m} (h_i x) = \mathbf{k}_{m+1} \notin \mathsf{iso}\,\mathbb{V}$ by Lemma 2.6, so that the variety $\mathbb{V}$ satisfies a nontrivial identity of the form $x \prod_{i=1}^{m} (h_i x) \approx \mathbf{w}$. Since $\mathbb{V}$ is neither commutative nor idempotent, the simple variables of the word $\mathbf{w}$, in order of appearance, are $h_1, h_2, \ldots, h_m$. It follows that $\mathbb{V}$ satisfies a nontrivial rigid identity (13.4) for some $e_0, e_1, \ldots, e_m \geq 0$. Let $\mathbf{u} = x \prod_{i=1}^{m}(h_i x)$ and $\mathbf{v} = x^{e_0} \prod_{i=1}^{m}(h_i x^{e_i})$, so that $\mathbf{u} \approx \mathbf{v}$ is the identity (13.4).

Suppose that the word $\mathbf{v}$ is square-free, that is, $e_0, e_1, \ldots, e_m \leq 1$. Then the nontriviality of $\mathbf{u} \approx \mathbf{v}$ implies that at least one of the exponents $e_0, e_1, \ldots, e_m$ is 0. Let φ denote the substitution

$$h_i \mapsto \begin{cases} h_i & \text{if } e_i = 1, \\ h_i x & \text{if } e_i = 0. \end{cases}$$

Then since $x \prod_{i=1}^{m}(h_i x) = x^{1-e_0}(\mathbf{v}\varphi) \overset{(13.4)}{\approx} x^{1-e_0}(\mathbf{u}\varphi)$, where $x^{1-e_0}(\mathbf{u}\varphi)$ is a rigid word of level m that is not square-free, the variety $\mathbb{V}$ satisfies some rigid identity of the form (13.4) with $e_j \geq 2$ for some $j \in \{0, 1, \ldots, m\}$. Therefore, generality is not lost by assuming that $e_j \geq 2$ to begin with.

Hence, the variety $\mathbb{V}$ satisfies a rigid identity $\sigma : x \prod_{i=1}^{m}(h_i x) \approx \mathbf{p}x^r\mathbf{q}$ with $r \geq 2$ and $\mathbf{p}, \mathbf{q} \in \mathscr{A}_{\varnothing}^{+}$. The identity σ is of the same form as the rigid identity (12.3) in the proof of Lemma 12.8. Following the argument in this proof, the rigid identity σ can be used to deduce another rigid identity of the form $\sigma' : x \prod_{i=1}^{m'}(h_i x) \approx \mathbf{p}'x^{r'}\mathbf{q}'$ with $r' \geq n$. The variety $\mathbb{V}$ thus satisfies the identity σ' and the lemma is established. □

Lemma 13.12 *Let $\mathbb{V}$ be any variety that satisfies the identity $\wp_n : x^{2n} \approx x^n$ for some $n \geq 2$. Then $\mathbb{K} \nsubseteq \mathbb{V}$ if and only if $\mathbb{V}$ satisfies a rigid identity of the form* (13.4), *where precisely one of the exponents $e_0, e_1, \ldots, e_m$ is equal to $n + 1$, while every other exponent is equal to* 1.

Proof Suppose that $\mathbb{K} \nsubseteq \mathbb{V}$. Then by Lemma 13.11, the variety $\mathbb{V}$ satisfies the rigid identity (13.4) with $e_j \geq n$ for some $j \in \{0, 1, \ldots, m\}$. There are two cases.

CASE 1: $0 < j \leq m$. Then the identity (13.4) is

$$x \prod_{i=1}^{m}(h_i x) \approx x^{e_0}\left(\prod_{i=1}^{j-1}(h_i x^{e_i})\right) h_j x^{e_j} \left(\prod_{i=j+1}^{m}(h_i x^{e_i})\right).$$

Let φ denote the substitution $h_j \mapsto h_j x^n$. Since

$$\begin{aligned} x \prod_{i=1}^{m}(h_i x) &\overset{(13.4)}{\approx} x^{e_0}\left(\prod_{i=1}^{j-1}(h_i x^{e_i})\right) h_j x^{e_j} \left(\prod_{i=j+1}^{m}(h_i x^{e_i})\right) \\ &\overset{\wp_n}{\approx} x^{e_0}\left(\prod_{i=1}^{j-1}(h_i x^{e_i})\right) h_j x^{e_j+n} \left(\prod_{i=j+1}^{m}(h_i x^{e_i})\right) \end{aligned}$$

$$= \left(x^{e_0}\left(\prod_{i=1}^{j-1}(h_i x^{e_i})\right)h_j x^{e_j}\left(\prod_{i=j+1}^{m}(h_i x^{e_i})\right)\right)\varphi$$

$$\overset{(13.4)}{\approx} \left(x\prod_{i=1}^{m}(h_i x)\right)\varphi$$

$$= x\left(\prod_{i=1}^{j-1}(h_i x)\right)h_j x^{n+1}\left(\prod_{i=j+1}^{m}(h_i x)\right),$$

the variety $\mathbb{V}$ satisfies the identity (13.4) with the required exponents.

CASE 2: $j = 0$. Repeat the argument in Case 1 with φ being the substitution $h_1 \mapsto x^n h_1$.

Conversely, if $\mathbb{K} \subseteq \mathbb{V}$, then it follows from Lemma 12.13 that $\mathbb{V}$ violates the identity (13.4). □

Lemma 13.13 *Let $\mathbb{V}$ be any subvariety of $\mathbb{A}^{\mathsf{com}}\{\rightleftarrows\}$. Then $\mathbb{K} \nsubseteq \mathbb{V}$ if and only if $\mathbb{V}$ satisfies the identity $\rho_m^{1,0}$ for some $m \geq 0$.*

Proof By assumption, the variety $\mathbb{V}$ satisfies the identities $\blacklozenge_n$ for some $n \geq 2$. Suppose that $\mathbb{K} \nsubseteq \mathbb{V}$. Then since the variety $\mathbb{V}$ satisfies the identity $\rightleftarrows$, it follows from Lemma 13.11 that it also satisfies some rigid identity $\sigma : x\prod_{i=1}^{m}(h_i x) \approx x^{n+1}\prod_{i=1}^{m}(h_i x)$. Hence, $\mathbb{V}$ satisfies the identity $\rho_m^{1,0}$ because

$$x^2\prod_{i=1}^{m}(h_i x) \overset{\sigma}{\approx} x^{n+2}\prod_{i=1}^{m}(h_i x)$$

$$\overset{\blacklozenge_n}{\approx} x^{n+1}\prod_{i=1}^{m}(h_i x)$$

$$\overset{\sigma}{\approx} x\prod_{i=1}^{m}(h_i x).$$

Conversely, if $\mathbb{V}$ satisfies the identity $\rho_m^{1,0}$, then $\mathbb{K} \nsubseteq \mathbb{V}$ by Lemma 12.13. □

13.4 Proof of Theorem 13.1

Lemma 13.14 *Any subvariety of $\mathbb{A}^{\mathsf{com}}\{\rightleftarrows\}$ that excludes the monoid $\mathcal{R}_{\mathsf{Q}}\{xyx\}$ is Cross.*

Proof Let $\mathbb{V}$ be any subvariety of $\mathbb{A}_n^{\mathsf{com}}\{\rightleftarrows\}$ such that $\mathcal{R}_{\mathsf{Q}}\{xyx\} \notin \mathbb{V}$. Then by Lemma 2.6, the variety $\mathbb{V}$ satisfies a nontrivial identity of the form $xyx \approx \mathbf{w}$. It is easily shown that if either $\mathsf{occ}(x, \mathbf{w}) \leq 1$ or $\mathsf{occ}(y, \mathbf{w}) \neq 1$, then $\mathbb{V}$ is idempotent and commutative, whence $\mathbb{V} \subseteq \mathscr{V}_{\mathsf{mon}}\{\mathcal{S}\ell_2\}$ is Cross; see Theorem 1.56. Therefore, assume that $\mathsf{occ}(x, \mathbf{w}) \geq 2$ and $\mathsf{occ}(y, \mathbf{w}) = 1$, so that the identity $xyx \approx \mathbf{w}$ is

$$xyx \approx x^p y x^q \tag{13.5}$$

for some $p, q \geq 0$ such that $p + q \geq 2$ and $(p, q) \neq (1, 1)$. There are two cases.

CASE 1: $p + q = 2$. Then the identity (13.5) is $xyx \approx x^2y$ or $xyx \approx yx^2$; by symmetry, it suffices to assume that (13.5) is $xyx \approx x^2y$. It is easily seen that the identities (11.1) are deducible from (13.5), so that $\mathbb{V} \subseteq \mathbb{O}$. Since

$$\begin{aligned} x\prod_{i=1}^{n}(h_i x) &\overset{(13.5)}{\approx} x^{n+1}\prod_{i=1}^{n} h_i \\ &\overset{\blacklozenge_n}{\approx} x^n \prod_{i=1}^{n} h_i, \end{aligned}$$

the subvariety $\mathbb{V}$ of $\mathbb{O}$ satisfies the limiting identity $x\prod_{i=1}^{n}(h_i x) \approx x^n \prod_{i=1}^{n} h_i$ and so is Cross by Proposition 12.11.

CASE 2: $p + q \geq 3$. Then $\mathbb{V}$ satisfies the identity $x^2 \approx x^{p+q}$, so that $\mathbb{V} \subseteq \mathbb{A}_2^{\mathsf{com}}$. Since

$$\begin{aligned} xyx &\overset{(13.5)}{\approx} x^p y x^q \\ &\overset{\rightleftarrows}{\approx} x^{p+q-1} yx \\ &\overset{\blacklozenge_2}{\approx} x^{p+q} yx \\ &\overset{\rightleftarrows}{\approx} x^{p+1} y x^q \\ &\overset{(13.5)}{\approx} x^2 yx, \end{aligned}$$

the variety $\mathbb{V}$ satisfies the identity $x^2yx \approx xyx$. Since $\mathbb{V} \subseteq \mathbb{A}_2^{\mathsf{com}}\{\rightleftarrows, x^2yx \approx xyx\} = \mathbb{Q}^1$ by (13.1), the variety $\mathbb{V}$ is Cross by Proposition 13.5. □

Lemma 13.15 *For each $n \geq 1$ and $r \geq 0$, the variety $(\mathbb{A}_n^{\mathsf{com}} \cap \mathbb{O})\{\rho_r^{1,0}, (13.2)\}$ is Cross.*

Proof It is clear that $(\mathbb{A}_n^{\mathsf{com}} \cap \mathbb{O})\{\rho_r^{1,0}, (13.2)\}$ is a subvariety of $\mathbb{O}$. Since

$$x\prod_{i=1}^{r+3}(h_i x) \overset{\rho_r^{1,0}}{\approx} x^n h_1 x h_2 x^n \prod_{i=3}^{r+3}(h_i x)$$

$$\overset{(13.2)}{\approx} x^n h_1 h_2 x^n \prod_{i=3}^{r+3}(h_i x)$$

$$\overset{\rho_r^{1,0}}{\approx} x h_1 h_2 x \prod_{i=3}^{r+3}(h_i x),$$

the variety $\mathbb{V}$ satisfies the limiting identity $x \prod_{i=1}^{r+3}(h_i x) \approx x h_1 h_2 x \prod_{i=3}^{r+3}(h_i x)$ and so is Cross by Proposition 12.11. □

Let $\mathbb{V}$ be any subvariety of $\mathbb{A}^{\mathsf{com}}\{\rightleftarrows\}$, so that $\mathbb{V} \subseteq \mathbb{A}_n^{\mathsf{com}}\{\rightleftarrows\}$ for some $n \geq 2$. It is clear that if $\mathbb{V}$ is Cross, then

(A) $\mathbb{J}_1, \mathbb{J}_2 \nsubseteq \mathbb{V}$;
(B) $\mathbb{K} \nsubseteq \mathbb{V}$.

Conversely, suppose that (A) and (B) hold. If $\mathcal{R}_{\mathcal{Q}}\{xyx\} \notin \mathbb{V}$, then the variety $\mathbb{V}$ is Cross by Lemma 13.14. Therefore, assume that $\mathcal{R}_{\mathcal{Q}}\{xyx\} \in \mathbb{V}$, so that $\mathbb{V}$ is distinguished. Hence, by (A) and Theorem 11.16, either

(C) $\mathbb{V} \subseteq \mathbb{O}$

or $\mathbb{V} \subseteq \mathbb{O}^{\triangleleft}$; by symmetry, it suffices to assume that (C) holds. Further, by (B) and Lemma 13.13,

(D) $\mathbb{V}$ satisfies the identity $\rho_r^{1,0}$ for some $r \geq 0$.

Now partition the subvarieties of $\mathbb{V}$ into the sets

$$\mathfrak{V}_{\in} = \{\mathbb{U} \in \mathfrak{L}(\mathbb{V}) \mid \mathcal{Q}^1 \in \mathbb{U}\} \text{ and } \mathfrak{V}_{\notin} = \{\mathbb{U} \in \mathfrak{L}(\mathbb{V}) \mid \mathcal{Q}^1 \notin \mathbb{U}\}.$$

In what follows, the sets $\mathfrak{V}_{\in}$ and $\mathfrak{V}_{\notin}$ are shown to contain only finitely many varieties. Consequently, the variety $\mathbb{V}$ contains only finitely many subvarieties and so is Cross by (C) and Lemma 12.10.

Lemma 13.16 *There exist only finitely many varieties in* $\mathfrak{V}_{\in}$.

Proof Since the monoid $\mathcal{Q}^1$ is noncommutative, every variety in $\mathfrak{V}_{\in}$ is noncommutative. Therefore, by (C), (D), Lemma 12.1, and Remark 12.3, each variety $\mathbb{U}$ in $\mathfrak{V}_{\in}$ possesses some identity basis of the form $\{(11.1), \blacklozenge_n, \rightleftarrows, \rho_r^{1,0}\} \cup \Theta \cup \Lambda$, where Θ is some set of efficient rigid identities satisfied by $\mathcal{Q}^1$ and Λ is some set of efficient identities from (13.3). Hence,

$$\mathbb{U} = \mathbb{O} \cap (\mathbb{A}_n^{\mathsf{com}}(\{\rightleftarrows, \rho_0^{1,0}\} \cup \Theta)) \cap (\mathbb{A}_n^{\mathsf{com}}(\{\rightleftarrows, \rho_0^{1,0}\} \cup \Lambda)).$$

By Lemmas 13.8 and 13.10, there exist only finitely many varieties of the forms

$$\mathbb{A}_n^{\mathsf{com}}(\{\rightleftarrows, \rho_0^{1,0}\} \cup \Theta) \text{ and } \mathbb{A}_n^{\mathsf{com}}(\{\rightleftarrows, \rho_0^{1,0}\} \cup \Lambda),$$

respectively. The result thus follows. □

Lemma 13.17 *There exist only finitely many varieties in* $\mathfrak{V}_{\notin}$.

Proof By Lemma 13.4, each variety in $\mathfrak{V}_{\notin}$ satisfies the identity (13.2). Hence, by (C) and (D), each variety in $\mathfrak{V}_{\notin}$ is a subvariety of $(\mathbb{A}_n^{\mathsf{com}} \cap \mathbb{O})\{\rho_r^{1,0}, (13.2)\}$. But the variety $(\mathbb{A}_n^{\mathsf{com}} \cap \mathbb{O})\{\rho_r^{1,0}, (13.2)\}$ is Cross by Lemma 13.15, so there exist only finitely many varieties in $\mathfrak{V}_{\notin}$. □

13.5 Summary

Theorem 13.1, the main result of the present chapter, is concerned with subvarieties of $\mathbb{A}^{\mathsf{com}}$ that satisfy the identity $x^2yx \approx xyx^2$: such a variety is Cross if and only if it excludes the almost Cross varieties $\mathbb{J}_1$, $\mathbb{J}_2$, and $\mathbb{K}$. This result was published in Lee (2014c).

Counterintuitive Examples of Monoids 14

The present chapter establishes the results stated in Sect. 1.6.4. Recall that a semigroup S is *conformable* if S is non-finitely based, while the monoid S^1 is finitely based.

In Sect. 14.1, the direct product of $\mathcal{R}_{\mathcal{Q}}\{xyx\}$ with any noncommutative group of finite exponent is shown to be non-finitely based. In Sect. 14.2, a method is presented from which finite conformable semigroups can be constructed. A conformable semigroup of order 20 is then exhibited, and it is shown that any conformable semigroup contains at least seven elements.

14.1 The Direct Product of $\mathcal{R}_{\mathcal{Q}}\{xyx\}$ with Noncommutative Groups of Finite Exponent

Theorem 14.1 *The direct product of the monoid* $\mathcal{R}_{\mathcal{Q}}\{xyx\}$ *with any noncommutative group of finite exponent is non-finitely based.*

As observed in Sect. 1.6.4, it follows from Proposition 1.63(ii) that for any group G of finite exponent, the direct product $\mathcal{R}_{\mathcal{Q}}\{xyx\} \times G$ is finitely based if and only if G is commutative.

After some restrictions on identities satisfied by the monoid $\mathcal{R}_{\mathcal{Q}}\{xyx\}$ and by noncommutative groups are established in Sect. 14.1.1, the proof of Theorem 14.1 is given in Sect. 14.1.2.

14.1.1 Identities Satisfied by $\mathcal{R}_{\mathcal{Q}}\{xyx\}$ and by Noncommutative Groups

Lemma 14.2 *Suppose that* $\mathbf{u} \approx \mathbf{v}$ *is any identity satisfied by some noncommutative group with* $\mathbf{u}[x, y] = xyxy$. *Then* $\mathbf{v}[x, y] \notin \{x^2y^2, xy^2x, yx^2y\}$.

E. W. H. Lee, *Advances in the Theory of Varieties of Semigroups*, Frontiers in Mathematics, https://doi.org/10.1007/978-3-031-16497-2_14

Proof Any group that satisfies $xyxy \approx \mathbf{w}$ for some $\mathbf{w} \in \{x^2y^2, xy^2x, yx^2y\}$ is commutative. □

Lemma 14.3 *Suppose that* $\mathbf{u} \approx \mathbf{v}$ *is any identity satisfied by* $\mathcal{R}_{\mathcal{Q}}\{xyx\}$. *Then*

(i) $\mathsf{sim}(\mathbf{u}) = \mathsf{sim}(\mathbf{v})$ *and* $\mathsf{con}(\mathbf{u}) = \mathsf{con}(\mathbf{v})$;
(ii) $\mathbf{u}[x, y] = xy$ *if and only if* $\mathbf{v}[x, y] = xy$;
(iii) $\mathbf{u}[x, y] = xyx$ *if and only if* $\mathbf{v}[x, y] = xyx$.

Proof These results are easily established since $xyx \in \mathsf{iso}\{\mathcal{R}_{\mathcal{Q}}\{xyx\}\}$. □

For each $n \geq 1$, define the word

$$\begin{aligned}\mathbf{x}_n &= \left(\prod_{i=1}^{n}(x_ih_i)\right)y\left(\prod_{i=1}^{n}(x_iz_i)\right)y\left(\prod_{i=1}^{n}(t_iz_i)\right)\\ &= x_1h_1x_2h_2\cdots x_nh_n \cdot y \cdot x_1z_1x_2z_2\cdots x_nz_n \cdot y \cdot t_1z_1t_2z_2\cdots t_nz_n.\end{aligned}$$

The word $\mathbf{x}_n$ was first employed by Jackson (2005b, proof of Lemma 5.5) to prove that the monoid $\mathcal{R}_{\mathcal{Q}}\{xhxyty\}$ is non-finitely based.

Lemma 14.4 *Suppose that* $\mathbf{x}_n \approx \mathbf{w}$ *is any identity satisfied by* $\mathcal{R}_{\mathcal{Q}}\{xyx\}$. *Then*

(i) $\mathbf{w}[x_i, h_j] = x_ih_jx_i$ *if and only if* $i \leq j$;
(ii) $\mathbf{w}[t_i, z_j] = z_jt_iz_j$ *if and only if* $i \leq j$;
(iii) $\mathbf{w}[x_i, t_j] \neq x_it_jx_i$ *and* $\mathbf{w}[h_i, z_j] \neq z_jh_iz_j$ *for all* i *and* j;
(iv) $\mathbf{w}[h_i, y] \neq yh_iy$ *and* $\mathbf{w}[y, t_i] \neq yt_iy$ *for all* i.

Proof These follow from Lemma 14.3(iii) because

(i) $\mathbf{x}_n[x_i, h_j] = x_ih_jx_i$ when $i \leq j$ and $\mathbf{x}_n[x_i, h_j] = h_jx_i^2$ otherwise;
(ii) $\mathbf{x}_n[t_i, z_j] = z_jt_iz_j$ when $i \leq j$ and $\mathbf{x}_n[t_i, z_j] = z_j^2t_i$ otherwise;
(iii) $\mathbf{x}_n[x_i, t_j] = x_i^2t_j$ and $\mathbf{x}_n[h_i, z_j] = h_iz_j^2$ for all i and j;
(iv) $\mathbf{x}_n[h_i, y] = h_iy^2$ and $\mathbf{x}_n[y, t_i] = y^2t_i$ for all i. □

Lemma 14.5 *Suppose that* $\mathbf{x}_n \approx \mathbf{w}$ *is any identity satisfied by some noncommutative group. Then*

(i) $\mathbf{w}[x_i, y] \neq x_i^2y^2$ *and* $\mathbf{w}[y, z_i] \neq y^2z_i^2$ *for all* i;
(ii) $\mathbf{w}[x_i, z_j] \neq x_iz_jx_iz_j$ *when* $i \leq j$;
(iii) $\mathbf{w}[x_i, z_j] \neq x_i^2z_j^2$ *when* $i > j$.

Proof These follow from Lemma 14.2 because

(i) $\mathbf{x}_n[x_i, y] = x_i y x_i y$ and $\mathbf{x}_n[y, z_i] = yz_i yz_i$ for all i;
(ii) $\mathbf{x}_n[x_i, z_j] = x_i^2 z_j^2$ when $i \leq j$;
(iii) $\mathbf{x}_n[x_i, z_j] = x_i z_j x_i z_j$ when $i > j$. □

Lemma 14.6 *Suppose that* $\mathbf{x}_n \approx \mathbf{w}$ *is any balanced identity satisfied by* $\mathcal{R}_Q\{xyx\} \times G$, *where G is some noncommutative group. Then* $\mathbf{x}_n = \mathbf{w}$.

Proof Since $\mathbf{x}_n[h_i, t_i \mid 1 \leq i \leq n] = h_1 h_2 \cdots h_n t_1 t_2 \cdots t_n$, it follows from Lemma 14.3(ii) that

$$\mathbf{w}[h_i, t_i \mid 1 \leq i \leq n] = h_1 h_2 \cdots h_n t_1 t_2 \cdots t_n.$$

Therefore, by parts (i)–(iii) of Lemma 14.4,

$$\text{(A)}\quad \mathbf{w}[x_i, h_i, t_i, z_i \mid 1 \leq i \leq n] = x_1 h_1 x_2 h_2 \cdots x_n h_n \mathbf{u} t_1 z_1 t_2 z_2 \cdots t_n z_n$$

for some $\mathbf{u} \in \{x_1, z_1, x_2, z_2, \ldots, x_n, z_n\}^+$ with $\mathsf{occ}(x_i, \mathbf{u}) = \mathsf{occ}(z_i, \mathbf{u}) = 1$ for all $i \in \{1, 2, \ldots, n\}$. If $\mathbf{u}[x_i, x_{i+1}] = x_{i+1} x_i$ for some i, so that $\mathbf{w}[x_i, x_{i+1}] = x_i x_{i+1}^2 x_i$, then Lemma 14.2 is violated due to $\mathbf{x}_n[x_i, x_{i+1}] = x_i x_{i+1} x_i x_{i+1}$. Therefore, $\mathbf{u}[x_i, x_{i+1}] = x_i x_{i+1}$ for all $i \in \{1, 2, \ldots, n\}$. By a similar argument, $\mathbf{u}[z_i, z_{i+1}] = z_i z_{i+1}$ for all $i \in \{1, 2, \ldots, n\}$. Hence,

$$\text{(B)}\quad \mathbf{u}[x_1, x_2, \ldots, x_n] = x_1 x_2 \cdots x_n \text{ and } \mathbf{u}[z_1, z_2, \ldots, z_n] = z_1 z_2 \cdots z_n.$$

By (A), (B), and parts (ii) and (iii) of Lemma 14.5,

$$\mathbf{w}[x_i, h_i, t_i, z_i] = x_1 h_1 x_2 h_2 \cdots x_n h_n x_1 z_1 x_2 z_2 \cdots x_n z_n t_1 z_1 t_2 z_2 \cdots t_n z_n.$$

Since the identity $\mathbf{x}_n \approx \mathbf{w}$ is balanced, $\mathsf{occ}(y, \mathbf{w}) = 2$. Consequently, $\mathbf{w} = \mathbf{x}_n$ by Lemmas 14.4(iv) and 14.5(i). □

14.1.2 Proof of Theorem 14.1

Suppose that G is any noncommutative group of finite exponent $m \geq 2$. Let $\mathbf{x}_n'$ denote the word obtained from $\mathbf{x}_n$ by replacing the first y with y^{m+1}, that is,

$$\mathbf{x}'_n = \Big(\prod_{i=1}^{n}(x_i h_i)\Big) y^{m+1} \Big(\prod_{i=1}^{n}(x_i z_i)\Big) y \Big(\prod_{i=1}^{n}(t_i z_i)\Big)$$
$$= x_1 h_1 x_2 h_2 \cdots x_n h_n \cdot y^{m+1} \cdot x_1 z_1 x_2 z_2 \cdots x_n z_n \cdot y \cdot t_1 z_1 t_2 z_2 \cdots t_n z_n.$$

Lemma 14.7 *For each $n \geq 1$, the identity $\mathbf{x}_n \approx \mathbf{x}'_n$ is satisfied by $\mathcal{R}_{\mathcal{Q}}\{xyx\} \times G$.*

Proof Since the group G is of exponent m, it satisfies the identity $y^{m+1} \approx y$ and so also the identity $\mathbf{x}_n \approx \mathbf{x}'_n$. It remains to verify that the monoid $\mathcal{R}_{\mathcal{Q}}\{xyx\}$ also satisfies the identity $\mathbf{x}_n \approx \mathbf{x}'_n$. Let $\varphi : \mathscr{A} \to \mathcal{R}_{\mathcal{Q}}\{xyx\}$ be any substitution. Since the first and last occurrences of y in $\mathbf{x}_n$ and in $\mathbf{x}'_n$ do not sandwich any simple variable, it is easily seen that if $y\varphi \neq \mathtt{1}$, then $\mathbf{x}_n\varphi$ and $\mathbf{x}'_n\varphi$ are not factors of xyx, whence $\mathbf{x}_n\varphi = \mathtt{0} = \mathbf{x}'_n\varphi$ in $\mathcal{R}_{\mathcal{Q}}\{xyx\}$. If $y\varphi = \mathtt{1}$, then evidently $\mathbf{x}_n\varphi = \mathbf{x}'_n\varphi$ in $\mathcal{R}_{\mathcal{Q}}\{xyx\}$. Therefore, the monoid $\mathcal{R}_{\mathcal{Q}}\{xyx\}$ satisfies the identity $\mathbf{x}_n \approx \mathbf{x}'_n$. □

Seeking a contradiction, suppose that the monoid $\mathcal{R}_{\mathcal{Q}}\{xyx\} \times G$ is finitely based, say with finite identity basis Σ. Then there exists some finite $n \geq 1$ such that $\#\Sigma = n$. It is shown below that no identity in Σ can be used to convert the word $\mathbf{x}_n$ into a different word. It follows that the identity $\mathbf{x}_n \approx \mathbf{x}'_n$ is not deducible from Σ and so is not satisfied by the monoid $\mathcal{R}_{\mathcal{Q}}\{xyx\} \times G$, whence Lemma 14.7 is contradicted.

Lemma 14.8 *Let $\mathbf{u} \approx \mathbf{v}$ be any identity satisfied by $\mathcal{R}_{\mathcal{Q}}\{xyx\} \times G$ with $|\mathsf{con}(\mathbf{uv})| \leq n$. Suppose there exist some words $\mathbf{e}, \mathbf{f} \in \mathscr{A}^+_\varnothing$ and a substitution $\varphi : \mathscr{A} \to \mathscr{A}^+_\varnothing$ such that $\mathbf{x}_n = \mathbf{e}(\mathbf{u}\varphi)\mathbf{f}$. Then $\mathbf{x}_n = \mathbf{e}(\mathbf{v}\varphi)\mathbf{f}$.*

Proof The lemma clearly holds if $\mathbf{u}\varphi = \mathtt{1}$, so assume that $\mathbf{u}\varphi \neq \mathtt{1}$. By Lemma 14.3(i),

(A) $\mathsf{sim}(\mathbf{u}) = \mathsf{sim}(\mathbf{v})$ and $\mathsf{con}(\mathbf{u}) = \mathsf{con}(\mathbf{v})$.

Further, the word $\mathbf{u}\varphi$ is a nonempty factor of $\mathbf{x}_n$ and $\mathsf{occ}(x, \mathbf{x}_n) \leq 2$ for all $x \in \mathscr{A}$. Therefore,

(B) if $x \in \mathsf{con}(\mathbf{u})$ and $x\varphi \neq \mathtt{1}$, then $\mathsf{occ}(x, \mathbf{u}) \leq 2$.

For the remainder of this proof, it is shown that

(◊) if $x \in \mathsf{con}(\mathbf{u})$ and $x\varphi \neq \mathtt{1}$, then $\mathsf{occ}(x, \mathbf{u}) = \mathsf{occ}(x, \mathbf{v})$.

It follows that $\mathbf{e}(\mathbf{u}\varphi)\mathbf{f} \approx \mathbf{e}(\mathbf{v}\varphi)\mathbf{f}$ is a balanced identity satisfied by $\mathcal{R}_{\mathcal{Q}}\{xyx\} \times G$. Consequently, $\mathbf{x}_n = \mathbf{e}(\mathbf{v}\varphi)\mathbf{f}$ by Lemma 14.6.

Let $x \in \mathsf{con}(\mathbf{u})$ be such that $x\varphi \neq \mathtt{1}$. If $\mathsf{occ}(x, \mathbf{u}) = 1$, then $\mathsf{occ}(x, \mathbf{u}) = 1 = \mathsf{occ}(x, \mathbf{v})$ by (A), so that (◊) holds. Therefore, further assume that $\mathsf{occ}(x, \mathbf{u}) \geq 2$, whence $\mathsf{occ}(x, \mathbf{u}) = 2$ by (B). Then $\mathbf{u} = \mathbf{u}_1 x \mathbf{u}_2 x \mathbf{u}_3$ for some $\mathbf{u}_1, \mathbf{u}_2, \mathbf{u}_3 \in \mathscr{A}_\varnothing^+$ with $x \notin \mathsf{con}(\mathbf{u}_1\mathbf{u}_2\mathbf{u}_3)$, so that

$$\mathbf{x}_n = \mathbf{e}(\mathbf{u}_1\varphi)(x\varphi)(\mathbf{u}_2\varphi)(x\varphi)(\mathbf{u}_3\varphi)\mathbf{f}.$$

It is easily seen that

(C) if $\mathbf{x}_n = \cdots \mathbf{w} \cdots \mathbf{w} \cdots$ for some $\mathbf{w} \in \mathscr{A}^+$, then $\mathbf{w}$ is one of the variables

$$x_1, x_2, \ldots, x_n, y, z_1, z_2, \ldots, z_n.$$

Therefore, the factor $x\varphi$ of $\mathbf{x}_n$ is also one of the above variables. It follows that

(D) the factor $\mathbf{u}_2\varphi$ of $\mathbf{x}_n$ contains $2n$ distinct variables.

Since $|\,\mathsf{con}(\mathbf{u}_1 x \mathbf{u}_2 x \mathbf{u}_3)| = |\,\mathsf{con}(\mathbf{u})| \leq n$, the word $\mathbf{u}_2$ contains fewer than n distinct variables. If each variable in $\mathbf{u}_2$ is mapped by φ to either 1 or a single variable, then $|\,\mathsf{con}(\mathbf{u}_2\varphi)| \leq |\,\mathsf{con}(\mathbf{u})| \leq n$ violates (D). Therefore, there exists some variable in $\mathbf{u}_2$, say z, such that $z\varphi$ is neither 1 nor a single variable, that is,

(E) the factor $z\varphi$ of $\mathbf{x}_n$ contains at least two variables.

Now write $\mathbf{u}_2 = \mathbf{u}_2' z \mathbf{u}_2''$ for some $\mathbf{u}_2', \mathbf{u}_2'' \in \mathscr{A}_\varnothing^+$, so that $\mathbf{u} = \mathbf{u}_1 x \mathbf{u}_2' z \mathbf{u}_2'' x \mathbf{u}_3$. If the variable z is non-simple in $\mathbf{u}$, then $\mathbf{x}_n = \mathbf{e}(\mathbf{u}\varphi)\mathbf{f} = \mathbf{e} \cdots z\varphi \cdots z\varphi \cdots \mathbf{f}$, which is impossible in view of (C) and (E). The variable z is thus simple in $\mathbf{u}$, whence $\mathbf{u}[x, z] = xzx$. It then follows from Lemma 14.3(iii) that $\mathbf{v}[x, z] = xzx$. Hence, $\mathsf{occ}(x, \mathbf{u}) = 2 = \mathsf{occ}(x, \mathbf{v})$, so that ($\lozenge$) holds. □

14.2 Finitely Based Monoids from Non-Finitely Based Semigroups

The following result is required for the construction of conformable semigroups in Theorem 1.71.

Lemma 14.9 (Volkov 1984b) *The direct product $S \times N$ of a semigroup S with any nilpotent semigroup N is finitely based if and only if S is finitely based.*

Proof of Theorem 1.71 Suppose that S and N are any semigroups such that S^1 is non-finitely based, N is nilpotent, and $S^1 \times N^1$ is finitely based. Then the direct product

$$P = S^1 \times N$$

is non-finitely based by Lemma 14.9. Since P is a subsemigroup of $S^1 \times N^1$, the inclusion $\mathscr{V}_{\mathsf{sem}}\{P^1\} \subseteq \mathscr{V}_{\mathsf{sem}}\{S^1 \times N^1\}$ holds. But since the monoids S^1 and N^1 are embeddable in P^1, the equality $\mathscr{V}_{\mathsf{sem}}\{P^1\} = \mathscr{V}_{\mathsf{sem}}\{S^1 \times N^1\}$ holds. It follows that the monoid P^1 is finitely based, whence P is conformable. □

Proof of Corollary 1.72 There exist finite nilpotent semigroups $N_0 \subset N_1 \subset N_2 \subset \cdots$ such that N_n^1 is finitely based if and only if n is even (Jackson and Sapir 2000, Corollary 4.4). Then following the proof of Theorem 1.71, it is routinely checked that $P_n = N_n^1 \times N_{n+1}$ has the desired properties. □

Now since the conformable semigroup $P = S^1 \times N$ is a direct product, its order $|S^1||N|$ can be quite large in general. It turns out that the semigroup P contains a proper subsemigroup that is also conformable. Consider the following subsemigroups of P:

$$S^1_\triangledown = \{(a, 0) \mid a \in S^1\}, \quad N_\triangledown = \{(1, b) \mid b \in N\}, \quad \text{and} \quad P_\triangledown = S^1_\triangledown \cup N_\triangledown.$$

It is easily seen that $S^1_\triangledown \cap N_\triangledown = \{(1, 0)\}$, so that $|P_\triangledown| = |S^1| + |N| - 1$.

Proposition 14.10 *The semigroup $P_\triangledown$ is conformable.*

Proof It is clear that S^1 and N are isomorphic to $S^1_\triangledown$ and $N_\triangledown$, respectively. Since

$$\begin{aligned} \mathscr{V}_{\mathsf{sem}}\{P\} &= \mathscr{V}_{\mathsf{sem}}\{S^1_\triangledown, N_\triangledown\} \\ &\subseteq \mathscr{V}_{\mathsf{sem}}\{P_\triangledown\} \\ &\subseteq \mathscr{V}_{\mathsf{sem}}\{P\}, \end{aligned}$$

the semigroups P and $P_\triangledown$ generate the same variety and thus satisfy the same identities. □

The order $|S^1| + |N| - 1$ of the semigroup $P_\triangledown$ is often much smaller than the order $|S^1||N|$ of the semigroup P. For instance, if

$$S = \mathcal{R}_Q\{xyxy\}\backslash\{1\} \text{ and } N = \mathcal{R}_Q\{x^2y^2, xy^2x\}\backslash\{1\},$$

so that $|S| = 8$ and $|N| = 12$, then $|P_\triangledown| = 9 + 12 - 1 = 20$ and $|P| = 9 \times 12 = 108$.

Proposition 14.11 *Every conformable semigroup is of order at least seven.*

Proof Semigroups of order five or less are finitely based (Trahtman 1991) and so are not conformable. Among non-unital semigroups of order six, up to isomorphism, only $\mathcal{A}_2^{\mathsf{g}}$ and $\mathcal{L}_3$ are non-finitely based (Lee et al. 2012). But since the monoids $(\mathcal{A}_2^{\mathsf{g}})^1$ and $\mathcal{L}_3^1$ are non-finitely based (M.V. Sapir 1987b; Zhang 2013), semigroups of order six are also not conformable. □

14.3 Summary

In Sect. 14.1, the direct product of the monoid $\mathcal{R}_{\mathbb{Q}}\{xyx\}$ with any noncommutative group of finite exponent is shown to be non-finitely based. In Sect. 14.2, a method for constructing finite conformable semigroups is given, and an explicit conformable semigroup of order 20 is exhibited. These results were published in Lee (2013c,d).

References

C.L. Adair, Bands with an involution. J. Algebra **75**(2), 297–314 (1982)

A.Ya. Aĭzenshtat, B.K. Boguta, On a lattice of varieties of semigroups (in Russian), in *Semigroup Varieties and Semigroups of Endomorphisms*, ed. by E.S. Lyapin (Leningrad. Gos. Ped. Inst., Leningrad, 1979), pp. 3–46

J. Almeida, *Finite Semigroups and Universal Algebra* (World Scientific, Singapore, 1994)

J. Almeida, A. Escada, On the equation $\mathbf{V} * \mathbf{G} = \mathscr{E}\mathbf{V}$. J. Pure Appl. Algebra **166**(1–2), 1–28 (2002)

J. Araújo, J.P. Araújo, P.J. Cameron, E.W.H. Lee, J. Raminhos, A survey on varieties generated by small semigroups and a companion website (2019). https://doi.org/10.48550/arXiv.1911.05817

D.N. Ashikmin, M.V. Volkov, W.T. Zhang, The finite basis problem for Kiselman monoids. Demonstr. Math. **48**(4), 475–492 (2015)

K. Auinger, M.B. Szendrei, On identity bases of epigroup varieties. J. Algebra **220**(2), 437–448 (1999)

K. Auinger, I. Dolinka, M.V. Volkov, Equational theories of semigroups with involution. J. Algebra **369**, 203–225 (2012a)

K. Auinger, I. Dolinka, M.V. Volkov, Matrix identities involving multiplication and transposition. J. Eur. Math. Soc. **14**(3), 937–969 (2012b)

K. Auinger, I. Dolinka, T.V. Pervukhina, M.V. Volkov, Unary enhancements of inherently non-finitely based semigroups. Semigroup Forum **89**(1), 41–51 (2014)

K. Auinger, Y.Z. Chen, X. Hu, Y.F. Luo, M.V. Volkov, The finite basis problem for Kauffman monoids. Algebra Universalis **74**(3–4), 333–350 (2015)

Yu.A. Bahturin, A.Yu. Ol'shanskiĭ, Identical relations in finite Lie rings (in Russian). Mat. Sb. (N.S.) **96**(4), 543–559 (1975). English translation: Math. USSR-Sb. **25**(4), 507–523 (1975)

A.Ya. Belov, The local finite basis property and local representability of varieties of associative rings (in Russian). Izv. Ross. Akad. Nauk Ser. Mat. **74**(1), 3–134 (2010). English translation: Izv. Math. **74**(1), 1–126 (2010)

A.P. Birjukov, Varieties of idempotent semigroups (in Russian). Algebra i Logika **9**(3), 255–273 (1970). English translation: Algebra and Logic **9**(3), 153–164 (1970)

G. Birkhoff, On the structure of abstract algebras. Proc. Cambridge Philos. Soc. **31**(4), 433–454 (1935)

A.D. Bol'bot, Finite basing of identities of four-element semigroups. Siberian Math. J. **20**(2), 323 (1979)

S. Burris, H.P. Sankappanavar, *A Course in Universal Algebra* (Springer, New York, 1981)

Y.Z. Chen, X. Hu, Y.F. Luo, O.B. Sapir, The finite basis problem for the monoid of two-by-two upper triangular tropical matrices. Bull. Aust. Math. Soc. **94**(1), 54–64 (2016)

E. W. H. Lee, *Advances in the Theory of Varieties of Semigroups*, Frontiers in Mathematics, https://doi.org/10.1007/978-3-031-16497-2

A.H. Clifford, A system arising from a weakened set of group postulates. Ann. Math. (2) **34**(4), 865–871 (1933)

A.H. Clifford, Semigroups admitting relative inverses. Ann. Math. (2) **42**(4), 1037–1049 (1941)

D.F. Cowan, N.R. Reilly, Partial cross-sections of symmetric inverse semigroups. Internat. J. Algebra Comput. **5**(3), 259–287 (1995)

S. Crvenković, I. Dolinka, Varieties of involution semigroups and involution semirings: a survey. Bull. Soc. Math. Banja Luka **9**, 7–47 (2002)

S. Crvenković, I. Dolinka, M. Vinčić, Equational bases for some 0-direct unions of semigroups. Studia Sci. Math. Hungar. **36**(3–4), 423–431 (2000)

A. Distler, T. Kelsey, The monoids of orders eight, nine & ten. Ann. Math. Artif. Intell. **56**(1), 3–21 (2009)

A. Distler, T. Kelsey, The semigroups of order 9 and their automorphism groups. Semigroup Forum **88**(1), 93–112 (2014)

I. Dolinka, All varieties of normal bands with involution. Period. Math. Hungar. **40**(2), 109–122 (2000a)

I. Dolinka, Remarks on varieties of involution bands. Comm. Algebra **28**(6), 2837–2852 (2000b)

I. Dolinka, On the lattice of varieties of involution semigroups. Semigroup Forum **62**(3), 438–459 (2001)

I. Dolinka, On identities of finite involution semigroups. Semigroup Forum **80**(1), 105–120 (2010)

P. Dubreil, Contribution à la théorie des demi-groupes. Mém. Acad. Sci. Inst. France (2) **63**(3), 52 pp. (1941)

C.C. Edmunds, On certain finitely based varieties of semigroups. Semigroup Forum **15**(1), 21–39 (1977)

C.C. Edmunds, Varieties generated by semigroups of order four. Semigroup Forum **21**(1), 67–81 (1980)

C.C. Edmunds, E.W.H. Lee, K.W.K. Lee, Small semigroups generating varieties with continuum many subvarieties. Order **27**(1), 83–100 (2010)

A. Escada, The **G**-exponent of a pseudovariety of semigroups. J. Algebra **223**(1), 15–36 (2000)

T. Evans, The lattice of semigroups varieties. Semigroup Forum **2**(1), 1–43 (1971)

S. Fajtlowicz, Equationally complete semigroups with involution. Algebra Universalis **1**(1), 355–358 (1971)

C. Fennemore, All varieties of bands. Semigroup Forum **1**(2), 172–179 (1970)

M. Gao, W.T. Zhang, Y.F. Luo, A non-finitely based involution semigroup of order five. Algebra Universalis **81**(3), Article 31, 14 pp. (2020a)

M. Gao, W.T. Zhang, Y.F. Luo, The monoid of 2×2 triangular boolean matrices under skew transposition is non-finitely based. Semigroup Forum **100**(1), 153–168 (2020b)

M. Gao, W.T. Zhang, Y.F. Luo, Finite basis problem for Lee monoids with involution. Comm. Algebra **49**(10), 4258–4273 (2021)

M. Gao, W.T. Zhang, Y.F. Luo, Finite basis problem for Catalan monoids with involution. Internat. J. Algebra Comput. **32**(6), 1161–1177 (2022)

J.A. Gerhard, The lattice of equational classes of idempotent semigroups. J. Algebra **15**(2), 195–224 (1970)

I.A. Gol'dberg, On the finite basis problem for the monoids of extensive transformations, in *Semigroups and Formal Languages (Lisbon 2005)*, ed. by J.M. André, V.H. Fernandes, M.J.J. Branco, G.M.S. Gomes, J. Fountain, J.C. Meakin (World Scientific, Singapore, 2007), pp. 101–110

G. Grätzer, *Universal Algebra* (Springer, New York, 1979)

C.K. Gupta, A. Krasilnikov, The finite basis question for varieties of groups—some recent results. Illinois J. Math. **47**(1–2), 273–283 (2003)

S.V. Gusev, On the ascending and descending chain conditions in the lattice of monoid varieties. Sib. Èlektron. Mat. Izv. **16**, 983–997 (2019)

S.V. Gusev, A new example of a limit variety of monoids. Semigroup Forum **101**(1), 102–120 (2020a)

S.V. Gusev, Cross varieties of aperiodic monoids with commuting idempotents (2020b). https://doi.org/10.48550/arXiv.2004.03470

S.V. Gusev, E.W.H. Lee, Varieties of monoids with complex lattices of subvarieties. Bull. London Math. Soc. **52**(4), 762–775 (2020)

S.V. Gusev, O.B. Sapir, Classification of limit varieties of J-trivial monoids. Comm. Algebra **50**(7), 3007–3027 (2022)

S.V. Gusev, B.M. Vernikov, Chain varieties of monoids. Dissertationes Math. **534**, 73 pp. (2018)

S.V. Gusev, E.W.H. Lee, B.M. Vernikov, The lattice of varieties of monoids. Jpn. J. Math. **17**(2), 117–183 (2022)

J. Hage, T. Harju, On involutions arising from graphs, in *Algorithmic Bioprocesses (Leiden 2007)*, ed. by A. Condon, D. Harel, J.N. Kok, A. Salomaa, E. Winfree. Natural Computing Series (Springer, Berlin, 2009), pp. 623–630

T.E. Hall, S.I. Kublanovskiĭ, S. Margolis, M.V. Sapir, P.G. Trotter, Algorithmic problems for finite groups and finite 0-simple semigroups. J. Pure Appl. Algebra **119**(1), 75–96 (1997)

B.B. Han, W.T. Zhang, Y.F. Luo, Equational theories of upper triangular tropical matrix semigroups. Algebra Universalis **82**(3), Article 44, 21 pp. (2021)

T.J. Head, The varieties of commutative monoids. Nieuw Arch. Wisk. (3) **16**, 203–206 (1968)

G. Higman, Ordering by divisibility in abstract algebras. Proc. London Math. Soc. **2**(1), 326–336 (1952)

C. Hollings, The early development of the algebraic theory of semigroups. Arch. Hist. Exact Sci. **63**(5), 497–536 (2009)

J.M. Howie, *Fundamentals of Semigroup Theory* (Clarendon Press, Oxford, 1995)

X. Hu, Y.Z. Chen, Y.F. Luo, On the finite basis problem for the monoids of partial extensive injective transformations. Semigroup Forum **91**(2), 524–537 (2015)

J.R. Isbell, Two examples in varieties of monoids. Proc. Cambridge Philos. Soc. **68**(2), 265–266 (1970)

M. Jackson, *Small Semigroup Related Structures with Infinite Properties*. Ph.D. thesis (University of Tasmania, Australia, 1999)

M. Jackson, Finite semigroups whose varieties have uncountably many subvarieties. J. Algebra **228**(2), 512–535 (2000)

M. Jackson, On the finite basis problem for finite Rees quotients of free monoids. Acta Sci. Math. (Szeged) **67**(1–2), 121–159 (2001)

M. Jackson, Small inherently nonfinitely based finite semigroups. Semigroup Forum **64**(2), 297–324 (2002)

M. Jackson, Finite semigroups with infinite irredundant identity bases. Internat. J. Algebra Comput. **15**(3), 405–422 (2005a)

M. Jackson, Finiteness properties of varieties and the restriction to finite algebras. Semigroup Forum **70**(2), 159–187 (2005b)

M. Jackson, E.W.H. Lee, Monoid varieties with extreme properties. Trans. Am. Math. Soc. **370**(7), 4785–4812 (2018)

M. Jackson, R.N. McKenzie, Interpreting graph colorability in finite semigroups. Internat. J. Algebra Comput. **16**(1), 119–140 (2006)

M. Jackson, O.B. Sapir, Finitely based, finite sets of words. Internat. J. Algebra Comput. **10**(6), 683–708 (2000)

M. Jackson, M.V. Volkov, The algebra of adjacency patterns: Rees matrix semigroups with reversion, in *Fields of Logic and Computation (Brno 2010)*, ed. by A. Blass, N. Dershowitz, W. Reisig. Lecture Notes in Computer Science, vol. 6300 (Springer, Berlin, 2010), pp. 414–443

M. Jackson, W.T. Zhang, From A to B to Z. Semigroup Forum **103**(1), 165–190 (2021)

P.R. Jones, The semigroups B_2 and B_0 are inherently nonfinitely based, as restriction semigroups. Internat. J. Algebra Comput. **23**(6), 1289–1335 (2013)

J. Kad'ourek, On varieties of combinatorial inverse semigroups. II. Semigroup Forum **44**(1), 53–78 (1992)

J. Kad'ourek, Uncountably many varieties of semigroups satisfying $x^2y \approx xy$. Semigroup Forum **60**(1), 135–152 (2000)

J. Kad'ourek, On bases of identities of finite inverse semigroups with solvable subgroups. Semigroup Forum **67**(3), 317–343 (2003)

J. Kad'ourek, On finite completely simple semigroups having no finite basis of identities. Semigroup Forum **97**(1), 154–161 (2018)

J. Kad'ourek, On bases of identities of finite central locally orthodox completely regular semigroups. Semigroup Forum **102**(3), 697–724 (2021)

J. Kalicki, D. Scott, Equational completeness of abstract algebras. Nederl. Akad. Wetensch. Proc. Ser. A **58**, 650–659 (1955)

J. Karnofsky, Finite equational bases for semigroups. Notices Am. Math. Soc. **17**(5), 813–814 (1970)

A.R. Kemer, Solution of the problem as to whether associative algebras have a finite basis of identities (in Russian). Dokl. Akad. Nauk SSSR **298**(2), 273–277 (1988). English translation: Soviet Math. Dokl. **37**(1), 60–64 (1988)

O.G. Kharlampovich, M.V. Sapir, Algorithmic problems in varieties. Internat. J. Algebra Comput. **5**(4–5), 379–602 (1995)

A. Kisielewicz, Varieties of commutative semigroups. Trans. Am. Math. Soc. **342**(1), 275–306 (1994)

E.I. Kleĭman, On basis of identities of Brandt semigroups. Semigroup Forum **13**(3), 209–218 (1977)

E.I. Kleĭman, Bases of identities of varieties of inverse semigroups (in Russian). Sibirsk. Mat. Zh. **20**(4), 760–777 (1979). English translation: Siberian Math. J. **20**(4), 530–543 (1979)

E.I. Kleĭman, A pseudovariety generated by a finite semigroup (in Russian). Ural. Gos. Univ. Mat. Zap. **13**(1), 40–42 (1982)

R.L. Kruse, Identities satisfied by a finite ring. J. Algebra **26**(2), 298–318 (1973)

S.I. Kublanovskiĭ, On the rank of the Rees–Sushkevich varieties (in Russian). Algebra i Analiz **23**(4), 59–135 (2011). English translation: St. Petersburg Math. J. **23**(4), 679–730 (2012)

S.I. Kublanovskiĭ, E.W.H. Lee, N.R. Reilly, Some conditions related to the exactness of Rees–Sushkevich varieties. Semigroup Forum **76**(1), 87–94 (2008)

E.W.H. Lee, *On the Lattice of Rees–Sushkevich Varieties*. Ph.D. thesis (Simon Fraser University, Canada, 2002)

E.W.H. Lee, Identity bases for some non-exact varieties. Semigroup Forum **68**(3), 445–457 (2004)

E.W.H. Lee, Subvarieties of the variety generated by the five-element Brandt semigroup. Internat. J. Algebra Comput. **16**(2), 417–441 (2006)

E.W.H. Lee, Minimal semigroups generating varieties with complex subvariety lattices. Internat. J. Algebra Comput. **17**(8), 1553–1572 (2007a)

E.W.H. Lee, On a simpler basis for the pseudovariety **EDS**. Semigroup Forum **75**(2), 477–479 (2007b)

E.W.H. Lee, On identity bases of exclusion varieties for monoids. Comm. Algebra **35**(7), 2275–2280 (2007c)

E.W.H. Lee, On the complete join of permutative combinatorial Rees–Sushkevich varieties. Int. J. Algebra **1**(1–4), 1–9 (2007d)

E.W.H. Lee, Combinatorial Rees–Sushkevich varieties are finitely based. Internat. J. Algebra Comput. **18**(5), 957–978 (2008a)
E.W.H. Lee, On the variety generated by some monoid of order five. Acta Sci. Math. (Szeged) **74**(3–4), 509–537 (2008b)
E.W.H. Lee, Finitely generated limit varieties of aperiodic monoids with central idempotents. J. Algebra Appl. **8**(6), 779–796 (2009a)
E.W.H. Lee, Hereditarily finitely based monoids of extensive transformations. Algebra Universalis **61**(1), 31–58 (2009b)
E.W.H. Lee, Combinatorial Rees–Sushkevich varieties that are Cross, finitely generated, or small. Bull. Aust. Math. Soc. **81**(1), 64–84 (2010a)
E.W.H. Lee, On a semigroup variety of György Pollák. Novi Sad J. Math. **40**(3), 67–73 (2010b)
E.W.H. Lee, Cross varieties of aperiodic monoids with central idempotents. Port. Math. **68**(4), 425–429 (2011a)
E.W.H. Lee, Finite basis problem for 2-testable monoids. Cent. Eur. J. Math. **9**(1), 1–22 (2011b)
E.W.H. Lee, A sufficient condition for the non-finite basis property of semigroups. Monatsh. Math. **168**(3–4), 461–472 (2012a)
E.W.H. Lee, Maximal Specht varieties of monoids. Mosc. Math. J. **12**(4), 787–802 (2012b)
E.W.H. Lee, Varieties generated by 2-testable monoids. Studia Sci. Math. Hungar. **49**(3), 366–389 (2012c)
E.W.H. Lee, Almost Cross varieties of aperiodic monoids with central idempotents. Beitr. Algebra Geom. **54**(1), 121–129 (2013a)
E.W.H. Lee, Finite basis problem for semigroups of order five or less: generalization and revisitation. Studia Logica **101**(1), 95–115 (2013b)
E.W.H. Lee, Finite basis problem for the direct product of some J-trivial monoid with groups of finite exponent. Vestn. St-Peterbg. Univ., Ser. I, Mat. Mekh. Astron. 2013(4), 60–64 (2013c)
E.W.H. Lee, Finitely based monoids obtained from non-finitely based semigroups. Univ. Iagel. Acta Math. **51**, 45–49 (2013d)
E.W.H. Lee, Inherently non-finitely generated varieties of aperiodic monoids with central idempotents. Zap. Nauchn. Sem. POMI **423**, 166–182 (2014a). Reprint: J. Math. Sci. (N.Y.) **209**(4), 588–599 (2015)
E.W.H. Lee. On a question of Pollák and Volkov regarding hereditarily finitely based identities. Period. Math. Hungar. **68**(2), 128–134 (2014b)
E.W.H. Lee, On certain Cross varieties of aperiodic monoids with commuting idempotents. Results Math. **66**(3–4), 491–510 (2014c)
E.W.H. Lee, A class of finite semigroups without irredundant bases of identities. Yokohama Math. J. **61**, 1–28 (2015)
E.W.H. Lee, Finite involution semigroups with infinite irredundant bases of identities. Forum Math. **28**(3), 587–607 (2016a)
E.W.H. Lee, Finitely based finite involution semigroups with non-finitely based reducts. Quaest. Math. **39**(2), 217–243 (2016b)
E.W.H. Lee, Equational theories of unstable involution semigroups. Electron. Res. Announc. Math. Sci. **24**, 10–20 (2017a)
E.W.H. Lee, On a class of completely join prime J-trivial semigroups with unique involution. Algebra Universalis **78**(2), 131–145 (2017b)
E.W.H. Lee, A sufficient condition for the absence of irredundant bases. Houston J. Math. **44**(2), 399–411 (2018a)
E.W.H. Lee, Varieties generated by unstable involution semigroups with continuum many subvarieties. C. R. Math. Acad. Sci. Paris **356**(1), 44–51 (2018b)

E.W.H. Lee, Variety membership problem for two classes of non-finitely based semigroups. Wuhan Univ. J. Nat. Sci. **23**(4), 323–327 (2018c)

E.W.H. Lee, Non-finitely based finite involution semigroups with finitely based semigroup reducts. Korean J. Math. **27**(1), 53–62 (2019a)

E.W.H. Lee, Varieties of involution monoids with extreme properties. Q. J. Math. **70**(4), 1157–1180 (2019b)

E.W.H. Lee, *Contributions to the Theory of Varieties of Semigroups*. D.Sc. thesis (National Research University Higher School of Economics, Russia, 2020a)

E.W.H. Lee, Non-Specht variety generated by an involution semigroup of order five. Tr. Mosk. Mat. Obs. **81**(1), 105–115 (2020b). Reprint: Trans. Moscow Math. Soc. **81**, 87–95 (2020)

E.W.H. Lee, Intervals of varieties of involution semigroups with contrasting reduct intervals. Boll. Unione Mat. Ital. **15**(4), 527–540 (2022)

E.W.H. Lee, A minimal pseudo-complex monoid. Arch. Math. (Basel) **120**(1), 15–25 (2023)

E.W.H. Lee, J.R. Li, Minimal non-finitely based monoids. Dissertationes Math. **475**, 65 pp. (2011)

E.W.H. Lee, J.R. Li, The variety generated by all monoids of order four is finitely based. Glas. Mat. Ser. III **50**(2), 373–396 (2015)

E.W.H. Lee, N.R. Reilly, The intersection of pseudovarieties of central simple semigroups. Semigroup Forum **73**(1), 75–94 (2006)

E.W.H. Lee, N.R. Reilly, Centrality in Rees–Sushkevich varieties. Algebra Universalis **58**(2), 145–180 (2008)

E.W.H. Lee, M.V. Volkov, On the structure of the lattice of combinatorial Rees–Sushkevich varieties, in *Semigroups and Formal Languages (Lisbon 2005)*, ed. by J.M. André, V.H. Fernandes, M.J.J. Branco, G.M.S. Gomes, J. Fountain, J.C. Meakin (World Scientific, Singapore, 2007), pp. 164–187

E.W.H. Lee, M.V. Volkov, Limit varieties generated by completely 0-simple semigroups. Internat. J. Algebra Comput. **21**(1–2), 257–294 (2011)

E.W.H. Lee, W.T. Zhang, The smallest monoid that generates a non-Cross variety (in Chinese). Xiamen Daxue Xuebao Ziran Kexue Ban **53**(1), 1–4 (2014)

E.W.H. Lee, W.T. Zhang, Finite basis problem for semigroups of order six. LMS J. Comput. Math. **18**(1), 1–129 (2015)

E.W.H. Lee, J.R. Li, W.T. Zhang, Minimal non-finitely based semigroups. Semigroup Forum **85**(3), 577–580 (2012)

E.W.H. Lee, J. Rhodes, B. Steinberg, Join irreducible semigroups. Internat. J. Algebra Comput. **29**(7), 1249–1310 (2019)

E.W.H. Lee, J. Rhodes, B. Steinberg, On join irreducible J-trivial semigroups. Rend. Semin. Mat. Univ. Padova **147**, 43–78 (2022)

J.R. Li, Y.F. Luo, On the finite basis problem for the monoids of triangular boolean matrices. Algebra Universalis **65**(4), 353–362 (2011)

J.R. Li, Y.F. Luo, Classification of finitely based words in a class of words over a 3-letter alphabet. Semigroup Forum **91**(1), 200–212 (2015)

J.R. Li, W.T. Zhang, Y.F. Luo, On the finite basis problem for certain 2-limited words. Acta Math. Sin. (Engl. Ser.) **29**(3), 571–590 (2013)

I.V. L'vov, Varieties of associative rings. I (in Russian). Algebra i Logika **12**(3), 269–297 (1973). English translation: Algebra and Logic **12**(3), 150–167 (1973)

S.O. MacDonald, M.R. Vaughan-Lee, Varieties that make one Cross. J. Austral. Math. Soc. Ser. A **26**(3), 368–382 (1978)

S.A. Malyshev, Permutational varieties of semigroups whose lattice of subvarieties is finite (in Russian), in *Modern Algebra* (Leningrad University, Leningrad, 1981), pp. 71–76

G.I. Mashevitzky, On identities in varieties of completely simple semigroups over abelian groups (in Russian), in *Modern Algebra* (Leningrad University, Leningrad, 1978), pp. 81–89

G.I. Mashevitzky, An example of a finite semigroup without an irreducible basis of identities in the class of completely 0-simple semigroups (in Russian). Uspekhi Mat. Nauk **38**(2), 211–212 (1983). English translation: Russian Math. Surveys **38**(2), 192–193 (1983)

G.I. Mashevitzky, On bases of completely simple semigroup identities. Semigroup Forum **30**(1), 67–76 (1984)

G.I. Mashevitzky, Varieties generated by completely 0-simple semigroups (in Russian), in *Semigroups and Their Homomorphisms*, ed. by E.S. Lyapin (Ross. Gos. Ped. University, Leningrad, 1991), pp. 53–62

G.I. Mashevitzky, On the finite basis problem for completely 0-simple semigroup identities. Semigroup Forum **59**(2), 197–219 (1999)

G.I. Mashevitzky, A new method in the finite basis problem with applications to rank 2 transformation semigroups. Internat. J. Algebra Comput. **17**(7), 1431–1463 (2007)

G.I. Mashevitzky, Bases of identities for semigroups of bounded rank transformations of a set. Israel J. Math. **191**(1), 451–481 (2012)

R.N. McKenzie, Equational bases for lattice theories. Math. Scand. **27**, 24–38 (1970)

R.N. McKenzie, Tarski's finite basis problem is undecidable. Internat. J. Algebra Comput. **6**(1), 49–104 (1996)

R.N. McKenzie, G.F. McNulty, W.F. Taylor, *Algebras, Lattices, Varieties. I.* (Wadsworth and Brooks/Cole, Monterey, 1987)

I.I. Mel'nik, Varieties and lattices of varieties of semigroups (in Russian), in *Studies in Algebra 2*, ed. by V.V. Vagner (Izdat. Saratov Univ., Saratov, 1970), pp. 47–57

I. Mikhailova, O.B. Sapir, Lee monoid L_4^1 is non-finitely based. Algebra Universalis **79**(3), Article 56, 14 pp. (2018)

H. Neumann, *Varieties of Groups* (Springer, New York, 1967)

S. Oates, M.B. Powell, Identical relations in finite groups. J. Algebra **1**(1), 11–39 (1964)

P. Perkins, *Decision Problems for Equational Theories of Semigroups and General Algebras*. Ph.D. thesis (University of California, Berkeley, 1966)

P. Perkins, Bases for equational theories of semigroups. J. Algebra **11**(2), 298–314 (1969)

P. Perkins, Finite axiomatizability for equational theories of computable groupoids. J. Symbolic Logic **54**(3), 1018–1022 (1989)

M. Petrich, *Inverse Semigroups* (Wiley, New York, 1984)

M. Petrich, N.R. Reilly, *Completely Regular Semigroups* (Wiley, New York, 1999)

L. Polák, On varieties of completely regular semigroups. I. Semigroup Forum **32**(1), 97–123 (1985)

L. Polák, On varieties of completely regular semigroups. II. Semigroup Forum **36**(3), 253–284 (1987)

L. Polák, On varieties of completely regular semigroups. III. Semigroup Forum **37**(1), 1–30 (1988)

G. Pollák, On hereditarily finitely based varieties of semigroups. Acta Sci. Math. (Szeged) **37**(3–4), 339–348 (1975)

G. Pollák, A class of hereditarily finitely based varieties of semigroups, in *Algebraic Theory of Semigroups (Szeged 1976)*, ed. by G. Pollák (Colloquia Mathematica Societatis János Bolyai 20, Amsterdam, 1979a), pp. 433–445

G. Pollák, On identities which define hereditarily finitely based varieties of semigroups, in *Algebraic Theory of Semigroups (Szeged 1976)*, ed. by G. Pollák (Colloquia Mathematica Societatis János Bolyai 20, Amsterdam, 1979b), pp. 447–452

G. Pollák, On two classes of hereditarily finitely based semigroup identities. Semigroup Forum **25**(1–2), 9–33 (1982)

G. Pollák, Some sufficient conditions for hereditarily finitely based varieties of semigroups. Acta Sci. Math. (Szeged) **50**(3–4), 299–330 (1986)

G. Pollák, A new example of limit variety. Semigroup Forum **38**(3), 283–303 (1989)

G. Pollák, M.V. Volkov, On almost simple semigroup identities, in *Semigroups (Szeged 1981)*, ed. by G. Pollák, Št. Schwarz, O. Steinfeld (Colloquia Mathematica Societatis János Bolyai 39, Amsterdam, 1985), pp. 287–323

V.V. Rasin, On the lattice of varieties of completely simple semigroups. Semigroup Forum **17**(2), 113–122 (1979)

V.V. Rasin, Varieties of orthodox Clifford semigroups (in Russian). Izv. Vyssh. Uchebn. Zaved. Mat. 1982(11), 82–85 (1982). English translation: Soviet Math. (Iz. VUZ) **26**(11), 107–110 (1982)

D. Rees, On semi-groups. Proc. Cambridge Philos. Soc. **36**(4), 387–400 (1940)

N.R. Reilly, Complete congruences on the lattice of Rees–Sushkevich varieties. Comm. Algebra **35**(11), 3624–3659 (2007)

N.R. Reilly, The interval $[\mathbf{B_2}, \mathbf{NB_2}]$ in the lattice of Rees–Sushkevich varieties. Algebra Universalis **59**(3–4), 345–363 (2008a)

N.R. Reilly, Varieties generated by completely 0-simple semigroups. J. Aust. Math. Soc. **84**(3), 375–403 (2008b)

N.R. Reilly, Shades of orthodoxy in Rees–Sushkevich varieties. Semigroup Forum **78**(1), 157–182 (2009)

V.B. Repnitskiĭ, M.V. Volkov, The finite basis problem for pseudovariety $\mathscr{O}$. Proc. Roy. Soc. Edinburgh Sect. A **128**(3), 661–669 (1998)

J. Rhodes, B. Steinberg, Krohn–Rhodes complexity pseudovarieties are not finitely based. Theor. Inform. Appl. **39**(1), 279–296 (2005)

J. Rhodes, B. Steinberg, *The* $\mathfrak{q}$*-Theory of Finite Semigroups*. Springer Monographs in Mathematics (Springer, New York, 2009)

M.V. Sapir, Inherently non-finitely based finite semigroups (in Russian). Mat. Sb. (N.S.) **133**(2), 154–166 (1987a). English Translation: Math. USSR-Sb. **61**(1), 155–166 (1988)

M.V. Sapir, Problems of Burnside type and the finite basis property in varieties of semigroups (in Russian). Izv. Akad. Nauk SSSR Ser. Mat. **51**(2), 319–340 (1987b). English translation: Math. USSR-Izv. **30**(2), 295–314 (1988)

M.V. Sapir, On Cross semigroup varieties and related questions. Semigroup Forum **42**(3), 345–364 (1991)

M.V. Sapir, Identities of finite inverse semigroups. Internat. J. Algebra Comput. **3**(1), 115–124 (1993)

M.V. Sapir, *Combinatorial Algebra: Syntax and Semantics*, Springer Monographs in Mathematics (Springer, Cham, 2014)

M.V. Sapir, E.V. Sukhanov, Varieties of periodic semigroups (in Russian). Izv. Vyssh. Uchebn. Zaved. Mat. 1981(4), 48–55 (1981). English translation: Soviet Math. (Iz. VUZ) **25**(4), 53–63 (1981)

M.V. Sapir, M.V. Volkov, On the joins of semigroup varieties with the variety of commutative semigroups. Proc. Am. Math. Soc. **120**(2), 345–348 (1994)

O.B. Sapir, *Identities of Finite Semigroups and Related Questions*. Ph.D. thesis (University of Nebraska–Lincoln, USA, 1997)

O.B. Sapir, Finitely based words. Internat. J. Algebra Comput. **10**(4), 457–480 (2000)

O.B. Sapir, The variety of idempotent semigroups is inherently non-finitely generated. Semigroup Forum **71**(1), 140–146 (2005)

O.B. Sapir, Finitely based monoids. Semigroup Forum **90**(3), 587–614 (2015a)

O.B. Sapir, Non-finitely based monoids. Semigroup Forum **90**(3), 557–586 (2015b)

O.B. Sapir, The finite basis problem for words with at most two non-linear variables. Semigroup Forum **93**(1), 131–151 (2016)

O.B. Sapir, Lee monoids are nonfinitely based while the sets of their isoterms are finitely based. Bull. Aust. Math. Soc. **97**(3), 422–434 (2018)

O.B. Sapir, Finitely based sets of 2-limited block-2-simple words. Semigroup Forum **99**(3), 881–897 (2019)

L.N. Shevrin, L.M. Martynov, Attainability and solvability for classes of algebras, in *Semigroups (Szeged 1981)*, ed. by G. Pollák, Št. Schwarz, O. Steinfeld (Colloquia Mathematica Societatis János Bolyai 39, Amsterdam, 1985), pp. 397–459

L.N. Shevrin, E.V. Sukhanov, Structural aspects of the theory of varieties of semigroups (in Russian). Izv. Vyssh. Uchebn. Zaved. Mat. 1989(6), 3–39 (1989). English translation: Soviet Math. (Iz. VUZ) **33**(6), 1–34 (1989)

L.N. Shevrin, M.V. Volkov, Identities of semigroups (in Russian). Izv. Vyssh. Uchebn. Zaved. Mat. 1985(11), 3–47 (1985). English translation: Soviet Math. (Iz. VUZ) **29**(11), 1–64 (1985)

L.N. Shevrin, B.M. Vernikov, M.V. Volkov, Lattices of semigroup varieties (in Russian). Izv. Vyssh. Uchebn. Zaved. Mat. 2009(3), 3–36 (2009). English translation: Russian Math. (Iz. VUZ) **53**(3), 1–28 (2009)

L.M. Shneerson, On the axiomatic rank of varieties generated by a semigroup or monoid with one defining relation. Semigroup Forum **39**(1), 17–38 (1989)

E.P. Simel'gor, On identities of four-element semigroups (in Russian). Modern Algebra, in *Republican Collection of Scientific Works* (Leningrad. Gos. Ped. Inst., Leningrad, 1978), pp. 146–152

W. Specht, Gesetze in Ringen. I. Math. Z. **52**(5), 557–589 (1950)

H. Straubing, The variety generated by finite nilpotent monoids. Semigroup Forum **24**(1), 25–38 (1982)

A.K. Suschkewitsch, Über die endlichen Gruppen ohne das Gesetz der eindeutigen Umkehrbarkeit. Math. Ann. **99**(1), 30–50 (1928)

A. Tarski, Equational logic and equational theories of algebras, in *Contributions to Mathematical Logic (Hannover 1966)*, ed. by H.A. Schmidt, K. Schütte, H.J. Thiele (North-Holland, Amsterdam, 1968), pp. 275–288

W.F. Taylor, Equational logic. Houston J. Math., Survey (1979)

A.V. Tishchenko, The finiteness of a base of identities for five-element monoids. Semigroup Forum **20**(2), 171–186 (1980)

N.G. Torlopova, Varieties of quasiorthodox semigroups (in Russian). Acta Sci. Math. (Szeged) **47**(3–4), 297–301 (1984)

A.N. Trahtman, A basis of identities of the five-element Brandt semigroup (in Russian). Ural. Gos. Univ. Mat. Zap. **12**(3), 147–149 (1981a)

A.N. Trahtman, Graphs of identities of a completely 0-simple five-element semigroup (in Russian) (Ural Polytechnic Institute, Sverdlovsk, 1981b). Deposited at VINITI, Moscow, no. 5558-81

A.N. Trahtman, The finite basis question for semigroups of order less than six. Semigroup Forum **27**(1–4), 387–389 (1983)

A.N. Trahtman, Some finite infinitely basable semigroups (in Russian). Ural. Gos. Univ. Mat. Zap. **14**(2), 128–131 (1987)

A.N. Trahtman, A six-element semigroup that generates a variety with a continuum of subvarieties (in Russian). Ural. Gos. Univ. Mat. Zap. **14**(3), 138–143 (1988)

A.N. Trahtman, Finiteness of identity bases of five-element semigroups (in Russian), in *Semigroups and Their Homomorphisms*, ed. by E.S. Lyapin (Ross. Gos. Ped. Univ., Leningrad, 1991), pp. 76–97

A.N. Trahtman, Identities of a five-element 0-simple semigroup. Semigroup Forum **48**(3), 385–387 (1994)

A.N. Trahtman, Identities of locally testable semigroups. Comm. Algebra **27**(11), 5405–5412 (1999)

P.G. Trotter, M.V. Volkov, The finite basis problem in the pseudovariety joins of aperiodic semigroups with groups. Semigroup Forum **52**(1), 83–91 (1996)

B.M. Vernikov, Distributivity, modularity, and related conditions in lattices of overcommutative semigroup varieties, in *Semigroups with Applications, Including Semigroup Rings*, ed. by S.I. Kublanovskiĭ, A.V. Mikhalev, P.M. Higgins, J. Ponizovskii (St. Petersburg State Technical University, St. Petersburg, 1999), pp. 411–439

B.M. Vernikov, Semidistributive law and other quasi-identities in lattices of semigroup varieties. Proc. Steklov Inst. Math. 2001(suppl. 2), S241–S256 (2001)

B.M. Vernikov, Special elements in lattices of semigroup varieties. Acta Sci. Math. (Szeged) **81**(1–2), 79–109 (2015)

M.V. Volkov, An example of a limit variety of semigroups. Semigroup Forum **24**(4), 319–326 (1982)

M.V. Volkov, Finite basis theorem for systems of semigroup identities. Semigroup Forum **28**(1–3), 93–99 (1984a)

M.V. Volkov, On the join of varieties. Simon Stevin **58**(4), 311–317 (1984b)

M.V. Volkov, Bases of identities of Brandt semigroups (in Russian). Ural. Gos. Univ. Mat. Zap. **14**(1), 38–42 (1985)

M.V. Volkov, Semigroup varieties with a modular lattice of subvarieties (in Russian). Izv. Vyssh. Uchebn. Zaved. Mat. 1989(6), 51–60 (1989a). English translation: Soviet Math. (Iz. VUZ) **33**(6), 48–58 (1989)

M.V. Volkov, The finite basis question for varieties of semigroups (in Russian). Mat. Zametki **45**(3), 12–23 (1989b). English translation: Math. Notes **45**(3), 187–194 (1989)

M.V. Volkov, A general finite basis condition for systems of semigroup identities. Semigroup Forum **41**(2), 181–191 (1990)

M.V. Volkov, Young diagrams and the structure of the lattice of overcommutative semigroup varieties, in *Transformation Semigroups (Colchester 1993)*, ed. by P.M. Higgins (University of Essex, Colchester, 1994), pp. 99–110

M.V. Volkov, On a class of semigroup pseudovarieties without finite pseudoidentity basis. Internat. J. Algebra Comput. **5**(2), 127–135 (1995)

M.V. Volkov, Covers in the lattices of semigroup varieties and pseudovarieties, in *Semigroups, Automata and Languages (Porto 1994)*, ed. by J. Almeida, G.M.S. Gomes, P.V. Silva (World Scientific, Singapore, 1996), pp. 263–280

M.V. Volkov, The finite basis problem for the pseudovariety $\mathcal{PO}$, in *Semigroups and Applications (St. Andrews 1997)*, ed. by J.M. Howie, N. Ruškuc (World Scientific, Singapore, 1998), pp. 239–257

M.V. Volkov, The finite basis problem for finite semigroups: a survey, in *Semigroups (Braga 1999)*, ed. by P. Smith, E. Giraldes, P. Martins (World Scientific, River Edge, NJ, 2000), pp. 244–279

M.V. Volkov, The finite basis problem for finite semigroups. Sci. Math. Jpn. **53**(1), 171–199 (2001)

M.V. Volkov, György Pollák's work on the theory of semigroup varieties: its significance and its influence so far. Acta Sci. Math. (Szeged) **68**(3–4), 875–894 (2002)

M.V. Volkov, On a question by Edmond W. H. Lee. Izv. Ural. Gos. Univ. Mat. Mekh. **7**(36), 167–178 (2005)

M.V. Volkov, A nonfinitely based semigroup of triangular matrices, in *Semigroups, Algebras and Operator Theory (Kochi, India 2014)*, ed. by P.G. Romeo, J.C. Meakin, A.R. Rajan. Springer Proceedings in Mathematics and Statistics vol. 142 (Springer, New Delhi, 2015), pp. 27–38

M.V. Volkov, Identities in Brandt semigroups, revisited. Ural Math. J. **5**(2), 80–93 (2019)

M.V. Volkov, I.A. Gol'dberg, Identities of semigroups of triangular matrices over finite fields (in Russian). Mat. Zametki **73**(4), 502–510 (2003). English translation: Math. Notes **73**(3–4), 474–481 (2003)

M.V. Volkov, I.A. Gol'dberg, The finite basis problems for monoids of triangular boolean matrices, in *Algebraic Systems, Formal Languages and Conventional and Unconventional Computation Theory*, ed. by M. Ito. Kôkyûroku, vol. 1366 (Research Institute for Mathematical Sciences, Kyoto University, 2004), pp. 205–214

S.L. Wismath, The lattices of varieties and pseudovarieties of band monoids. Semigroup Forum **33**(2), 187–198 (1986)

W.T. Zhang, Existence of a new limit variety of aperiodic monoids. Semigroup Forum **86**(1), 212–220 (2013)

W.T. Zhang, Y.F. Luo, The variety generated by a certain transformation monoid. Internat. J. Algebra Comput. **18**(7), 1193–1201 (2008)

W.T. Zhang, Y.F. Luo, A new example of a minimal nonfinitely based semigroup. Bull. Aust. Math. Soc. **84**(3), 484–491 (2011)

W.T. Zhang, Y.F. Luo, A sufficient condition under which a semigroup is nonfinitely based. Bull. Aust. Math. Soc. **93**(3), 454–466 (2016)

W.T. Zhang, Y.F. Luo, The finite basis problem for involution semigroups of triangular 2×2 matrices. Bull. Aust. Math. Soc. **101**(1), 88–104 (2020)

W.T. Zhang, Y.D. Ji, Y.F. Luo, The finite basis problem for infinite involution semigroups of triangular 2×2 matrices. Semigroup Forum **94**(2), 426–441 (2017)

W.T. Zhang, Y.F. Luo, N. Wang, Finite basis problem for involution monoids of unitriangular boolean matrices. Algebra Universalis **81**(1), Article 7, 23 pp. (2020)

A.I. Zimin, Blocking sets of terms (in Russian). Mat. Sb. (N.S.) **119**(3), 363–375 (1982). English translation: Math. USSR-Sb. **47**(2), 353–364 (1984)

List of Symbols

General Symbols

E. W. H. Lee, *Advances in the Theory of Varieties of Semigroups*, Frontiers in Mathematics, https://doi.org/10.1007/978-3-031-16497-2

Aspects of a General Word w

Words

Identities

Finite Algebras

Varieties of Semigroups

Varieties or Classes of Monoids

Index

E. W. H. Lee, *Advances in the Theory of Varieties of Semigroups*, Frontiers in Mathematics, https://doi.org/10.1007/978-3-031-16497-2

Printed in the United States
by Baker & Taylor Publisher Services